Industrial
Hygiene

INDUSTRIAL HYGIENE

Robert W. Allen

Michael D. Ells

Andrew W. Hart

Ferris State College

Prentice-Hall, Inc., *Englewood Cliffs, New Jersey*

Library of Congress Cataloging in Publication Data

Allen, Robert W (date)
   Industrial hygiene.

   Bibliography: p.
   Includes index.
   1. Industrial hygiene.   I. Ells, Michael D.,
(date) joint author.   II. Hart, Andrew W.,
joint author.   III. Title.
RC967.A46   613.6'2   75-35564
ISBN   0-13-461202-7

299922

Printed in the United States of America

10   9   8   7   6   5   4   3

Prentice-Hall International, Inc., *London*
Prentice-Hall of Australia Pty. Limited, *Sydney*
Prentice-Hall of Canada, Ltd., *Toronto*
Prentice-Hall of India Private Limited, *New Delhi*
Prentice-Hall of Japan, Inc., *Toyko*
Prentice-Hall of Southeast Asia Pte. Ltd., *Singapore*

This book is dedicated
to Neoma Allen, Carol Ells,
and the memory of John Weston Hart

Contents

2

SAFETY TRAINING AND INVESTIGATING PROGRAMS   7

3

AIR CONTAMINANTS   36

4

NONIONIZING RADIATION   87

7

HEAT   179

8

NOISE   206

9

MICROBIOLOGICAL HAZARDS   240

10

VENTILATION   293

APPENDIX   319

INDEX   357

Preface

This book is intended both as an introductory textbook for students of industrial hygiene and as a reference book for manufacturers, public service organizations, and government agencies. Included is basic information needed to recognize, evaluate, and cope with the stresses that are of concern to industrial hygienists.

Three features set the book apart from other industrial hygiene texts:

1. The contents stress information that is of practical utility. Theoretical information and material of secondary importance is largely de-emphasized.
2. The book contains considerable information not included in other general industrial hygiene treatises. An example is the chapter on microbiological hazards.
3. The book includes those portions of the Occupational Safety and Health Act standards that pertain to industrial stresses. In addition, a special section discusses the overall significance of the act to employees. These materials will assist employees greatly in complying with the stipulations of the act.

Because of the foregoing features, this book should prove highly useful not only to students but also to every industrial employer, even those with extensive industrial hygiene libraries.

Acknowledgments

Like all authors of technical books, we are indebted to a number of organizations and persons for the assistance they have provided.

First and foremost, we would like to thank The Dow Chemical Company for generously allowing us to use, as a first draft, a manuscript prepared by one of the authors as an employee of the company. This manuscript, and thus our book, drew heavily upon the information, insights, and techniques developed by Dow during the nearly eight decades of its existence. During this period the company has established a reputation for its pioneering efforts in toxicology, as well as industrial hygiene, medicine, and safety; and many employees have earned national and international recognition for their accomplishments in these fields. Without the original manuscript, the book would never have been possible.

Special thanks are also due our employer, Ferris State College. This institution, which recently established one of the nation's first two-year programs in industrial hygiene, has given further evidence of its commitment by the generous and continuing encouragement its officials have given our project.

Adding considerably to the interest and utility of the book are its pictures of safety devices and equipment. Although individual acknowledgments accompany each of the pictures, we wish to add our general thanks to each of the manufacturers who furnished them to us.

Finally we are very much indebted to Lawrence G. Silverstein of Dow for his careful review of the manuscript in its final form.

Our thanks to all of you.

Industrial Hygiene

1

Industrial Hygiene: General Considerations

WHAT IS INDUSTRIAL HYGIENE?

Industrial hygiene is broadly concerned with the chemical and physical stresses that may impair the health and well-being of workers. Such stresses may include:

- Chemical dusts, liquids, gases, fumes, vapors, or mists that present dangers when they are swallowed, breathed, or contact the eyes or skin.
- Nonionizing and ionizing radiation.
- Noise, vibration, high and low temperatures, and pressures.
- Microbiological stresses.
- Ergonomic stresses that arise when an employee's physical and mental attributes do not match the requirements of his job.

An effective industrial hygiene program involves three separate factors. First, it must be possible to *recognize* the various stresses generated in a particular work environment. Second, the stress must be *evaluated* in terms of its magnitude, an operation that generally involves one or more measurements supplemented with the insights gained through past experience. Third, the stress must be *controlled*. This may be accomplished by substituting a less hazardous process, modifying the original process, utilizing special equipment such as ventilating systems and radiation screens, or employing personal protective equipment.

HISTORY AND IMPORTANCE OF INDUSTRIAL HYGIENE

The history of industrial hygiene can be said to begin with Hippocrates, who in the fourth century B.C. recognized and recorded the problem of lead toxicity in the mining industry. About 500 years later, the elder Pliny commented on the hazards involved in handling lead and silver and devised a crude face mask for persons working around large quantities of dust or lead fumes. Still later, in the second century A.D., the Greek physician Galen wrote at length on occupational diseases, recognizing the dangers of acid mists to copper miners.

After this promising beginning, however, interest in industrial hygiene languished until the publication in 1473 of a pamphlet by Ulrich Ellenbog, which discussed toxic fumes and vapors, and also described ways of coping with them. The following century saw the publication of Georgius Agricola's *De Re Metallica*, a 12-part treatise which described mining accidents, offered suggestions for mine ventilation and protective masks for miners, and discussed silicosis, a lung disease caused by inhaling silica dust. Other noteworthy treatises of the Renaissance and its immediate aftermath include those of Paracelsus (1567) and Bernardo Ramazzini (1700). The latter work, which outlined preventive measures for alleviating a wide variety of industrial hazards, remained the chief work in the industrial hygiene field for the next 100 years.

During the eighteenth and nineteenth centuries industrial hygiene became of increasing interest, and the period saw the publication of numerous works on the subject, as well as the passage in both England and the rest of Europe of the first effective legislation designed to protect the health of workers. As early as 1910 in the United States, the U.S. Public Health Service and the U.S. Bureau of Mines began exploratory studies of occupational diseases in the mining and steel industries. Two years later, the Public Health Service began investigating many industries, and in 1913 New York and Ohio established the first state programs in industrial hygiene.

Although these developments and the numerous others that followed them have contributed notably to improved working conditions and better worker health in the United States, the industrial health and safety problem remains enormous. Its magnitude is shown dramatically by a 1971 survey conducted by the

Bureau of Labor Statistics, including figures on both injury and illness. This survey was based on some 60,000 employers in private, nonfarm industry and covered the first six months of 1971. Sampling results showed that about 3.1 million occupational illnesses and injuries and nearly 4,300 deaths occurred during this period, and the overall injury and illness rate was 12.1 per 100 full-time employees. Assuming that about the same rate held for the second half of 1971, almost one out of every eight workers suffered a job-related illness or injury sometime during the year. In the reporting period, an estimated 12.2 million employee days were lost. Without counting the deaths, this figure translates into a loss to industry of about 50,000 employee years of work for the six-month period. As another disturbing note, Dr. Orlan Hampton, former chairperson of the Trauma Committee of the American College of Surgeons, has estimated that in 1968 occupational injuries consumed nearly 4.5 million hospital days. Since there is good evidence for thinking that occupational illnesses are at least as great a problem as injuries, it is possible that work-related injuries and illnesses accounted for 9 million hospital days.

Within recent years, a growing awareness of the dimensions of the industrial safety and health problem has led to the passage of three landmark pieces of legislation. These acts are the Metal and Nonmetallic Mine Safety Act of 1966, the Federal Coal Mine Health and Safety Act of 1969, and—most notably—the Occupational Safety Health Act of 1970, whose purpose is to assure so far as possible safe and healthful working conditions for every working person in the country.

Between 1970—when the Occupational Safety and Health Act was passed—and October 1974, the authorities responsible for carrying out the act's provisions conducted 185,099 inspections resulting in 124,029 citations alleging 642,000 violations. Proposed financial penalties totaled over $16 million. The results of this massive effort are reflected in statistics, which show a noticeable decrease in occupational injuries and illnesses. In 1973, for example, work-related fatalities were seven percent below those for 1972 (5,200 and 5,500 respectively), although nearly three million more workers were employed in 1973. The overall injury and illness rate for 1973 was 11.0 per 100 full-time workers, and about one out of 10 workers experienced a job-related illness or injury—both substantially lower than the figures reported in the 1971 Bureau of Labor Statistics study.

What of the future? Dr. Marcus M. Key, director of the National Institute for Occupational Safety and Health, has said, "The effectiveness of . . . Act of 1970 is directly dependent on NIOSH's ability to produce sufficient manpower to carry it out—that represents our greatest challenge." Unfortunately, the professional personnel situation leaves much to be desired. In 1972, for example, the following shortages were noted:

10,000	industrial hygienists
10,000	occupational health nurses
3,000	occupational health physicians
5-10,000	safety personnel
8,000	occupational health scientists of various types

Today, despite a massive array of governmental and academic training programs, backed by a variety of public and private grants, serious shortages still exist. Successful industrial hygiene students are thus assured a vital and personally satisfying role in the task of protecting and enhancing the well-being of our nation's greatest resource—its people.

PERSONNEL RESPONSIBLE FOR EFFECTIVE PROGRAMS

Effective industrial hygiene is dependent upon the efforts of many different persons. The specific responsibilities of the various types of personnel involved are discussed below.

Design Engineers

Design engineers must design new facilities and machines so that workers are not unnecessarily exposed to harmful stresses. The engineer should notify medical and safety personnel when new operations are about to be started and should ask for an industrial hygiene survey before permitting employees to operate new or altered equipment or carry out new procedures.

Medical Personnel

Medical personnel are responsible for carrying out health care programs, with primary emphasis upon preventing employee disability through proper medical investigations and recommendations. Such personnel participate in establishing and maintaining optimum facilities for medical services. They schedule and administer physical examinations, informing the individuals of the results as needed. They provide medical care for occupational injuries, acute or chronic

occupational illnesses, and—in emergencies—nonoccupational illnesses and injuries. They maintain appropriate treatment files, as well
as provide medical treatment forms, medical literature, and other
materials needed for an effective medical program. They appraise
environmental stresses as they affect individual employees, establish
health education programs, and conduct medical investigations where
applicable and appropriate. Periodically, they review existing medical
programs with management.

Supervisors

Supervisors must know the toxic, corrosive, chemical, and
physical properties of all materials in their areas, and understand the
health hazards resulting from their use. They must educate those in
their charge as to the proper methods for controlling environmental
health hazards. They must follow the schedule for environmental
measurements and physical examinations as laid down by medical
and safety personnel. They must assist in evaluating the design of
new, expanded, or altered facilities. They must remain alert at all
times for changes in the physical condition of their personnel.

Safety Personnel

These personnel have the task of being aggressively supportive
in assisting supervisors in identifying problems of environmental
exposure and seeking the appropriate company group for solutions.
They assist supervisors in training personnel to be alert to potential
hazards, to avoid exposure, and to know the indications of exposure.
They assist in reviewing and formulating operating procedures, as
well as help to review plans for new, expanded, or revised facilities.
They also must maintain an imaginative awareness of the potential
for hazardous exposures, and must plan and instigate actions to
neutralize or safeguard against such exposures.

Industrial Hygienists

The industrial hygienist has the primary function of reducing or
eliminating worker exposure to various chemical and physical stresses
encountered in industry. In carrying out this function, the industrial
hygienist alerts company managers to potential in-plant environmental hazards, measures the hazards, recommends proper engineering controls, and periodically monitors the controlled environment.
Qualifications for an industrial hygienist include extensive training in

physics, chemistry, and toxicology; a considerable background in engineering, medicine, and the biological sciences is highly desirable. Because of their specialized training in toxicology, industrial hygienists are especially helpful in assisting physicians and nurses treating cases of chemical exposure.

Employees

The individual employee is responsible for working safely. To do so he or she must know and faithfully carry out all operating and safety instructions and procedures; know and use the proper equipment for the job; keep alert and warn others of possible environmental health hazards; suggest improvements that will reduce or eliminate environmental health hazards; and report promptly any changes in physical condition or in the work environment that might result in increased exposure.

An effective industrial hygiene program cannot remain static. As new processes and materials appear, such programs, and the roles that various groups play in them, must be constantly revised to insure that the goal of a healthy work environment will continue to be met.

EXERCISES: CHAPTER 1

1. Industrial hygiene is devoted to the _____, _____, and _____ of environmental stresses or hazards in the work area.

2. Contrast the jobs of the industrial hygienist, safety professional, and medical professional.

3. Identify the following types of stresses:
   a. chemical forms
   b. physical forms
   c. biological forms
   d. ergonomic forms

4. Discuss the results of the 1971 Bureau of Labor Statistics survey of work-related health and safety problems, with particular attention to:
   a. the scope of the survey
   b. the overall injury and illness rate
   c. losses in employee hours and employee days

2

Safety Training
and
Investigating Programs

To be successful, any industrial hygiene plan must begin with a comprehensive safety and loss prevention training and accident investigation program. Without such a program, failure is far too often the result, regardless of the emphasis placed on recognizing, evaluating, and controlling physical and chemical stresses.

SAFETY POLICY

Each division, department, and service organization within a company should have a clearly stated and written safety policy signed by the manager. The policy should be communicated, understood, and applied to all levels of supervision so that every employee is aware of management's concern for safe operation. The safety policy should be part of a system that also provides for approving appropriate safety standards, reviewing accident experience, planning for accident prevention, measuring progress, and persistent education.

EMPLOYEE EDUCATION

From the time he or she is first hired, the employee must be made aware of possible hazards and must be responsible for his or her own, as well as coworker's, safety. Proper awareness can be fostered by making each individual's safety attitude, record, and contribution a part of his or her job performance review, and thus a factor in wage increases.

Each new employee should participate in a safety indoctrination session with his or her superior. Temporary employees should also be indoctrinated. Preferably, the department should maintain a safety indoctrination log, which should be signed and dated by both supervisor and employee on completion of the worker's indoctrination. The new employee must understand that "safety is first." He or she must be made to realize that a proper safety attitude, being alert and anticipating hazards—even in simple, routine operations—and always getting proper instruction before doing a job are of primary importance.

The following items are examples of what should be emphasized during the safety indoctrination program of each new full-time and temporary employee.

1. Organization of department and department safety program.
2. Detailed description of hazards existing in work area.
3. Restrictions on clothing and footwear.
4. Eye protection and respiratory protection.
5. Smoking regulations.
6. Procedures for handling dangerous materials and equipment, and for storing dangerous materials.
7. Procedures for disposal of various waste materials.
8. How to handle minor injuries, policy concerning trips to the medical facility.
9. Location and use of emergency equipment (fire shower, fire blanket, fire extinguisher, fire hose, eye shower, gas masks, and others).
10. Location of emergency exits.
11. Recognition and meaning of "alert", "evacuation," and "all-clear" signals on alarm system, and procedures to follow in each case.
12. Location and use of safety manuals and other literature.

SUPERVISOR TRAINING

Line supervisors play a greater role than any other person in insuring a safe work environment and safe employee work practices. To achieve the desired performance, each line supervisor must be thoroughly trained in the philosophy and skills of accident prevention. He or she must also learn to communicate this knowledge so that employees understand it and must take prompt action to insure control of hazards related to the equipment and work area. The training program should stress the fact that safety is an integral part of every job. The supervisor should be instructed to carry out the task of preventing accidents in the same manner as he or she attends to other aspects of the job.

The initial training session should clearly establish management's concern for the safety and loss prevention program. It should emphasize the supervisor's role in the program and make him or her understand that safety performance is a primary consideration in promotions. The instructor should also present an outline of the course.

The second segment of the program should focus on the total accident picture. This segment should begin with a general discussion of accidents, including definitions of an accident, a disabling injury, a potentially serious accident, and a potentially serious injury. The relationship of disabling injuries to workers' compensation should be stressed. During the second half of the segment, performance measurement methods should be reviewed. The instructor should give reasons for measurement and discuss in detail specific measurement methods, such as frequency rate, severity rate, loss index, and serious injury rate.

Once the accident picture is understood, attention can be turned to employee motivation and training. The instructor should impress upon supervisors the importance of providing employees with facts that will enable them to avoid accidents. In addition, he or she should stress the importance of convincing employees that the supervisor is concerned with and involved in their safety. Following this, the various techniques by which these aims can be furthered— job safety analyses, accident prevention key point card programs, and safety meetings (all discussed in this chapter), as well as such things as written operating procedures and new employee indoctrination programs—should be taken up in detail.

Investigation of accidents (discussed in this chapter) is an important factor in job safety and should not be neglected. As a learning exercise, an actual accident case history may be employed advantageously. The history may involve injury, property damage, and/or business interruptions, alone or in any combination. The supervisors should simulate an accident investigation, fill out a complete accident investigation report, and discuss causes and preventive measures. At the conclusion of this course segment, supervisors should have a clear understanding of why it is important to report and investigate accidents.

Inspections and emergency planning must receive consideration. After establishing the importance of inspections in the supervisor's mind, the instructor should take up point-by-point the things to check in any inspection. These may include employee practices, manufacturing procedures, adherence to safety regulations, condition of tools and equipment, and adequacy of housekeeping practices.

Other matters that should receive attention include the importance of checklists, use of such techniques as team inspections and photographs, and the necessity for follow-up inspections. Emergency planning involves establishing a comprehensive plan for each production and rescue unit, as well as a coordinated plan for each location comprised of two or more separate production units. In addition, such planning involves instructing and training employees in how to cope with storms, power failures, fires, and other emergencies. At the conclusion of this segment of the program, supervisors should be fully cognizant of all these factors.

The final portion of the program should be concerned with operating procedures for hazard control. Disucssion should center around such specific topics as operations manuals, work permit systems, personal protective equipment, the design, layout, and modification of operating equipment, and prestart-up surveys.

Whenever possible, the instructor should utilize up-to-date visual aids (overhead projection, slides, movies, charts, graphs). Demonstrations and workshops should also be utilized. Appropriate literature by the National Safety Council, National Fire Protection Association, and similar groups should be discussed.

The course should conclude with a written examination (preferably true/false and multiple choice) covering the basic facts and ideas presented. The test should not be designed to pass or fail participants but to reinforce learning and provide instructors with a measurement of the success or failure of the course.

JOB SAFETY ANALYSIS

Job Safety Analysis is a procedure that identifies the hazards or potential accidents associated with each step of a job, and develops solutions that will either eliminate or guard against such hazards. Once completed it can be the beginning of, or can supplement, the operating procedure. It then may be used in personalized job instruction or education of groups of workers doing the same task.

This method involves the supervisor and the person on the job working together to identify and record all the possibilities for accidents that the job poses. Careful observation of both the worker and the job environment is critical to a successful analysis. Each possible cause of an accident must be dealt with separately as a correction problem. Only a thorough, meticulous approach will avoid missing the less obvious hazards.

How To Use the Job Safety Analysis Method

Define the Job

Defining the job is important, for it helps the supervisor and employee to know its scope. Accidents have occurred on a portion of a job that the supervisor was not aware the employee was doing.

Separate the Job Into Steps

The supervisor should begin by observing the employee doing the job. Before doing so, however, the supervisor should contact the employee to:

1. Explain to the worker what he or she is doing—what the Job Safety Analysis is and why it is being done.
2. Instruct the worker to operate normally.

During the observation, each step of the job procedure should be listed and checked to see if it is still being done or if a new procedure is being used.

If this is the first analysis, it is important that all basic steps of the job be recorded. Avoid, however, making the breakdown so detailed that there is an unnecessary large number of steps. The supervisor must know the job thoroughly to make a proper analysis. Once the procedure is complete, examine each step for hazards.

Identify the Hazards

Basically, there are two important areas to check for hazards. They are: (1) work habits or practices of the employees and (2) work environment or the conditions found in or about the job location.

1. Work habits cover the full range of what the employee does or how he or she goes about doing the job. Habits may involve lifting, holding, pushing, walking, reaching; use of protective equipment and clothing; and attention to established work rules and health hazards.

2. Working conditions (environment) are often checked by using types of accidents to examine possible injury or property damage systematically. Those types may include:

  caught in or between.
  falls to same or other level.
  chemical releases (liquid, solid, or gas).
  sprains or strains.
  struck by or against.
  electrical shock or ignition.

A review of accident reports may also reveal a number of proven hazards.

Safety Control for the Hazards

Once a hazard is identified, it should be recorded and dealt with separately as a correction problem. Usually, two alternatives are possible: (1) changing the design of machinery or equipment or providing mechanical safeguards, or (2) providing a safe work procedure and instructions in following it. Obviously, the first alternative is preferable.

Reviewing Results with Employee

When the job procedure steps, identified hazards, and safety controls are recorded, review them with the employee. He or she may suggest a better way of describing a job procedure or add something that has been overlooked. It is advisable to let the operator help by reviewing each step listed for other accident possibilities. Have the employee think in terms of what could happen to a new or inexperienced worker. If the employee feels that the supervisor is interested in his or her opinion on safety matters, most will react by trying to be helpful. Supervisors should not forget to thank employees for any contributions. Even if no suggestions were made, the fact that the supervisor reviewed the safety analysis with the workers will give them a sense of participating in the analysis.

Whenever an accident occurs on a job covered by Job Safety Analysis, the analysis should be reviewed to determine the need for revision. All employees concerned should then be informed of any changes and instructed in the new procedures.

Consideration of New Ways to do the Job

A thorough review of the job by Job Safety Analysis can set the stage for studying methods of doing the job in a completely new way. The work goal of the job should be detemined, then new ways to reach the goal safely should be considered. Improving efficiency and the quality of the product can be considered at the same time, thus enhancing the value of Job Safety Analysis.

Use of the Form

The form illustrated below is satisfactory for the Job Safety Analysis. It is simple and easy to follow. As the "job procedure steps" are listed, any hazard identified should be listed directly opposite the step at which it was found, in the "hazard present" column. Likewise, the "safety control" should appear opposite the hazard it is controlling.

Some variations in the form, depending on the plant or operation, may be desired. They can be designed to suit the needs of the location.

ACCIDENT PREVENTION KEY POINTS PROGRAM

The development of safe work habits by the employee is a major key to safety. This conclusion is based on the conviction that when a skillful worker has moments of mental lapse, he or she will follow safe work-habit patterns and avoid the deviations that lead to an accident. Accidents can be avoided if employees: (1) understand the hazards of their jobs, (2) know how to circumvent those hazards, and (3) put that knowledge into practice so that it becomes a pattern of work behavior.

The Key Points Card

The Accident Prevention Key Points Program is designed to help employees meet the above listed goals. This program, an original development of The Dow Chemical Company, complements and follows the Job Safety Analysis. The program employs a card on which the key accident prevention elements of the job are listed. The cards are divided into two major sections: (1) a "safe practices" section detailing the procedures and practices that must be followed to control the hazards of the job, and (2) a "safe conditions" section outlining the conditions necessary for safe operations.

Use of the Key Points Card

In carrying out the program, the employee and the supervisor consult and reach agreement on the safest way to perform the work

FIGURE 2.1 Job Safety Analysis

Classification	Grinder Operator	Job: LZ Grinder, X Building		
Employee:	John H. Jones	Supervisor: W. P. Smith		
Date Analysis Completed:		Revised: (1)	(2) (3)	(4)

Job Procedure Steps	Hazard Present	Safety Control
1. Start grinder	Possible loose parts being thrown from grinder	All employees clear of grinder
2. Start conveyor belt *after* grinder	Plugged grinder—reaching into grinder/ hopper to unplug	Proper sequence of starting grinder before conveyor
3. Dump scrap material from packs onto sorting table	a. Standing on table—losing balance—falling (slipping on granules) b. Back strains c. Cutting fingers and hand on pack rim d. Bruises from falling patties	a. Clear space to stand. Railings in place b. Proper lifting methods c. Gloves d. Amount of scrap poured on table—stand back when emptying packs
4. Feed material on conveyor	Cuts from sharp edges of plastic	Gloves
5. Breaking oversize patties	Flying pieces of plastic Hammer deflected off pattie striking employee	Eye protection and face shield Employee clear of swinging arc of hammer
6. Cleaning around grinder	a. Flying granules—using air b. Slipping on granules c. Dangers of compressed air	a. Eye and face protection b. Walk on cleaned areas—rubber soled shoes preferred c. Nozzle never directed at body. Other employees clear of flying material.
a. Use of vacuum cleaner for cleaning	a. Electric shock b. Strains from pulling vacuum cleaner	a. Check plug wires and tension clamp on cable before using b. Proper pulling. Path clear of pieces of scrap
7. Removing and replacing pack covers	Cutting fingers and hands on pack rims	Gloves required
8. Moving packs and material	a. Strains to back b. Foot injuries c. Fingers and hands caught in or between	a. Use of power lift truck, portable conveyors and hand trucks wherever possible b. Safety shoes c. Clear of packs and materials being lifted
9. Maintenance on grinder	Catching hands or fingers in moving parts	Use lockout on grinder switch box
10. Emergencies	Possibility of fires Dust most combustible	a. Prevent plug-ups b. Maintain low dust accumulation by frequent cleanup c. Know location of nearest fire hose and extinguisher

14

under discussion. The conclusions are then noted on the card, after which the supervisor notes the employee's name, the date, and any follow-up on the back of the card, and files it in an active safety file.

The supervisor then uses the card to provide personalized safety instruction for employees. Instruction continues until the employee can demonstrate skill in carrying out every facet of the job safely. Under no circumstances should employees be asked to sign the card to certify they have received instruction. Experience has shown that employees resent signing because they fear the signature may be used against them on their record or to disclaim company liability in the event they become involved in an accident.

Types of Key Points Cards

Five specific key points cards are in use at Dow, each designed for a particular job or situation.

1. Job Planning Card. The job planning card is adapted to service field work. On arrival at a proposed job, the field crew surveys the job site and discusses with the local plant supervisors the safety practices and conditions to be employed. The safety key points of the proposed field job are then recorded on the card.

2. Project Planning Card. The project planning card is one used by laboratories, particularly for larger projects of some duration.

3. General Department Key Point Card. These cards are not specific to a single job but cover a wide range of safety key points to be considered in a plant or operation. They are often used to cover general items not covered on the specific job cards.

4. Special Equipment and/or Technique Card. This type of card has been used widely by all major Dow plant groups. It is excellent for covering the specific safety key points to be considered on a single machine, piece of laboratory equipment, or in jobs requiring handling of hazardous chemicals. It is particularly recommended on types of work where employees use such equipment or materials intermittently.

5. Out-Plant–Field-Work Card. The "Out-Plant or Field-Work" type of card has been developed and used to some degree in technical service, sales, and engineering groups. A prior listing of the safety key points to be considered when visiting customer plants or field operations is the primary objective of this type of card.

Following are examples of the first four types of these key points cards.

FIGURE 2.2

Job Planning Card

JOB TITLE				BUILDING	DOOR
CHARGE	W.O.	A.V.O.	B.M.	CREW	

JOB PLANNING — SAFE PRACTICES TO BE FOLLOWED PROTECTIVE EQUIPMENT, WORK HABITS, ETC.	JOB AREA AND BUILDING HAZARDS CHEMICALS, VENTILATING EQUIPMENT, ETC.	*(LISTED BY OR WITH AID OF PRODUCTION SUPERVISOR OR OPERATOR.)*
LOAN TOOLS		

SIGNED *(MAINTENANCE FOREMAN)*	DATE	SIGNED *(SUPERVISOR OR OPERATOR)*	DATE

NAME OF EMPLOYEE	DATE	FOLLOW UP BY SUPERVISOR

EMERGENCY ALARM SYSTEMS	CHECK ONE:		ALERT	ALL CLEAR	EVACUATE
	SIREN ☐		ONE RISE TO MAXIMUM PITCH	TWO RISES TO MAXIMUM PITCH	CONTINUOUS
	HORN ☐ BELL ☐	WHISTLE ☐ BUZZER ☐	INTERMITTENT BLASTS FOR TWO MINUTES	TWO BLASTS INTERMITTENTLY	CONTINUOUS

PHONES: Ambulance   Fire

16

FIGURE 2.3
Project Planning Card

JOB TITLE	BUILDING	NEAREST DOOR FOR AMBULANCE

JOB PLANNING — SAFE PRACTICES TO BE FOLLOWED PROTECTIVE EQUIPMENT, WORK HABITS, ETC.	JOB AREA AND BUILDING HAZARDS CHEMICALS, VENTILATING, EQUIPMENT, ETC.

SIGNED (SUPERVISOR)	DATE

NAME OF EMPLOYEE (SIGNATURE)	DATE	NAME OF EMPLOYEE (SIGNATURE)	DATE

PHONES: AMBULANCE   FIRE

FIGURE 2.4

General Department Key Point Card

JOB TITLE		SECTION	
SAFE PRACTICES		**SAFE CONDITIONS**	
WORK HABITS, PROTECTIVE EQUIPMENT, HYGIENE, ETC.		*CHEMICALS, GUARDING, ELECTRICAL CONTROLS, TOOLS, ETC.*	
1.		1.	
2.		2.	
3.		3.	
4.		4.	
5.		5.	
6.		6.	
7.		7.	
8.		8.	
9.		9.	
10.		10.	
11.		11.	

ACCIDENT PREVENTION KEY POINTS
THE DOW CHEMICAL COMPANY   SIGNED_____   DATE _____
STOCK FORM 23090 PRINTED IN U.S.A. R2-63   *SUPERVISOR*

RECORD OF DISCUSSIONS WITH EMPLOYEES

	NAME OF EMPLOYEE	DATE	FOLLOW-UP REQUIRED
1			
2			
3			
4			
5			
6			
7			
8			
9			
10			
11			
12			
13			
14			
15			
16			

18

FIGURE 2.5 Special Equipment and/or Technique Card

TITLE	Circulation Loop	SECTION	Lab

1. Only personnel familiar with the loop should operate it.
2. Do not enter loop room when red light is on.
3. Inspect loop for leaks before putting into operation. This can be done by pressurizing with cold water.
4. Flush loop before and after use.
5. Wear goggles when handling chemicals. This includes taking samples.
6. Wear protective clothing, chemical goggles and full face shield when entering loop room if chemicals are under pressure. Goggles must be used whenever chemicals are in the loop.
7. Inspect loop from glass windows before entering room.
8. Lower fluid temperature to $120°$ F and vent system before draining.
9. Apply 2 psi nitrogen to the loop if system is being cooled below $212°$ F.
10. Wear rubber gloves and goggles when removing and flushing tube.
11. Immediately clean up any chemical spills.

PRESTART-UP INSPECTION

SECTION HEAD _____ COMMITTEE _____ NOT APPLICABLE _____ DATE _____

ACCIDENT PREVENTION KEY POINTS
THE DOW CHEMICAL COMPANY

PREPARED BY _____ DATE _____

I, THE SUPERVISOR, HAVE DISCUSSED AND OBSERVED THIS OPERATION WITH THE EMPLOYEES NAMED BELOW.

EMPLOYEE	SUPERVISOR SIGNATURE	DATE	EMPLOYEE	SUPERVISOR SIGNATURE	DATE

19

SAFETY MEETINGS

Good safety meetings are the result of careful planning, thorough preparation and involvement of the participants. The key factors that must be taken into account for any safety program to be successful include establishing objectives, selecting topics, involving participants in the proceedings, and communicating information effectively. These factors are discussed below.

Establishing Objectives

The objectives of a program of safety meetings are the changes sought through the meetings. There is great value in establishing objectives: the overview puts meaning in the program and helps establish controls that will assure continuity.

Objectives must be as clear and realistic as possible to guide the planners in reaching them. It should be realized that no objective probably can be met by a single meeting. The safety meeting should be considered as a channel of communication providing information that will gradually change behavior, knowledge, and skills, and thus eventually improve safety performance.

Supervisors, technical experts, and safety and fire protection engineers normally establish objectives and plan safety meetings. Often overlooked by planners, however, is a session involving employees, who discuss objectives and problems as they see them. Such a session helps employees to identify with the objectives and makes them feel that they are part of the program. Once objectives are established, the planners can plot their attack to reach them. It is well to remember that mere scheduling is not planning. Planning is the mental process of selecting topics, ideas, and methods to reach the established objectives.

Selecting Topics

Selecting proper accident-prevention topics is of prime importance. The topics must relate to the long-range objective.

Pertinency of the topic can be established from accident-cause summary data. The description and cause of accidents may be obtained from cause analysis, incident case histories, and job-hazard analysis surveys.

Subcommittees of participants can develop topics for discussion. They should know best what is concerning them. They are familiar with the practices, conditions, and near-misses occurring in their work areas.

All too often safety meetings limit discussions of accidents to those involving personal injuries. However, loss prevention topics are equally important to an effective accident control program. An accident must be thought of as an unplanned event that could have been avoided and that results in (1) personal injury, (2) dollar losses (property damage), or (3) has a potential of either of the first two.

Cause analysis summaries will reveal a marked similarity of events leading to personal injury, losses, or both.

Involving Participants

The extent to which participants in a meeting become actively involved largely determines how much they gain from the meeting. In planning and leading a meeting, the aim should be to involve all members in the discussion. A number of discussion patterns have proved effective in achieving this aim.

The Lecture Forum

The chairperson introduces the speaker, who gives a speech or lecture on a topic. He or she then encourages the audience to discuss the topic with the speaker. If possible, the chairperson should seat the speaker among the group to encourage participation in the forum period. The chairperson—not the lecturer—conducts the program.

The Buzz Session

Discussion may be opened with a short talk, film strip, or case study. The large group is then broken down into small groups which meet simultaneously. Each group continues discussion on a specific question arising out of the film, or whatever. Speakers from the smaller groups report group thinking and an open discussion period follows.

Cooperative Investigation

The leader picks several members beforehand to prepare brief subtopic reports on the topic under study. The group adds information after reports are given. The reports occupy half the total time allowed. The remaining time is for general discussion. This format insures collaboration prior to decision making.

The Panel

The chairperson guides the conversation. An outline is prepared by the chairperson and panelists in advance. If the program is planned for an hour, the panel portion is concluded after 30 or 40 minutes. The remaining time is for audience participation. There are no set speeches and the panel develops a conversational quality well adapted to small audiences.

The Dialogue

Two people take part in a dialogue or conversation. The audience listens, then takes part in a planned discussion period. In a dialogue one person questions, the second person is the respondent. The respondent is an expert on the topic under study. The questioner questions in such a way that the respondent will not make lengthy speeches.

Role Playing

The discussion leader guides a few members of the group in acting out examples of the problem under discussion. Other members of the group observe the performance. Multirole playing involving all present is also possible. When the role playing is over, all members join in a discussion of the specific example and the general problem. To be effective, the leader needs experience with role playing.

Brainstorming Conference

The goal may not be problem solving or accumulating facts, but creative idea finding. The group accumulates and refines as many ideas as possible in a short session. A leader announces a topic for consideration. The participants suggest ideas for proposed improvements as rapidly as they can. As ideas begin to flow, the group is interstimulated. An incomplete idea suggested by one member is usually completed by another.

Whatever the technique, the leader has the task of guiding the discussion and involving group members. Skillful leadership will encourage all members to participate and discuss the topic thoroughly and thereby help to insure that the objectives of the meeting will be achieved.

Safety meetings are often used for problem solving. To avoid the pitfalls of straight lecture sessions and unplanned group discussions, this five-point plan can be followed.

1. A simple, specific safety problem is stated.
2. Discussion is focused on a single hazard. The positive and

negative factors are listed. Next, methods by which the problem can be solved are listed.

3. A plan of action is formulated.
4. Follow-up schedules are set.
5. The discussion and agreed upon action is summarized.

With problem-solving meetings there is danger that the responsibility for decision making will be shifted away from the supervisor. Employees may not know all the factors required to make an intelligent decision.

Communicating Information Effectively

Those selected to lead or address a safety meeting must communicate effectively if their message is to be understood and retained. A common problem is the person who is impressed with his or her command of the language or is lost in technical jargon. The speaker with a point to make should do so without burying it under a mumbo-jumbo of words. A speaker should decide what he or she is trying to say and how to say it concisely, using, whenever possible, simple words instead of technical language.

As we have already noted, a good safety meeting is the product of careful planning and hard work. The success of meetings is measured by results, and results are dependent on each participant being committed to doing his or her job safely.

ACCIDENT INVESTIGATION

Accident investigation is a scientific approach to accident prevention. It is a systematic method of collecting information that is used to reconstruct the accident accurately and determine why it happened. Once this has been done, preventive measures can be developed and applied.

The main goal of accident investigation is to prevent recurrence. The extent to which this goal is met, however, largely depends on how well the information concerning causes is communicated to all operations having identical or similar accident-causing situations.

All accidents must be investigated promptly regardless of severity; otherwise vital evidence may be lost. The accident may range from a large-scale disaster involving multiple fatalities and extensive property damage to a potentially severe incident where no injury or property loss actually occurs. The type of investigation depends on the nature and magnitude of the accident.

Disabling Injuries/Reportable Losses

Who Should Investigate

The investigation should be made by the immediate supervisor of the individual involved or of the area or equipment connected with the accident. This does not mean other investigations should not be made. There are, however, several reasons why the immediate supervisor is the first choice.

1. The supervisor is responsible for the welfare of his or her employees and their work environment.
2. The supervisor knows the employee—particularly his or her training and work habits—better than anyone else.
3. The supervisor knows the job, the operating instructions, the equipment, and the work area.
4. By demonstrating personal concern for their well-being, the supervisor promotes better relations with employees.
5. When directly involved in the investigation, the supervisor is more apt to put corrective measures into effect.

Disabling injuries and reportable property damage accidents ($10,000 and over) are usually investigated by special committees appointed for that purpose. In all cases the supervisor of the employee, equipment, or area involved should be included, as well as a safety and loss prevention specialist.

When such committees do the investigating, all disciplines involved in the accident or the probable corrective actions should be represented in the committee.

Employee Considerations

The first concern should be for the comfort and treatment of the injured employee. Immediate questioning should be avoided. Normally questioning can be carried out within a reasonable period after the person has received treatment and is comfortable. It is better that the investigator go directly to the scene of the accident as a first step in the investigation.

To be truly effective, the investigator must convince employees being questioned that the aim of the investigation is to find facts, not fault. Most employees will cooperate when they are convinced that they may be helpful in preventing others from becoming injured in a similar accident. Some, however, will hesitate to talk for fear of self-incrimination. Sarcasm or any attempt to assess blame heightens the tension and often makes people withdraw or become belligerent. To avoid hostile worker reactions the investigator should never attempt to put words in a witness's mouth or ask leading questions. Instead, the investigator should identify with the worker by acting friendly and concerned and by seeking worker suggestions. Fairness in all cases is absolutely essential. Decisions made following the investigation must be fair and impartial to assure future cooperation in that work group.

Investigational Methods

The investigation must be carried out promptly since conditions at the accident scene may soon change—unintentionally or intentionally. Evidence may be destroyed or removed and details forgotten. Moreover, witnesses are more apt to describe circumstances accurately right after an accident, without the added conjecture that comes later from discussion of the accident with other employees. Checking the scene promptly also assures employees that management is concerned for their well-being and determined to prevent future accidents.

To get the facts, the investigator must (1) obtain a description of the accident, and (2) determine any unsafe acts or conditions and why they occurred or existed. To obtain complete information requires a well-organized approach. In addition, the investigator must avoid the tendency to draw conclusions before all the facts are obtained and considered.

Following are suggestions on how to get the facts:

1. Check the site and circumstances of the accident before anything has been changed. Photographs are useful in many instances.
2. Secure pertinent facts from witnesses who saw the accident.
3. Reconstruct the events that resulted in the accident. Proceed slowly so that an accident is not repeated during any re-enactment. Begin as far back into the history of the event as it is practical to explore, particularly to determine what activities were being carried out before the accident. Study the operating procedures to determine deviations from safe practices.
4. If help is needed in determining the cause, ask for it. Technical personnel and/or safety and loss prevention specialists are sources of assistance.

5. Discuss the accident with injured employee(s), but only after treatment and under favorable circumstances.

6. Encourage other employees to contribute ideas for prevention.

Once the facts are gathered, studied, and carefully weighed, causes can be determined. For each cause a suitable remedy should be adopted and formally recorded. Preventive measures usually involve: (1) employee education, (2) design improvements (mechanical guards, for example), and/or (3) improved supervision.

Most accident causes indicate either (1) lack of knowledge of safe operating procedures, (2) failure to learn and exercise hazard-control information received in job training, (3) violations of safety rules, or (4) failure to enforce rules. Once the cause of an accident has been determined, the persons involved in the accident may have to be re-educated. If the hazard information is new, all employees involved in similar operations will require instruction in the new safe-work procedures.

Design improvements on equipment may range from a simple addition of a cover guard to a complex change in the actual design of the equipment. The tendency may be to make a minor modification and continue to operate. A follow-up inspection will help to assure permanent improvement. When the design cannot be improved, greater emphasis must be placed on employee safe-work procedures to control the hazard.

Supervisory methods must also be examined when an accident occurs, and the kind and quality of employee safety education must be carefully assessed. Frequent observation of employees at work to determine if they are practicing what has been taught or are deviating from safe procedures is a necessary part of any accident follow-up.

Maintenance of safe equipment and facilities also are the responsibility of supervision. Failure to provide or practice procedures to detect unsafe conditions must be corrected.

Disasters

Successful investigation of accidents involving extensive property damage and possible multiple fatalities requires advance planning. To coordinate the investigators required and cope with the great amount of information that will be uncovered, an investigating committee is desirable.

Make-Up of Investigating Committee

The chairperson of the committee should have the authority and responsibility to obtain the facts required. He or she should have a good technical background on the plant or operation involved but should not be directly responsible for the operation or be a friend of the concerned personnel. Of course, the chairperson should be able to spend the time necessary for a thorough investigation.

Committee members should include personnel with expert knowledge of the equipment or processes in question, as well as the supervisor directly responsible for the unit involved in the accident. Outside personnel may be used to supplement company personnel when such people possess expert knowledge not available within the organization. All disciplines involved in the accident or likely to be involved in the corrective actions should be represented.

A legal advisor should be appointed to assist the committee members. This person, who may be a company attorney, the insuror's attorney, or a lawyer hired especially for the investigation, informs the committee chairperson of pertinent statutes and the company's legal obligations before and during the investigation. In addition, the advisor studies the legal consequences of committee actions to prevent unnecessary liability. Other assistants to the committee may include such people as stenographers, messengers, workers.

Functions of Investigating Committee

This committee has three prime functions: interviewing persons who were involved in or who witnessed the accident, collecting evidence, and reporting findings. Interviewing should begin immediately, before rumors can spread and become inextricably entangled with fact. Interviews should be conducted individually, and should, of course, be completely free of attempts to harass the interviewee or to put words in his or her mouth. Stenographic records, which inhibit employer comments less than tapes, are preferable as records of the interview. In conducting this phase of the investigation, casual observers of the disaster should not be overlooked, nor should persons who may have photographed it. Whoever is conducting interviews should always bear in mind that eyewitness accounts may be unreliable.

The first step in collecting evidence is to rope off the disaster area and post guards if necessary. Any evidence must be photographed before it is disturbed. The actual collection procedure may involve collecting instrument charts and logbooks; collecting, numbering, and mapping missiles and fragments, including weighing or estimating their weights to aid in energy calculations; sending items for chemical and metallurgical examinations; and disassembling equipment, sifting debris, and cleaning up. The chairperson should appoint individuals or teams to carry out each phase of the investigation.

Daily meetings of the committee are vital in order to keep members aware of the progress of the investigation and to plan further action. At these meetings, stenographic records should be reviewed, reports of individual members or investigating teams read, and findings discussed. Before the meeting is adjourned, further action should be determined. All theories should be investigated until completely ruled out.

Prior to issuing reports, the committee should seek the advice of the corporation's public relations staff and management concerning reporting techniques. Medical bulletins and technical reports should be issued promptly whenever possible. Adopt a consistent policy for newspapers, radio, and television. Remember, these media will publish and comment anyway, so it is better to supply them with all the available facts. If the investigation is lengthy, interim and consultants' reports may be released as they become available. The final report, issued over the chairperson's signature, may include separate reports from members and consultants.

Accident Report Forms

The Disabling Injury and/or Loss Report form (Figure 2.6) is designed primarily for reporting disabling injuries as per ANSI Z16.1-1967, and property damage and business interruption losses of $10,000 or more. Potentially severe injuries or property damage incidents can also be reported on this form if they are properly identified.

Completeness of the report is essential to a meaningful understanding of what really happened, why it happened, and what the investigator recommends for prevention. It is particularly important that the investigator seek for the record the reason or reasons for the causes listed. The report should be as brief as possible. However, it must have all significant information if it is to be useful. The use of a general term such as "carelessness" is unacceptable because it has little meaning.

Examples of forms to be used by superintendents are the Supervisor's Investigation of Accident (Figure 2.7) and the Supervisor's Accident Report (Figure 2.8). (Note that Figure 2.7 also includes an assessment by higher supervisory levels of supervisors' responsibilities that may have contributed to the accident.) Copies of these reports should go to the department head and the safety group. The Supervisor's Accident Report may also be used for potentially severe injuries, property damage incidents, and/or off-the-job accidents.

Follow-Up

Conclusions on preventive measures or corrective action to be taken are recorded with good intent to implement them. Time, however, has a way of eroding good intentions, and a second accident may occur before anything is done about the first one. Specific dates for follow-up checks and names of those who are responsible for doing so are extremely important. Indication of intended follow-up should be an integral part of the Disabling Injury and/or Loss Report form.

Communications

To prevent accidents from recurring, it is vital that the accident report be communicated to all operations having identical or similar accident potentials. Unless this is done, any lessons learned are fruitless.

Insurance Considerations

The results of disaster investigations should not be disclosed to anyone other than appropriate personnel without first securing proper clearances. Investigators for possible third-party claims should not be permitted to interview company personnel without prior clearance by local management, the company's legal department, and the insurance department.

FIGURE 2.6

Disabling Injury and/or Loss Report

DISABLING INJURY AND/OR LOSS REPORT

THE DOW CHEMICAL COMPANY
LOSS PREVENTION

'D. I.: U.S.A. S.I. Z16 ... LOSS: Property damage and/or business interruption – fire, explosion, storm, boiler, machinery and leaks, breaks and spills)

	(CHECK ONE OR BOTH)	DIVISION OR DEPARTMENT		LOCATION OR PLANT				
	☐ DISABLING INJURY							
		ACCIDENT LOCATION						
1	☐ LOSS							
	DATE OF ACCIDENT	TIME OF ACCIDENT	TYPE OF OPERATION					
			☐ PRODUCTION ☐ PILOT PLANT ☐ LABS ☐ SERVICES (POWER, SHOPS) ☐ WAREHOUSE ☐ OFFICES AND MISC.					

	NAME OF INJURED EMPLOYEE		AGE	CLASSIFICATION		TIME WITH COMPANY	TIME IN CLASSIFIC.
2	INJURY						EST. TIME LOST
	PERSONAL DATA ON INJURED EMPLOYEE:	① FAMILY			② OTHER		

	ESTIMATED COMPENSATION AND MEDICAL COST $	ESTIMATED PROPERTY DAMAGE (Replacement cost) $	ESTIMATED BUSINESS INTERRUPTION LOSS $	NUMBER OF DAYS
3	PRODUCT OR EQUIPMENT INVOLVED			

DESCRIPTION (CONTINUE ON PAGE 2 IF NECESSARY)

4 CAUSE

1. **UNSAFE ACT OR PRACTICE** (INCL. - REASON FOR)

2. **UNSAFE CONDITION** (INCL. - REASON FOR)

PREVENTIVE MEASURES TAKEN OR PLANNED

5	FOLLOW UP	PLANT, DEPARTMENT OR OTHER	BY	DATE TO BE CHECKED	CHECKED BY

COPIES TO:

(SEE REVERSE) | REPORTED BY | DATE REPORTED

FORM C-23360 PRINTED IN U.S.A. R2-67

FIGURE 2.6 (cont.)

Disabling Injury and/or Loss Report

DISABLING INJURY AND/OR LOSS REPORT

ANALYSIS

A. TYPE

1. DISABLING INJURIES

☐ STRUCK AGAINST	☐ FALL ON SAME LEVEL	☐ CAUGHT IN, UNDER OR BETWEEN	☐ TEMPERATURE EXTREMES-CONTACT
☐ STRUCK BY	☐ BODILY REACTION	☐ RUBBED, ABRADED, SCRATCHED	☐ CHEMICALS, RADIATION-CONTACT
☐ FALL FROM ELEVATION	☐ OVEREXERTION	☐ ELECTRICAL CURRENT-CONTACT	☐ MOTOR VEHICLE ACCIDENTS

2. PROPERTY LOSS

☐ FIRE	☐ STORM	☐ MACHINERY	☐ OTHER (SHOW TYPE)
☐ EXPLOSION	☐ BOILER	☐ LEAK, BREAK, SPILL	

B. CAUSE

1. ADMINISTRATION

A. WRITTEN OPERATING PROCEDURES	B. EMPLOYEE TRAINING	C. SAFETY RULE OR POLICY	D. SUPERVISION
☐ COMPLETE	☐ ADEQUATE	☐ PROVIDED	
☐ PARTIAL	☐ INADEQUATE	☐ NOT PROVIDED	☐ ADEQUATE
☐ NONE	☐ NONE	☐ ENFORCED	☐ INADEQUATE
		☐ NOT ENFORCED	

6 **2. UNSAFE ACT   (CHECK ONE - PRIMARY)**

☐ WORKING ON MOVING, RUNNING, ELECTRICALLY ENERGIZED OR PRESSURED EQUIPMENT	☐ IMPROPER USE OF EQUIPMENT	☐ OPERATING OR WORKING AT UNSAFE SPEEDS	☐ USING UNSAFE EQUIPMENT
☐ FAILURE TO USE AVAILABLE PERSONAL PROTECTIVE EQUIP.	☐ IMPROPER USE OF HANDS OR BODY PARTS	☐ TAKING UNSAFE POSITION OR POSTURE	☐ IMPROPER WELDING OR BURNING PROCEDURE
☐ FAILURE TO SECURE OR WARN	☐ INATTENTION TO FOOTINGS, OR SURROUNDINGS	☐ DRIVING ERRORS	☐ OTHER (SPECIFY)
☐ HORSEPLAY, DISTRACTING, QUARRELING	☐ MAKING SAFETY DEVICES INOPERATIVE	☐ UNSAFE PLACING, MIXING, COMBINING	

3. UNSAFE CONDITION   (CHECK ONE - PRIMARY)

☐ IMPROPERLY GUARDED	☐ IMPROPER MAINTENANCE	☐ VENTILATION
☐ POOR DESIGN OR LAYOUT	☐ HOUSEKEEPING	☐ DRESS OR APPAREL
☐ DEFECTIVE FABRICATION OR INSTALLATION	☐ ILLUMINATION	☐ OTHER (SPECIFY)

C. CONTRIBUTING FACTORS

1. FIRE OR OTHER PROTECTIVE SYSTEMS		2. EMERGENCY PROCEDURES	
☐ EFFECTIVE	☐ INEFFECTIVE	☐ ADEQUATE	☐ INADEQUATE

(CONTINUED FROM PAGE 1)

FIGURE 2.7

Supervisor's Investigation of Accident

The Dow Chemical Company

Date of
Accident: _____

Time: _____   Shift: A B C D

Date Accident
Investigated: _____

Classification of Accident

  Accident-Not Potentially Serious   ☐ Accident - Potentially Serious

Block Name and Location In Block Where Accident Occurred: _____

Name of Injured (If Any): _____

Classification of Employee: _____   Contractor Involved (If Any): _____

Accident Description: _____

_____

Caused by: ☐ Equipment Failure   ☐ Human Failure   ☐ Design Feature   ☐ Improper Maint./Installation

Explain: _____

_____

Supervisory Action Resulting From Accident: _____

_____

(To Be Completed By Supt.)

Area of Supervisory Responsibility Which Contributed To This Accident:

☐ Maint. of Work Environment
☐ Training of Personnel
☐ Attitude — Involvement Prior to Accident
☐ Enforcement of Safety Discipline
☐ Not Applicable

$ _____   Estimated Cost to _____ of Property Damage or Spill (If Any)

FIGURE 2.7 (cont.)

Supervisor's Investigation of Accident

☐ This PSA will be investigated further by APC and Safety Board Action.

Superintendent's Signature	Investigated By	Signature of Injured (If Any)

Distribution: Safety Department; Department Reporting; Department Head

Scope: To secure the reporting of ALL accidents which resulted in, or had the potential to result in, injury, property loss, or hazard to the community.

1. Purpose:

   a) To investigate, in depth, each accident to determine the cause, reason for the cause, and resulting supervisory action so that it will not re-occur.

2. Reporting Procedure:

   a) All accidents will be reported on the attached form. "Supervisor's Investigation of Accident". This form replaces both the "Supervisor's Investigation of Injury", and the "Incident Report" forms.

   b) Please type or print legibly on this form.

   c) Routing of the form will be (1) Safety Department Copy (2) Department (Reporting) Copy and (3) Department Head Copy.

   d) The investigating supervisor will classify all accidents as provided on the form.

   1) Definitions:

      a. Accident - Not Potentially Serious - Include in this classification all accidents where the conditions cause & result of the accident were minor in nature & did not have the potential to result in a major injury, property damage, or product release.

      b. Accident - Potentially Serious - Include in this classification all accidents where the conditions of the accident resulted in or had the potential to result in a major injury, or property damage, or product release.

3. Special Investigations:

   All potentially serious accidents (PSA) that require further attention will be investigated with an APC member, Safety Board Member, and Safety Engineer at the request of the plant superintendent; a separate report will be written by the responsible supervision concerned. This is mandatory for disabling injuries, and for property damage in excess of $10,000.

FIGURE 2.8

Supervisor's Accident Report

ACCIDENT REPORT

☐ *Off-the-job accident 1, 2, 3 and 4
☐ Industrial accident 1, 2, 3, 4, 5 and 6

*SEE REVERSE
OF COPY 3

POTENTIAL SEVERITY OF ACCIDENT

☐ MINOR ☐ MODERATE ☐ EXTREME

1 | NAME | OCCUPATION | DEPARTMENT

2 | LOCATION OF ACCIDENT | DATE | TIME ☐ DAY ☐ P.M. ☐ MID-NIGHT ☐ OVER-TIME | *DATE OF FIRST FULL DAY LOST OFF-THE-JOB

3 | NATURE OF INJURY

WHAT HAPPENED?

4

DATE RETURNED TO WORK?

STATE SPECIFIC ACT OR CONDITION

CHECK PRIMARY CAUSE

☐ PERSONAL ☐ MECHANICAL

5

PREVENTIVE MEASURES TAKEN

6

SIGNED | REPORTED BY (IMMEDIATE SUPERVISOR) | PHONE NO. | APPROVED: DEPARTMENT SUPT. | DATE

STOCK FORM 16760
PRINTED IN USA R8-66

☐ PLEASE CHECK IF ACCIDENT COULD BE SAFETY MEETING TOPIC

IMMEDIATE SUPERVISOR

NOTE: Report due in Safety Department Office not later than the fifth (5th) day of succeeding month.

EXERCISES: CHAPTER 2

1. Indicate the points that should be stressed in safety indoctrination programs for new full-time and temporary employees.

2. Discuss the factors involved in supervisor training and the types of teaching aids appropriate for instructors to utilize.

3. Name the purpose of a Job Safety Analysis and the steps involved in carrying one out.

4. Define a key points card, indicate the types of information it includes, and discuss its use.

5. Compare the following formats for safety meetings:
   a. the buzz session
   b. the panel
   c. role playing
   d. the brainstorming conference

6. Summarize the procedure for investigating a disabling injury accident, with special attention to the investigation methods that should be followed.

7. Summarize the procedure for investigating accidents involving extensive property damage and/or one or more fatalities, with special attention to the makeup of the investigating committee and its functions.

3

Air Contaminants

Air contaminants include chemical dusts, gases, vapors, fumes, or mists that present dangers when they are breathed or contact the eyes or skin. For convenience, ingestion will also be covered in this chapter.

The various types of air contaminants can be defined as follows:

Dusts: Solid particles that have been reduced to small size by some mechanical process.

Gases: Materials that are in a physical state such that they will diffuse and occupy the space in which they are enclosed. These materials do not appear in the solid or liquid state at ordinary temperatures or pressures.

Vapors: The gaseous form of a substance that is normally a liquid or a solid.

Fumes: Solid, microscopically small particles formed by condensation from the gaseous state.

Mists: Suspensions in air of very small drops, usually formed by mechanical means (atomization) or condensation from the gaseous state.

Paragraph 1910.93 of the OSHA standards deals with air contaminants. It consists primarily of a series of tables that set threshold limit values for a large number of materials. The paragraph stipulates that exposures by inhalation, ingestion, skin absorption, or contact to concentrations above those set in the tables should be avoided. If this is not possible, protective equipment must be provided and used. The paragraph further states that to achieve compliance with this requirement:

> . . . feasible administrative or engineering controls must first be determined and implemented in all cases. In cases where protective equipment, or protective equipment in addition to other measures, is used as the method of protecting the employee, such protection must be approved for each specific application by a competent industrial hygiene or other technically qualified source.

The OSHA regulations are shown in Tables 1, 2, and 3.

DEFINITIONS

To have an intelligent understanding of the evaluation of air contamination hazards, it is necessary to know the meaning of several terms.

Toxicity: The capacity of a substance to produce injury by other than physical means.

Toxic Hazard: The possibility of toxic injury occurring under stated circumstances.

Acute: Having the character of being single, of short duration, or of high intensity. An exposure of short duration, at high concentrations of material, and with fairly prompt response is thought of as being acute. The term may also include intermittent exposure over several days with the response delayed for several hours or even one or two days.

Ceiling Concentration: The maximum concentration to which workers can be exposed. There are a number of substances for which ceiling concentrations have not as yet been established.

Chronic: Having the character of being prolonged and/or repeated. A chronic condition is, thus, one that persists over a period of months or years, and a chronic dosage is one that is repeated or continuous over a similarly long period. Frequently, the effects of chronic dosages come on gradually.

Subacute: Having a character intermediate between acute and chronic, such as a repeated dosage over a period of days or weeks. It should be noted that the terms "acute," "chronic,"

TABLE 1

Substance	p.p.m.[a]	mg./M³ [b]
Acetaldehyde	200	360
Acetic acid	10	25
Acetic anhydride	5	20
Acetone	1,000	2,400
Acetonitrile	40	70
Acetylene dichloride, see 1, 2-Dichloroethylene		
Acetylene tetrabromide	1	14
Acrolein	0.1	0.25
Acrylamide—Skin		0.3
Acrylonitrile—Skin	20	45
Aldrin—Skin		0.25
Allyl alcohol—Skin	2	5
Allyl chloride	1	3
**C Allylglycidyl ether (AGE)	10	45
Allyl propyl disulfide	2	12
2-Aminoethanol, see Ethanolamine		
2-Aminopyridine	0.5	2
**Ammonia	50	35
Ammonium sulfamate (Ammate)		15
n-Amyl acetate	100	525
sec-Amyl acetate	125	650
Aniline—Skin	5	19
Anisidine (o, p-isomers)—Skin		0.5
Antimony and compounds (as Sb)		0.5
ANTU (alpha naphthyl thiourea)		0.3
Arsenic and compounds (as As)		0.5
Arsine	0.05	0.2
Azinphos-methyl—Skin		0.2
Barium (soluble compounds		0.5
p-Benzoquinone, see Quinone		
Benzoyl peroxide		5
Benzyl chloride	1	5
Biphenyl, see Diphenyl		
Bisphenol A, see Diglycidyl ether		
Boron oxide		15
C Boron trifluoride	1	3
Bromine	0.1	0.7
Bromoform—Skin	0.5	5
Butadiene (1, 3-butadiene)	1,000	2,200
Butanethiol, see Butyl mercaptan		
2-Butanone	200	590
2-Butoxy ethanol (Butyl Cellosolve)—Skin	50	240
Butyl acetate (n-butyl acetate)	150	710
sec-Butyl acetate	200	950
tert-Butyl acetate	200	950
Butyl alcohol	100	300
sec-Butyl alcohol	150	450
tert-Butyl alcohol	100	300
C Butylamine—Skin	5	15
C tert-Butyl chromate (as CrO₃)—Skin		0.1
n-Butyl glycidyl ether (BGE)	50	270
*Butyl mercaptan	10	35
p-tert-Butyltoluene	10	60
Calcium arsenate		1
Calcium oxide		5
**Camphor	2	
Carbaryl (Sevin®)		5
Carbon black		3.5

TABLE 1

Substance	p.p.m.[a]	mg./M³ [b]
Carbon dioxide	5,000	9,000
Carbon monoxide	50	55
Chlordane—Skin		0.5
Chlorinated camphene—Skin		0.5
Chlorinated diphenyl oxide		0.5
*Chlorine	1	3
Chlorine dioxide	0.1	0.3
C Chlorine trifluoride	0.1	0.4
C Chloroacetaldehyde	1	3
a-Chloroacetophenone (phenacylchloride)	0.05	0.3
Chlorobenzene (monochlorobenzene)	75	350
o-Chlorobenzylidene malononitrile (OCBM)	0.05	0.4
Chlorobromomethane	200	1,050
2-Chloro-1, 3-butadiene, see Chloroprene		
Chlorodiphenyl (42 percent Chlorine)—Skin		1
Chlorodiphenyl (54 percent Chlorine)—Skin		0.5
1-Chloro,2,3-epoxypropane, see Epichlorhydrin		
2-Chloroethanol, see Ethylene chlorohydrin		
Chloroethylene, see Vinyl chloride		
C Chloroform (trichloromethane)	50	240
1-Chloro-1-nitropropane	20	100
Chloropicrin	0.1	0.7
Chloroprene (2-chloro-1, 3-butadiene)—Skin	25	90
Chromium, sol. chromic, chromous salts as Cr		0.5
Metal and insol. salts		1
Coal tar pitch volatiles (benzene soluble fraction) anthracene, BaP, phenanthrene, acridine, chrysene, pyrene		0.2
Cobalt, metal fume and dust		0.1
Copper fume		0.1
Dusts and Mists		1
Cotton dust (raw)		1
Crag® herbicide		15
Cresol (all isomers)—Skin	5	22
Crotonaldehyde	2	6
Cumene—Skin	50	245
Cyanide (as CN)—Skin		5
Cyclohexane	300	1,050
Cyclohexanol	50	200
Cyclohexanone	50	200
Cyclohexene	300	1,015
Cyclopentadiene	75	200
2, 4-D		10
DDT—Skin		1
DDVP, see Dichlorvos		
Decaborane—Skin	0.05	0.3
Demeton®—Skin		0.1
Diacetone alcohol (4-hydroxy-4-methyl-2-pentanone)	50	240
1,2-diaminoethane, see Ethylenediamine		
Diazomethane	0.2	0.4
Diborane	0.1	0.1
Dibutylphthalate		5
C o-Dichlorobenzene	50	300
p-Dichlorobenzene	75	450

Substance	p.p.m.[a]	mg./M³ [b]
Dichlorodifluoromethane	1,000	4,950
1,3-Dichloro-5,5-dimethyl hydantoin		0.2
1,1-Dichloroethane	100	400
1,2-Dichloroethylene	200	790
C Dichloroethyl ether —Skin	15	90
Dichloromethane, see Methylenechloride		
Dichloromonofluoromethane	1,000	4,200
C 1,1-Dichloro-1-nitroethane	10	60
1,2-Dichloropropane, see Propylenedichloride		
Dichlorotetrafluoroethane	1,000	7,000
Dichlorvos (DDVP)—Skin		1
Dieldrin—Skin		0.25
Diethylamine	25	75
Diethylamino ethanol—Skin	10	50
Diethylether, see Ethyl ether		
Difluorodibromomethane	100	860
C Diglycidyl ether (DGE)	0.5	2.8
Dihydroxybenzene, see Hydroquinone		
Diisobutyl ketone	50	290
Diisopropylamine—Skin	5	20
Dimethoxymethane, see Methylal		
Dimethyl acetamide—Skin	10	35
Dimethylamine	10	18
Dimethylaminobenzene, see Xylidene		
Dimethylaniline (N-dimethylaniline)—Skin	5	25
Dimethylbenzene, see Xylene		
Dimethyl 1,2-dibromo-2, 2-dichloroethyl phosphate, (Dibrom)		3
Dimethylformamide—Skin	10	30
2,6-Dimethylheptanone, see Diisobutyl ketone		
1,1-Dimethylhydrazine—Skin	0.5	1
Dimethylphthalate		5
Dimethylsulfate—Skin	1	5
Dinitrobenzene (all isomers)—Skin		1
Dinitro-o-cresol—Skin		0.2
Dinitrotoluene—Skin		1.5
Dioxane (Diethylene dioxide)—Skin	100	360
Diphenyl	0.2	1
Diphenylmethane diisocyanate (see Methylene bisphenyl isocyanate (MDI)		
Dipropylene glycol methyl ether—Skin	100	600
Di-sec, octyl phthalate (Di-2-ethylhexylphthalate)		5
Endrin—Skin		0.1
Epichlorhydrin—Skin	5	19
EPN—Skin		0.5
1,2-Epoxypropane, see Propyleneoxide		
2,3-Epoxy-1-propanol, see Glycidol		
Ethanethiol, see Ethylmercaptan		
Ethanolamine	3	6
2-Ethoxyethanol—Skin	200	740

Substance	p.p.m.[a]	mg./M³ [b]
2-Ethoxyethylacetate (Cellosolve acetate)—Skin	100	540
Ethyl acetate	400	1,400
Ethyl acrylate—Skin	25	100
Ethyl alcohol (ethanol)	1,000	1,900
Ethylamine	10	18
Ethyl sec-amyl ketone (5-methyl-3-heptanone)	25	130
Ethyl benzene	100	485
Ethyl bromide	200	890
Ethyl butyl ketone (3-Heptanone)	50	230
Ethyl chloride	1,000	2,600
Ethyl ether	400	1,200
Ethyl formate	100	300
C Ethylmercaptan	10	25
Ethyl silicate	100	850
Ethylene chlorohydrin—Skin	5	16
Ethylenediamine	10	25
Ethylene dibromide, see 1,2-Dibromoethane		
Ethylene dichloride, see 1,2-Dichloroethane		
C Ethylene glycol dinitrate and/or Nitroglycerin—Skin	[d]0.2	1
Ethylene glycol monomethyl ether acetate, see Methyl cellosolve acetate		
Ethylene imine—Skin	0.5	1
Ethylene oxide	50	90
Ethylidene chloride, see 1,1-Dichloroethane		
N-Ethylmorpholine—Skin	20	94
Ferbam		15
Ferrovanadium dust		1
Fluoride (as F)		2.5
Fluorine	0.1	0.2
Fluorotrichloromethane	1,000	5,600
Formic acid	5	9
Furfural—Skin	5	20
Furfuryl alcohol	50	200
Glycidol (2,3-Epoxy-1-propanol)	50	150
Glycol monoethyl ether, see 2-Ethoxyethanol		
Guthion®, see Azinphosmethyl		
Hafnium		0.5
Heptachlor—Skin		0.5
Heptane (n-heptane)	500	2,000
Hexachloroethane—Skin	1	10
Hexachloronaphthalene—Skin		0.2
Hexane (n-hexane)	500	1,800
2-Hexanone	100	410
Hexone (Methyl isobutyl ketone)	100	410
sec-Hexyl acetate	50	300
Hydrazine—Skin	1	1.3
Hydrogen bromide	3	10
C Hydrogen chloride	5	7
Hydrogen cyanide—Skin	10	11
Hydrogen peroxide (90%)	1	1.4
Hydrogen selenide	0.05	0.2
Hydroquinone		2
C Iodine	0.1	1
Iron oxide fume		10
Isoamyl acetate	100	525
Isoamyl alcohol	100	360
Isobutyl acetate	150	700
Isobutyl alcohol	100	300

39

Substance	p.p.m.[a]	mg./M³ [b]
Isophorone	25	140
Isopropyl acetate	250	950
Isopropyl alcohol	400	980
Isopropylamine	5	12
Isopropylether	500	2,100
Isopropyl glycidyl ether (IGE)	50	240
Ketene	0.5	0.9
Lead arsenate		0.15
Lindane—Skin		0.5
Lithium hydride		0.025
L.P.G. (liquified petroleum gas)	1,000	1,800
Magnesium oxide fume		15
Malathion—Skin		15
Maleic anhydride	0.25	1
C Manganese		5
Mesityl oxide	25	100
Methanethiol, see Methyl mercaptan		
Methoxychlor		15
2-Methoxyethanol, see Methyl cellosolve		
Methyl acetate	200	610
Methyl acetylene (propyne)	1,000	1,650
Methyl acetylene-propadiene mixture (MAPP)	1,000	1,800
Methyl acrylate—Skin	10	35
Methylal (dimethoxymethane)	1,000	3,100
Methyl alcohol (methanol)	200	260
Methylamine	10	12
Methyl amyl alcohol, see Methyl isobutyl carbinol		
Methyl (n-amyl) ketone (2-Heptanone)	100	465
C Methyl bromide—Skin	20	80
Methyl butyl ketone, see 2-Hexanone		
Methyl cellosolve—Skin	25	80
Methyl cellosolve acetate—Skin	25	120
Methyl chloroform	350	1,900
Methylcyclohexane	500	2,000
Methylcyclohexanol	100	470
o-Methylcyclohexanone—Skin	100	460
Methyl ethyl ketone (MEK), see 2-Butanone		
Methyl formate	100	250
Methyl iodide—Skin	5	28
Methyl isobutyl carbinol—Skin	25	100
Methyl isobutyl ketone, see Hexone		
Methyl isocyanate—Skin	0.02	0.05
C Methyl mercaptan	10	20
Methyl methacrylate	100	410
Methyl propyl ketone, see 2-Pentanone		
C a Methyl styrene	100	480
C Methylene bisphenyl isocyanate (MDI)	0.02	0.2
Molybdenum:		
Soluble compounds		5
Insoluble compounds		15
Monomethyl aniline—Skin	2	9
C Monomethyl hydrazine—Skin	0.2	0.35
Morpholine—Skin	20	70

Substance	p.p.m.[a]	mg./M³ [b]
Naphtha (coaltar)	100	400
Naphthalene	10	50
Nickel carbonyl	0.001	0.007
Nickel, metal and soluble cmpds, as Ni		1
Nicotine—Skin		0.5
Nitric acid	2	5
Nitric oxide	25	30
p-Nitroaniline—Skin	1	6
Nitrobenzene—Skin	1	5
p-Nitrochlorobenzene—Skin		1
Nitroethane	100	310
Nitrogen dioxide	5	9
Nitrogen trifluoride	10	29
Nitroglycerin—Skin	0.2	2
Nitromethane	100	250
1-Nitropropane	25	90
2-Nitropropane	25	90
Nitrotoluene—Skin	5	30
Nitrotrichloromethane, see Chloropicrin		
Octachloronaphthalene—Skin		0.1
*Octane	500	2,350
*Oil mist, mineral		5[e]
Osmium tetroxide		0.002
Oxalic acid		1
Oxygen difluoride	0.05	0.1
Ozone	0.1	0.2
Paraquat—Skin		0.5
Parathion—Skin		0.11
Pentaborane	0.005	0.01
Pentachloronaphthalene—Skin		0.5
Pentachlorophenol—Skin		0.5
*Pentane	1,000	2,950
2-Pentanone	200	700
Perchloromethyl mercaptan	0.1	0.8
Perchloryl fluoride	3	13.5
Petroleum distillates (naphtha)	500	2,000
Phenol—Skin	5	19
p-Phenylene diamine—Skin		0.1
Phenyl ether (vapor)	1	7
Phenyl ether-biphenyl mixture (vapor)	1	7
Phenylethylene, see Styrene		
Phenylglycidyl ether (PGE)	10	60
Phenylhydrazine—Skin	5	22
Phosdrin (Mevinphos®)—Skin		0.1
Phosgene (carbonyl chloride)	0.1	0.4
Phosphine	0.3	0.4
Phosphoric acid		1
Phosphorus (yellow)		0.1
Phosphorus pentachloride		1
Phosphorus pentasulfide		1
Phosphorus trichloride	0.5	3
Phthalic anhydride	2	12
Picric acid—Skin		0.1
Pival® (2-Pivalyl-1,3-indandione)		0.1
Platinum (Soluble Salts) as Pt		0.002
Propargyl alcohol—Skin	1	
Propane	1,000	1,800
n-Propyl acetate	200	840
Propyl alcohol	200	500
n-Propyl nitrate	25	110

Substance	p.p.m.[a]	mg./M³ [b]
Propylene dichloride	75	350
Propylene imine—Skin	2	5
Propylene oxide	100	240
Propyne, see Methyl-		
acetylene		
Pyrethrum		5
Pyridine	5	15
Quinone	0.1	0.4
RDX—Skin		1.5
Rhodium, Metal fume		
and dusts, as Rh		0.1
Soluble salts		0.001
Ronnel		10
Rotenone (commercial)		5
Selenium compounds		
(as Se)		0.2
Selenium hexafluoride	0.05	0.4
Silver, metal and soluble		
compounds		0.01
Sodium fluoroacetate		
(1080)—Skin		0.05
Sodium hydroxide		2
Stibine	0.1	0.5
*Stoddard solvent	500	2,950
Strychnine		0.15
Sulfur dioxide	5	13
Sulfur hexafluoride	1,000	6,000
Sulfuric acid		1
Sulfur monochloride	1	6
Sulfur pentafluoride	0.025	0.25
Sulfuryl fluoride	5	20
Systox, see Demeton®		
2,4,5T		10
Tantalum		5
TEDP—Skin		0.2
Tellurium		0.1
Tellurium hexafluoride	0.02	0.2
TEPP—Skin		0.05
C Terphenyls	1	9
1,1,1,2-Tetrachloro-2,		
2-difluoroethane	500	4,170
1,1,2,2-Tetrachloro-1,		
2-difluoroethane	500	4,170
1,1,2,2-Tetrachloro-		
ethane—Skin	5	35
Tetrachloroethylene,		
see Perchloroethylene		
Tetrachloromethane, see		
Carbon tetrachloride		
Tetrachloronaphthalene—		
Skin		2
Tetraethyl lead (as Pb)—		
Skin		0.075
Tetrahydrofuran	200	590
Tetramethyl lead (as		
Pb)—Skin		0.07
Tetramethyl succinonitrile—		
Skin	0.5	3

Substance	p.p.m.[a]	mg./M³ [b]
Tetranitromethane	1	8
Tetryl (2,4,6,-trinitrophenyl-		
methylnitramine)—Skin		1.5
Thallium (soluble com-		
pounds)		0.1
Thiram		5
Tin (inorganic cmpds,		
except oxides)		2
Tin (organic cmpds)		0.1
C Toluene-2,4-diisocyanate	0.02	0.14
o-Toluidine—Skin	5	22
Toxaphene, see Chlorinated		
camphene		
Tributyl phosphate		5
1,1,1-Trichloroethane, see		
Methyl chloroform		
1,1,2-Trichloroethane—		
Skin	10	45
Titaniumdioxide		15
Trichloromethane, see		
Chloroform		
Trichloronaphthalene—		
Skin		5
1,2,3-Trichloropropane	50	300
1,1,2-Trichloro 1,2,2-		
trifluoroethane	1,000	7,600
Triethylamine	25	100
Trifluoromonobromo-		
methane	1,000	6,100
2,4,6-Trinitrophenol,		
see Picric acid		
2,4,6-Trinitrophenyl-		
methylnitramine, see		
Tetryl		
Trinitrotoluene—Skin		1.5
Triorthocresyl phosphate		0.1
Triphenyl phosphate		3
Turpentine	100	560
Uranium (soluble		
compounds)		0.05
Uranium (insoluble		
compounds)		0.25
C Vanadium:		
V₂O₅ dust		0.5
V₂O₅ fume		0.1
Vinyl benzene, see		
Styrene		
**C Vinyl chloride	500	1,300
Vinylcyanide, see		
Acrylonitrile		
Vinyl toluene	100	480
Warfarin		0.1
Xylene (xylol)	100	435
Xylidine—Skin	5	25
Yttrium		1
Zinc chloride fume		1
Zinc oxide fume		5
Zirconium compounds		
(as Zr)		5

*1970 Addition.

**1972 Addition.

[a]Parts of vapor or gas per million parts of contaminated air by volume at 25°C. and 760 mm. Hg pressure.

[b]Approximate milligrams of particulate per cubic meter of air.

(No footnote "c" is used to avoid confusion with ceiling value notations.)

[d]An atmospheric concentration of not more than 0.002 p.p.m., or personal protection may be necessary to avoid headache.

[e]As sampled by method that does not collect vapor.

C: Ceiling concentration.

TABLE 2

Material	8-hour time weighted average	Acceptable ceiling concentration	Acceptable maximum peak above the acceptable ceiling concentration for an 8-hour shift.	
			Concentration	Maximum duration
Benzene (Z37.4-1969)	10 p.p.m.	25 p.p.m.	50 p.p.m.	10 minutes.
Beryllium and beryllium compounds (Z37.29-1970)	2 μg./M^3	5 μg./M^3	25 μg./M^3	30 minutes.
Cadmium fume (Z37.5-1970)	0.1 mg./M^3	3 mg./M^3		
Cadmium dust (Z37.5-1970)	0.2 mg./M^3	0.6 mg./M^3		
Carbon disulfide (Z37.3-1968)	20 p.p.m.	30 p.p.m.	100 p.p.m.	30 minutes.
Carbon tetrachloride (Z37.17-1967)	10 p.p.m.	25 p.p.m.	200 p.p.m.	5 minutes in any 4 hours.
Ethylene dibromide (Z37.31-1970)	20 p.p.m.	30 p.p.m.	50 p.p.m.	5 minutes.
Ethylene dichloride (Z37.21-1969)	50 p.p.m.	100 p.p.m.	200 p.p.m.	5 minutes in any 3 hours.
Formaldehyde (Z37.16-1967)	3 p.p.m.	5 p.p.m.	10 p.p.m.	30 minutes.
Hydrogen fluoride (Z37.28-1969)	3 p.p.m.			
Fluoride as dust (Z37.28-1969)	2.5 mg./M^3			
Lead and its inorganic compounds (Z37.11-1969)	0.2 mg./M^3			
Methyl chloride (Z37.18-1969)	100 p.p.m.	200 p.p.m.	300 p.p.m.	5 minutes in any 3 hours.
Methylene chloride (Z37.3-1969)	500 p.p.m.	1,000 p.p.m.	2,000 p.p.m.	5 minutes in any 2 hours.
Organo (alkyl) mercury (Z37.30-1969)	0.01 mg./M^3	0.04 mg./M^3		
Styrene (Z37.15-1969)	100 p.p.m.	200 p.p.m.	600 p.p.m.	5 minutes in any 3 hours.
Trichloroethylene (Z37.19-1967)	100 p.p.m.	200 p.p.m.	300 p.p.m.	5 minutes in any 2 hours.
Tetrachloroethylene (Z37.22-1967)	100 p.p.m.	200 p.p.m.	300 p.p.m.	5 minutes in any 3 hours.
Toluene (Z37.12-1967)	200 p.p.m.	300 p.p.m.	500 p.p.m.	10 minutes.
Hydrogen sulfide (Z37.2-1966)		20 p.p.m.	50 p.p.m.	10 minutes once only if no other measurable exposure occurs.
Mercury (Z37.8-1971)		1 mg./10M^3		
Chromic acid and chromates (Z37.7-1971)		1 mg./10M^3		

TABLE 3

Mineral Dusts

Substance	$Mppcf^e$	Mg/M^3
Silica:		
Crystalline:		
Quartz (respirable)	250^f	$10mg/M^{3\,m}$
Quartz (total dust)	$\dfrac{\%SiO_2+5}{}$	$\dfrac{\%SiO_2+2}{30mg/M^3}$
		$\dfrac{}{\%S_2O_2+2}$
Cristobalite: Use 1/2 the value calculated from the count or mass formulae for quartz		
Tridymite: Use 1/2 the value calculated from the formulae for quartz.		
Amorphous, including natural diatomaceous earth	20	$80mg/M^3$
		$\dfrac{}{\%SiO_2}$
Silicates (less than 1% crystalline silica):		
Mica	20	
Soapstone	20	
Talc (non-asbestos-form)	20^n	
Talc (fibrous). Use asbestos limit		
Tremolite (see talc, fibrous)		
Portland cement	50	
Graphite (natural)	15	
Coal dust (respirable fraction less than 5% SiO_2)		$2.4mg/M^3$ or
For more than 5% SiO_2		$10mg/M^3$
		$\dfrac{}{\%SiO_2+2}$
Inert or Nuisance Dust:		
Respirable fraction	15	$5mg/M^3$
Total dust	50	$15mg/M^3$

Note: Conversion factors:

mppcf $\times 35.3$ = million particles per cubic meter
= particles per c.c.

[e]Millions of particles per cubic foot of air, based on impinger samples counted by light-field technics.

[f]The percentage of crystalline silica in the formula is the amount determined from air-borne samples, except in those instances in which other methods have been shown to be applicable.

[m]Both concentration and percent quartz for the application of this limit are to be determined from the fraction passing a size-selector with the following characteristics:

Aerodynamic diameter (unit density sphere)	Percent passing selector
2	90
2.5	75
3.5	50
5.0	25
10	0

[n]Containing < 1% quartz; if > 1% quartz, use quartz limit.

and "subacute" are only comparative; it is perfectly possible to get chronic effects from acute exposures and vice versa. For this reason it is desirable to use definitive descriptions of dosages, time periods, and effects whenever possible.

LC_{50}: Statistically derived figure representing the concentration in air of a material that would be expected to kill 50 percent of the test animals at a specified concentration or upon exposure for a specified time.

LD_{50}: Statistically derived figure representing the amount of ingested or absorbed material expected to kill 50 percent of the test animals.

T.L.V.: "Threshold limit value" a term developed by the American Conference of Governmental Industrial Hygienists. Basically, it is a time-weighted average atmospheric concentration that is not expected to cause injury to workers exposed up to eight hours per day for their working lifetimes. T.L.V.'s are guides by definition and do not represent a fine line between safe and unsafe exposure levels. It should be noted that there is no common denominator to these values—some are based on toxicity, others on disagreeable odor or irritation.

T.W.A.: "Time-weighted average concentration" (atmospheric) is a figure the industrial hygienist reaches when he or she calculates the results of a job study. Mathematically:

$$\text{T.W.A.} = \frac{C_1 T_1 + C_2 T_2 + \cdots + C_x T_x}{480}$$

where C = concentration of material found in sample taken over a given time interval,

T = time interval in minutes.

The sum of the CT products in the numerator represents the total job exposure. When this sum is divided by the minutes (480) in an eight-hour day, the T.W.A. is obtained.

The following example illustrates the procedure utilized in determining the T.W.A. for a particular environment in which trichloroethylene, a chemical degreasing solvent, is being utilized. Let us assume that a worker spends the day as follows:

6.0 hours at degreaser (average exposure concentration: 90 parts per million-ppm).

0.5 hour cleaning degreaser (average exposure concentration: 150 ppm).

1.0 hour sorting parts (average exposure concentration: 40 ppm).

0.5 hour lunchroom (average exposure concentration: < 5 ppm).

In this example:

$$\text{T.W.A.} = \frac{90 \times 360 + 150 \times 30 + 40 \times 60 + 5 \times 30}{480} = 82 \text{ ppm.}$$

For this material, OSHA standards specify a maximum T.W.A. exposure of 100 ppm, with an acceptable ceiling concentration of 200 ppm and maximum excursion peaks of up to 300 ppm for a duration of no more than five minutes in any two-hour period. The results obtained in the above example represent an acceptable exposure, since they meet the OSHA criteria. If the T.W.A. had exceeded 100 ppm, the employer would have been in violation of OSHA.

On May 8, 1975, OSHA and NIOSH published proposed standards for six toxic ketones and included in the proposal the concept of "action levels." This is the first model toxic substances standard, and 400 more will be developed in the next few years.

According to the proposal, the action level will be one-half the T.L.V. At or above the action level, the employer will be responsible for initiating employee exposure measurements, employee training, and medical surveillance. If the result of the measurement indicates that the employee is exposed to concentrations in excess of the action level, all other employees similarly exposed must also have their exposures monitored. If the concentrations are above the action level but below the T.L.V., measurements of the employee's exposure must be made every two months. If the concentrations are above the T.L.V., the employer must measure the exposure monthly, inform the employee of his or her overexposure, and institute control measures.

When two consecutive measurements at least one week apart show the employee is no longer being exposed above the action level, the sampling program may be ended for that person.

EVALUATING HAZARDS

Responsibility for Conducting Evaluations

In large organizations, the responsibility for evaluating the hazards associated with air contaminants falls primarily on two groups of personnel: toxicologists and industrial hygienists.

The role of the toxicologist is to develop basic toxicological information on every chemical or chemical composition of significant interest to his or her organization. The extent to which these properties are determined will depend upon the importance of the material and upon the probability of and degree of exposure likely to be encountered either by workers, by customers, or by the public.

Additionally, the toxicologist is obligated to keep up with current developments in the field through literature as well as through contacts with other professionals. This aspect of his or her duties may include the evaluation of new information and new techniques. When appropriate, the toxicologist should publish the results of his or her own toxicological research. Other duties include maintaining a close liaison with his or her organization's medical and safety departments to serve their needs properly for toxicological information, to distribute to the personnel in the organization the results of his or her work, and upon occasion to provide expert consultation to customers.

The industrial hygienist is charged with developing techniques appropriate for the sampling, analysis, and control of environmental contamination, including the evaluation and adaptation of commercially available equipment to his or her organization's needs. When such information is requested, he or she must develop data that describe the environmental conditions in any plant.

Like the toxicologist, the industrial hygienist must keep up with developments in the field, publish when appropriate the results of his or her own work, and maintain the proper liaison with other groups, such as medical personnel, to serve their needs properly. Occasionally, the industrial hygienist may serve as a customer consultant.

Toxicological Evaluations

Sources of Toxicological Information

The toxicologist can turn to several sources for information when a question arises about the toxicity and hazard of a material in a certain process or usage. When the material has been made for an extended period in another plant of the company, plant experience may well have provided valuable insights concerning the levels of material that can be tolerated. Thus, experience in one location can be transferred to another location.

The toxicologist can consult published literature as a source of data. Such literature includes publications of federal agencies—for example, the National Institute of Occupational Safety and Health—as well as of state agencies. The data may describe someone else's experience with the material. However, if the material is a newly developed chemical or formulation, obviously no one has had any experience in handling it and obviously nothing will have been published.

A third source is the supplier, who may provide the toxicologist with what data he or she has. This can guide engineers in designing or controlling the plant or process. A more likely source of information, however, is based upon animal studies. Using various types of animals, the toxicologist tries to simulate possible exposures in the plant by exposing animals to different concentrations and observing the effects upon their health. Based upon the severity of injury to the animals, he or she can draw conclusions as to the likelihood of injury to workers. In these studies, the toxicologist can expose animals once to determine the effects of acute exposure or can give them repeated exposures and get some idea of the chronic toxicity. The effects resulting from acute exposure may be considerably different from those resulting from repeated exposure.

Unfortunately, there is no assurance that the rat or monkey, or any animal species, will respond in the same way as a human, and therefore any results obtained must be interpreted with caution. However, it appears that the similarities are far greater than the dissimilarities, and it is possible to get a good approximation of what the likely effect will be on humans.

Inhalation Toxicity

Two types of inhaled materials are of concern in toxicology:

1. *Vapors or gases*, which behave as single molecules, are of extremely small size, and tend to obey the gas laws; and
2. *Particles* (dust, mists, fumes), which may be considerably bigger than individual molecules. There is a fuzzy area of overlap in size, since many of the extremely fine dusts are so small that they almost attain molecular size and behave more like gases than particles.

With both gases and particles, three things must be known before the toxicologist can speak with any intelligence about an inhalation exposure:

1. The concentration inhaled; *how much?*
2. The duration of exposure; *how long?*
3. The frequency with which the exposure occurs; *how often?*

If any one of these three factors is missing, the toxicologist has only a partial picture of the total problem. The effects of concentration and duration on the seriousness of exposure are obvious; but it is not so obvious that inhalation may consist of single exposures or repeated exposures, and the seriousness of the response may be quite different depending on the exposure frequency.

We often think of single exposures as accidental, or certainly occurring sporadically. For example, the person working at home is apt to have a single exposure to a vapor or dust since he or she is not likely to work with a material for more than a few hours at a time, and probably has several days between exposures. The home gardener may get an exposure while dusting crops. Transporation of chemicals may mean a single exposure; the ruptured railroad car or broken bottle in a carton may result in a single exposure of the people in the area of the accident.

In an industrial establishment, we think of single exposures when an employee handles a chemical for a short period of time, or is exposed when mishandling or an accident occurs. This is not totally correct, however, for the worker can receive repeated exposures.

Repeated exposures often accompany industrial processing. Workers generally stay with the process for months or even years. In some cases, a lifetime may be spent working in the production plant with exposure to the same materials. Every plant, of course, has its own potential repeated exposures. For example, a degreasing operation for cleaning metal parts may be done repeatedly. A dry cleaning operation may result in daily repeated exposures. Plastic fabricating and sawing are other examples. Professional pest control operators who apply weed killers or insect sprays may be receiving repeated exposures, as do painters and other tradespeople. These examples should not be considered exhaustive; they merely represent the many types of single and repeated vapor exposures that can result from usage of a material.

One other major problem that influences toxicological studies is the probable end-use of a material. When a material is used within a single organization, the problem is relatively simple because the company is generally well aware of its own capabilities and is ordinarily reasonably well qualified to handle the material. When a product is sold to another company, however, its personnel may or may not have the know-how and capability of handling it. Furthermore, the seller has lost control and can only advise, not enforce, safe practices.

If the material becomes a consumer item and ends up in the household, the problems are even greater. "If a material *can* be misused, it *will* be misused" is a basic tenet of all toxicological studies involving household items. Furthermore, a household may contain all stages of age and health, ranging from the newborn to the old and infirm. We are dealing with men, women, and children who have various chronic illnesses and disabilities. In the vast majority of cases, these persons have had absolutely no experience in handling hazardous materials, and they are likely to assume that any material they purchase is absolutely safe or it would not be sold. Needless to say, these considerations have a profound effect on the studies undertaken.

Range-Finding Studies. In the early stages of its development, only small quantities of a material are utilized and few workers are involved. Work on the material is often carried out in a chemical hood, and the exposures, therefore, are relatively minor. But what about accidental exposure from a broken or spilled container? What should be done in case of such an accident? Is it necessary to evacuate the building? Do workers have to take special precautions in cleaning it up, or is the material such that it can be cleaned up without fear of inhaling too much vapor? To answer these questions, animals are exposed to essentially saturated vapors. This is done by blowing air through a sample and passing the saturated air over a group of five rats. The maximum exposure period is seven hours.

If nothing happens when animals are exposed to a saturated atmosphere for seven hours, it can probably be concluded that there is little acute hazard to the workers. If there is an effect from the seven-hour exposure, animals are exposed for shorter periods of time—four, two, one hour or less, down to six minutes. If adverse effects occur from a six-minute exposure to a saturated vapor, there is a definite acute hazard. Thus, even though the material may not have a high toxicity, if it has adequate toxicity and enough vapor pressure, hazardous concentrations can occur. Such a material may be of concern to a laboratory worker who must plan on several minutes to clean up a potential spill.

If ill effects occur between the seven-hour exposure and the six-minute exposure, interpretation is required. This is where the art of the toxicologist is needed to interpret the results. To make absolutely certain that he or she is exposing animals severely enough to rule out the likelihood of acute hazard, the toxicologist may heat the test sample to some predetermined temperature—around $100°C$—and expose animals to these supersaturated vapors. Such an exposure can actually occur in a production plant when a worker puts his head into an open manhole of a heated kettle or when a large amount of heated material is exposed to the atmosphere. If animals survive seven hours of exposure to the supersaturated atmosphere, the toxicologist can feel rather secure in the belief that there is little or no acute hazard.

Extended Studies On Single Exposures. When the next stage of development of a material is reached and specific uses emerge, it is necessary to have quantitative data indicating what will happen from various durations of exposure to various concentrations. The actual procedure for making exposure concentrations is relatively complicated. In essence, however, it consists of metering in a chemical at a known rate and diluting it with a known rate of air flow. Of course, all concentrations must be verified by analytical techniques, which involve the use of such devices as infrared, gas chromatographic, and combustion conductivity analyzers.

Repeated Exposures. Repeated exposure tests represent yet another type of inhalation study. In these experiments, the toxicologist selects large groups of animals, exposes them several hours a day, five days a week for months, and determines the effects upon the health of the animals. The results of these experiments give the maximum concentration that the animals can inhale without adverse effect upon their health. This no-effect concentration becomes the basis for selecting guides for controlling the atmosphere in the plant.

Particle Inhalation. Particles can be generated by many methods; for example, sawing, shaking, sieving, grinding, or breaking of materials can produce fine particles or dusts. Air bubbling through a liquid will often give off mists as a result of the breaking of the bubbles. These fine mists may evaporate to form solid particles or they may stay as discrete droplets in the air.

When speaking of particle inhalation, one must consider two rather broad, ill-defined categories. The first is composed of those materials for which the respiratory tract merely serves as a portal of entry, and the toxic effects are not different from those that would be attained by ingestion. The second category consists of those materials that produce their toxic effects in the respiratory tract.

For materials in the first category, the size of the particle is not as important as the total quantity of material inhaled. This is because all the dust in the air that enters the nose may be trapped, swallowed, and eventually end up in the bloodstream. The larger particles will be trapped by the nose and upper parts of the respiratory tract; the smaller particles will reach the lungs. Since the larger particles make up the greatest share of the weight of the material inhaled, they will be most influential in determining the toxicity. A reasonably good approximation of the toxic hazard can be obtained by determining the total quantity of material suspended in a volume of air.

Acid mists, silica, and beryllium are examples of the second type of materials, which have their effect in the respiratory tract and for which particle size is of extreme importance. Particles larger than 10 microns tend to fall out of the air rather rapidly and do not stay suspended. Particles larger than one to two microns will enter the respiratory tract but may be trapped in the nose and trachea; hence only those particles of less than one or two microns are likely to enter the lungs. Acid mists commonly cause injury in the nose and upper respiratory tract. The particles of interest in silicosis are in the range of a few microns. These obviously are not visible to the naked eye, and thus, general observation is not adequate to assure that no hazard exists.

Effects of Interest to the Toxicologist. What sort of effects does the toxicologist look for as the result of exposure to a certain material? The most severe effect, of course, is death. However, obviously death is not the end-point desired when dealing with human workers; rather the toxicologist wants to determine the maximum exposure conditions that permit total survival. In addition, he or she may want to find out what exposure conditions are likely to cause no organic injury, no blood changes, no irritation, no lacrymation, no anesthesia, and no other manifestations of toxicity. The organs observed most carefully in inhalation studies are the central nervous system, respiratory tract, the liver, and the kidneys. Observations also include other major organs of test animals. These organs can either be looked at grossly or with the microscope to detect more subtle changes in the condition of the animal organs. For such examination, the pathologist is needed to tell whether a certain exposure condition produced adverse effects in exposed animals.

Another criteria that might influence the level the toxicologist recommends as safe for humans is comfort. Some materials produce so much pain that no one will willfully tolerate injurious concentrations. It is fairly safe to say, for example, that a chemical such as chloropicrin (tear gas) will not be willfully tolerated by a worker if the concentrations are injurious. Chloropicrin, therefore, has what we call an "adequate warning property" to prevent injurious exposure. Notice that the term "willful" is involved. If a person is trapped and has to stay, he or she can become injured from chloropicrin, as well as from many other materials considered to have warning properties.

It is not good to depend too much on warning properties to prevent excessive exposures. Warning properties are not dependable; the sense of smell and pain can fatigue, and hence thresholds of preception can change. Some materials may have adequate warnings to prevent injury from a single exposure but not from repeated exposures.

Eye-Contact Toxicity

Eye-contact toxicity can occur with gases, liquids, or solids. However, the effect of gases upon the eyes is usually obtained during inhalation studies; therefore, the following section will deal with eye exposure from liquids or solids.

Toxicological literature repeatedly asserts the principle that possible eye effects can be evaluated only by eye-contact tests. To evaluate eye effects from skin contact tests would be misleading or even disastrous.

In industrial operations, eye exposure is most likely to result from spilling or splashing of either liquids or solids. The usual eye exposure is likely to be a single exposure, and routine eye tests will likely consist of a single application.

Since eye exposure can result in serious vision impairment, or even blindness, one of the chief purposes of any eye-contact toxicity study is to establish the best procedure for decontamination should eye exposure occur. In many eye tests, therefore, the effect of flushing the contaminated eye with water is studied.

Eye-contact tests are increasingly being required by law at the local, state, and federal levels. Federal laws are primarily concerned with consumer products, pesticides, drugs and cosmetics, food additives, and industrial chemicals. Hence, essentially all products are covered by one law or another requiring or recommending eye testing.

Range-Finding Studies. Range-finding studies seek to determine the type as well as the severity of adverse effects. Experience has shown that for the most part the rabbit eye serves as a reasonable model for evaluating effects on the human eye. To simulate possible human exposure, liquids are placed between the eyelid and the eyeball. Solids are usually ground to a powder and a small amount is introduced into the eye. One eye is observed for adverse effects while another eye (usually on the same animal) is flushed with flowing tap water to evaluate the benefit of flushing the eye.

Extended Studies. Further studies usually are designed to evaluate the same type of effects. However, they generally involve more animals per test, and sometimes other animals, such as the dog or monkey.

Effects of Interest to the Toxicologist. To evaluate fully the adverse effects of chemicals upon the eye, the toxicologist studies the following:

1. **The effect on the eyelids.** Because the rabbit has a large nictitating membrane (a thin membrane that can extend across

the eyeball), which is easily examined, both it and the eyelids are looked at for such things as irritation, edema, and necrosis.

2. **Corneal effects.** The surface of the eyeball is examined for adverse effects such as haziness, pitting, edema, and necrosis.

3. **Internal effects.** These effects can often be observed without mechanical aids, but there are times when a slit lamp or other tools are utilized. The effects in question involve the anterior chamber, the iris, the lens, and the posterior chamber, including the retina.

4. **Systemic effects.** The general well-being of the test animal is observed to help evaluate the possibility of toxic amounts of material being absorbed through the eye.

5. **Reaction to Pain.** During the actual application the reaction of the animal is observed to get some idea of the degree of pain that may occur.

Skin-Contact Toxicity

Disability from skin-contact toxicity accounts for a large number of chemical exposure problems in industry. The problems can arise from a single contact, from gross, frequently repeated exposures, or from any intermediate degree. Hence, it becomes necessary in many instances to study both single and repeated exposures.

Several factors must be taken into account in any test program. They include the effects on unbroken skin and on scratched, abraded, or wounded skin, the absorption of the material through the skin, and the potential of the material to cause allergenic reactions or skin cancer. To complicate the picture even further, exposure can involve uncovered skin or covered skin, as well as dry skin or skin wet with perspiration, and these parameters must also be considered in the tests.

Skin exposure can occur with solids, liquids, gases, mists, and aerosols. Usually the effects of gases, mists, or aerosols are observed during inhalation studies. However, evaluating the effects of solids or liquids can best be carried out in tests especially designed to simulate possible skin exposure. Special tests other than inhalation studies may be needed to evaluate the effect of mists or aerosols on the skin.

Skin-contact toxicity has caused as much legislative concern as eye-contact toxicity. Thus, almost every chemical product manufactured is covered by one or more local, state, or federal laws recommending or requiring skin testing.

The type of study required for any material is greatly

dependent upon its stage of development as well as its possible uses. It is seldom that any one material will require every type of study described below.

Range-Finding Studies. The object of this study is to obtain information suitable for evaluating some of the problems that may occur when the product is handled. Such tests may use only a single exposure or several repeated exposures. They may include application to the ear of the test animal, which simulates uncovered exposure; to shaved and healed belly skin under a cloth bandage, which simulates situations where the material is confined to the skin; to an abraded and covered area of the skin; and—if the material is a solid—to dry skin as well as skin that has been wet with water to simulate perspiration. Usually such exposures are repeated over several days or even a week or two. In addition, the behavior and condition of the test animal is observed to judge if the material is being absorbed through the skin in toxic amounts. The usual animal for these tests is the rabbit; however, other species may also be tested.

Patch Test. If a material causes a severe skin reaction, it is highly desirable to establish the time required to bring about the start of the effect. This is usually accomplished by applying the material under a pad of cotton for predetermined lengths of time. This type of testing is also used to define the skin corrosivity of a product as required by certain laws.

Skin Absorption. Often it is not enough to show that a material may be absorbed through the skin in toxic amounts; some idea of the quantity of material capable of causing the observed effect is also required. Such information is usually obtained by applying to the skin surface known doses of the material, usually in grams or milliliters per kilogram of animal body weight. By utilizing suitable numbers of animals, the range of toxicity or the estimated level of test material likely to cause various effects (usually death, survival, or LD_{50}—see definitions at beginning of this chapter—as well as systemic effects) may be determined. Unless body tissues are actually analyzed for the material or its metabolites, all that can be evaluated is the material's ability to penetrate the skin in sufficient amounts to cause toxic effects. Thus when discussing the toxicity by absorption, the statement is usually made that a material is or is not absorbed through the skin in toxic amounts.

The test program may involve single or repeated exposures, depending upon the proposed use of the material and its stage of development.

Skin Hypersensitivity. Depending again upon the proposed use or uses of a specific material, it may be necessary to evaluate its potential to sensitize skin under use conditions. Numerous techniques and types of animals may be used for such tests. Most programs use guinea pigs in the initial screening procedure. The material is applied one or more times to the clipped skin or is injected. After a rest period of one to three weeks, during which time no additional material is applied or injected, a final or challenge application is made at the original, or at another, site. Based upon the response to the final application, a judgment can be made as to the sensitization potential. A strong guinea pig response is usually considered predictive, but a negative response cannot always be assumed to indicate that the material will not sensitize human skin. In fact, further testing upon humans may be desirable, and even required.

Skin Cancer. Cancer in any form is of grave concern, and skin cancer is known to be produced by repeated skin contact with certain chemicals. This aspect of skin exposure, therefore, has been the subject of much thought, discussion, and experimentation. The problems involved in testing are many and serious. They include such factors as mode of application, animal of choice, sex of animal, significant number of animals to use, and the use of positive and negative controls. Finally, there is the very real problem of evaluating the results in terms of humans.

The method usually employed involves making several applications per week of the test material to large numbers of animals, generally mice. The application, called skin painting, is continued for the lifetime of the animal. The toxicologist looks for number and type of tumors, the time it takes them to develop, and their permanence.

Effects of Interest to the Toxicologist. Range-Finding Test. As this is the first test to determine the effects of skin contact, the toxicologist is on the alert for many things. First, he or she is interested in the type of effect on the skin itself. Specifically he or she wishes to learn if the material causes irritation, edema, exfoliation, or necrosis, and—if there is necrosis—whether scabs and scars form. Second, the toxicologist seeks to discover how quickly these effects occur. In addition, he or she looks for differences between the reactions at the various sites. The effects are compared on abraded and intact skin, covered and uncovered skin. Finally, the toxicologist sees if there are any special reactions, such as hyperplasia, chloracne-type responses, and signs of toxicity due to absorption through the skin.

Patch Test. The main purpose of this test is to establish the length of time that will cause beginning irritation or the beginning evidence of a burn. Hence, the nature of the ultimate reaction is secondary.

Skin Absorption. This test is designed to determine the amount of the product that causes toxic signs due to absorption; the skin reaction of the animal is of secondary importance. The main object is to establish the dose-response relationship so that an evaluation can be made of the health hazard to humans.

Skin Allergy and Skin Cancer. Because of the specialized nature of the information sought through these tests, and the difficulty of interpreting the results, it would be difficult to outline all the observations made. The particular observations in any individual case are those of significance for the evaluation desired. Somewhat specialized medical personnel are needed.

Ingestion Toxicity

The problem of ingesting chemicals is not widespread in industry—most workers have the common sense not to swallow materials they handle. Occasionally, however, a material must be handled which is so toxic that accidental swallowing of hazardous amounts may occur. Therefore, ingestion toxicity studies are usually carried out when the health hazard of a product is evaluated.

Ingestion toxicity tests are also valuable because the data obtained provide a convenient means of determining the relative toxicities of a number of materials. The information sought is usually obtained by comparing the LD_{50} values by ingestion. The LD_{50} value, an estimate of the dose response, is one of the more precise toxicological measurements that can be made.

Ingestion hazard may result from a single or acute exposure as well as from repeated or even continuous exposures. Thus, the hazard from ingestion is usually qualified by stating that it is either a single-dose or acute oral toxicity or chronic oral toxicity. To be meaningful, any discussion of chronic oral toxicity should state the length of the test period; for instance, 90-day feeding tests, two-year feeding tests, or lifetime studies.

Although chronic ingestion of materials in industrial operations is a fairly unlikely occurrence, there is often a need to know the difference in toxicity between a single exposure and repeated exposures. Many times the repeated exposure studies will provide a feel for the mechanisms of detoxication, excretion, and metabolism of a material that is most helpful in the overall evaluation of the health hazard from using a compound. In addition, such chronic studies pinpoint the "target" organ(s) or tissue(s) that may be adversely affected.

Acute and chronic ingestion tests are also required by many state and federal laws. The latter include such laws as the Federal Hazardous Substances Act, the Food and Drug Act, the Rodenticide, Fungicide, and Insecticide Act, and the Occupational Safety and Health Act, as well as numerous regulations governing the transportation of chemicals.

Chronic ingestion tests are required, of course, where chemicals are used as intentional or incidental food additives. They are also required for drugs, cosmetics, and other materials that may be administered to humans or economic animals. There is also increasing concern as to the chronic toxicity of materials that get into the environment, and eventually into human food and drinking water.

To evaluate the potential chronic hazard to health from handling and using a material, the toxicologist must establish how much is too much. Thus, as with single-dose oral toxicity, the goal of chronic testing is to establish a dose-response relationship, using the best criteria of adverse or pathological body changes. The changes must be distinguished from ones that occur as a result of stresses arising from the test situation, but that are not adverse or pathologic in nature.

Range-Finding or Single-Dose Oral-Toxicity Tests. These tests are designed to provide a basis for evaluating the hazard from an accidental or willful single ingestion. Depending upon the needs, one or more species of warm-blooded animals are used. The number of animals is influenced by the desired preciseness of the dose-response relationship; the more precise the figure, the more test animals should be used and the smaller the difference between doses should be. The dose levels usually chosen differ logarithmically, not arithmetically, as this permits better use of statistics in estimating an LD_{50} value.

At least two, but more frequently as many as five or more dosage levels are chosen. Hopefully one of the doses will be without an adverse effect, and another dose will have a detectable adverse effect. In estimating an LD_{50} value, it is desirable to feed one dose that will cause the death of all the test animals.

Doses normally are administered by stomach tube or hypodermic syringe to small laboratory animals; stomach tubes or feeding of capsulated material may be used with larger animals, such as the dog, sheep, or cow.

Chronic or Repeated Oral-Ingestion Toxicity Tests. Although the goal of these tests—establishing a dose-response relationship—is the same as for the single-dose oral toxicity tests, the testing techniques are different. Except for special studies, the material under test is administered to the animals by mixing it into their food or water. The material can be fed in any one of a number of dosage rates; milligrams/kilogram of body weight/day, parts per million, or percentage in diet.

Factors that must be taken into account in setting up any test program include the duration of the test; the type, number, and sex of the test animals used; the types of criteria to be studied; the control material (if any) to be used; analysis of materials fed; and perhaps other special factors.

The duration of the study can range from 14 days to the entire life-span of the animals. The type of animals employed depends upon the results desired and may include anything from laboratory animals such as rats, rabbits, and dogs, to large animals such as sheep, hogs, and cows. The number may range from several per dose for the large animals, to as many as 20 or more of the small laboratory animals for each dosage level. Usually both male and female animals are tested, and from three to six dosage levels evaluated. A negative control group is always run, and on some occasions a positive control group as well.

The effects usually studied include growth rate, behavior, appearance, gross and microscopic pathology of tissues, and changes in the blood and urine, as shown by clinical and chemical studies. Other effects may be studied as required. Because beginning effects are of prime interest in the evaluation, the validity of any observed fact is often best assessed by statistical evaluation of the data. This necessitates selecting and using the individual animals so that the best use of the statistical analysis of the results can be made.

Effects of Interest to the Toxicologist. Range-Finding Oral Studies. In addition to establishing dose-response relationships, it is helpful to obtain as much information as possible with respect to adverse effects. The toxicologist looks for toxic signs and, if there are any, clearly records the time of onset, the organs affected, and the clinical or chemical effects. If death does not occur but illness does, the time it takes for the apparent recovery of the test animal is recorded.

Chronic Oral Studies. Again, although the main goal is to establish a dose-response relationship, other information is gathered and reported as suggested in the discussion of factors involved in setting up the study.

Special Toxicological Studies

The tests so far described are the most helpful in providing a basis for evaluating the health hazards in the handling and use of chemicals and materials. But they do not answer all the questions that may be of concern. More and more, government groups are suggesting such studies as teratology, mutagenicity, effects on reproduction, and metabolism studies. These areas have always been of concern, but more recently, with the greatly increased emphasis on the environmental effects of chemicals and materials, the urgency for answers has become greater. At the federal level, there is increasing pressure to include one or more of these special studies in health-hazard evaluations.

Although cancer study has been a part of most chronic-type studies, special tests are being suggested or even required to evaluate more fully the ability of a chemical or material to cause cancer.

Conclusions

Obviously, the studies designed to provide suitable data for evaluating the health hazards encountered when chemicals and other materials are handled and used are highly specialized and require a number of technical skills. Involved are chemists to do the analysis and the clinical chemical studies; medical specialists to conduct the clinical studies; animal pathologists to evaluate gross and microscopic pathology; pharmacologists to study metabolism, excretion, and detoxication; specially trained animal handlers and caretakers; and—finally—a team to correlate the findings with those that may be expected in human beings or economic animals. It is equally obvious that once the toxicological team has provided its information, special skills (including analytical and engineering skills) may be needed for an industrial hygiene evaluation of the working environment. A high degree of inquisitiveness and initiative is also helpful in ferreting out the facts with regard to a material, and in putting them together to establish a workable, economically feasible, and acceptable solution to industrial hygiene problems.

Monitoring Air Contaminants

Air monitoring must be carried out to determine whether or not concentrations of air contaminants are below acceptable levels, as defined by OSHA or established by toxicological investigations. Such monitoring is the responsibility of the industrial hygienist, who uses a variety of air sampling and analysis equipment to obtain the needed data.

Factors Involved in Monitoring

Monitoring may be carried out for a number of reasons. It may be utilized in emergency situations; for example, when a worker has been overcome. Unfortunately, monitoring under such circumstances provides only an estimate of the conditions that existed at the exact time the exposure occurred. Monitoring is frequently used to check areas in which emergencies are likely to occur. There monitoring confirms or refutes the existence of a hazard and accurately defines hazard if found. Monitoring is also used to check employee complaints and to determine whether atmospheric concentrations of contaminants meet standards.

To monitor an individual's exposure it is necessary to know what plant locations he or she occupies and how the work day is distributed among them. This kind of information would be expected to apply to any other individuals who hold the same job classification. One source is the worker's supervisor; another is the worker. Both should have a fair idea how the worker spends his or her time, but they may not agree; and both may be surprised at the amount of time that cannot easily be accounted for. The industrial hygienist probably will need to make some personal observations to resolve any differences and to be convinced that he or she has a good idea of the worker's work pattern. At the same time the hygienist may be able to detect undesirable or unnecessary chemical contacts or personal habits that may contribute to exposure.

Where, when, and how much to sample will depend to an extent upon the questions to be answered. The intent may be to survey an entire plant for documentation or to examine a single job because of suspected conditions or presence of symptoms. The purpose also may be to confirm the effectiveness of a control measure. The amount of sampling sufficient to answer a specific question will vary considerably.

The level desired to control a contaminant determines the sensitivity needed from an analytical method. Interfering substances that are likely to be present will also influence the choice. If peak concentrations are important, a direct sampling device is in order, and if average values are more pertinent, sampling may be carried out over an extended period with subsequent analysis. Continuous monitoring may be indicated for a reliable picture of changing conditions. Allowing the worker to wear a sampling device may be the best way to measure his or her exposure in some cases.

The industrial hygienist will need some means of handling the occasional operation—the one that is not performed every day. How long does it take? What is the peak concentration? Or what does it average over the time period? Many operations or tasks (either occasional or repetitive) involve exposure that is related more to what the worker is doing than to where he or she is. To describe the exposure, air in the breathing zone must be sampled while the operation is being performed. If there is more than one health-important agent in the environment, it will be necessary to sample for each and to consider possible synergistic effects.

Care is needed to see that operating conditions are representative when the sampling is done. If the operating rate, or material variety, or weather conditions will likely alter the exposure, then sufficient samples must be taken to describe all the conditions. The process or work pattern or both may change on the night shift, if there is one. In that case, night samples might be needed. Rotating shifts, of course, will smooth out changes in the work pattern.

In many process layouts there will be spots where contaminant levels are high but where operating personnel do not go. Samples from such spots are not pertinent to a study designed to describe exposure, neither are samples that represent the worker's breathing zone in locations or during operations for which he or she normally wears respiratory protective equipment. Despite the levels of contamination, such conditions do not contribute to the exposure.

There may also be some conditions which are not obvious but which contribute to the worker's exposure—time spent in transit from one work site to another, chats with fellow workers, time spent outside the work area (offices, for instance), coffee breaks, lunchtime. Such things could add up to a considerable portion of the day and may be supported by air samples.

Types of Samples

There are two major types of samples, the *short-term*, or *grab*, sample and the *long-term* sample. The grab sample is one taken over a short period of time, generally less than five minutes. Because of the short sampling time, the atmospheric concentration is assumed to be constant during the entire period. To account for changes in concentration, a series of grab samples is usually taken. Long-term samples are taken over a long enough time period so that any variations in concentration average out. Ordinarily, however, this type of sample does not allow the industrial hygienist to detect any peak exposure concentrations that might occur. Because of the drawbacks of each type of sample, neither by itself is sufficient. To obtain representative information, a combination of both must be used.

Sampling Equipment

A wide variety of portable air sampling and analysis equipment is available to the industrial hygienist. Some devices are used not only to determine the concentration of air contaminants, but also to detect the presence of combustible atmospheres. and to ascertain sufficient or insufficient oxygen.

Components of air analyzers often include: (1) an air mover, (2) some means for determining the volume of air that has been sampled, (3) a sample collector, and (4) a sample analyzer. Sometimes only the collection portions of the system are portable, with the actual analysis carried out in the laboratory. The available battery of portable air sampling equipment includes devices that simply move air, those that collect a sample, and some that sample and analyze on the spot. Some samplers give readings directly in terms of concentrations; those that provide a measure of the quantity of a contaminant must also determine the volume of air sampled so that a concentration may be calculated. Because of the great variety of air contaminants that can produce undesirable effects, there is no single air sampler or sampling system to handle all industrial hygiene problems.

The major types of air samplers and analyzers available to industrial hygienists are described below. Devices for detecting flammable ranges of combustible gases and concentrations of oxygen are also included in the list.

Air Movers. This category includes hand-operated pumps, battery-operated pumps, and electric-circuit-driven devices.

Squeeze bulbs and glass valve syringes are examples of hand-operated pumps. They are used primarily to move air to the collecting medium or detector. They can be used only when precise flow rate or volume is not critical. Hand-operated piston pumps provide an accurate measurement of air flow and volume because of their precision machined cylinders and pistons. Such pumps are ordinarily operated by a hand crank or similar device, and nonfluctuating suction is maintained regardless of variations in the rate of cranking.

Battery-operated pumps usually operate off one or more small batteries, such as Class C flashlight batteries. These devices provide reasonably constant sampling rates and are adaptable for use with air-sampling pipettes, bubblers, and filter papers. Generally, such a pump cannot be used with high-resistance collectors.

Diaphragm pumps, vacuum rotary pumps, centrifugal fans, and aspirators are among the air-moving devices that run off electric circuits. These devices deliver large volumes of air at constant rates and are effective for moving air through sources of high resistance. The aspirator-type devices utilize a vapor jet of a fluoridated hydrocarbon or other material, which passes through a small nozzle at high velocity and produces the vacuum needed to draw air samples through a collection system.

Common problems causing errors with pumps are wear, temperature variations, pressure differences, plugged orifices, and leaking valves. Figures 3.1 and 3.2 show a syringe-type and a battery-operated air pump sampling unit and analyzer.

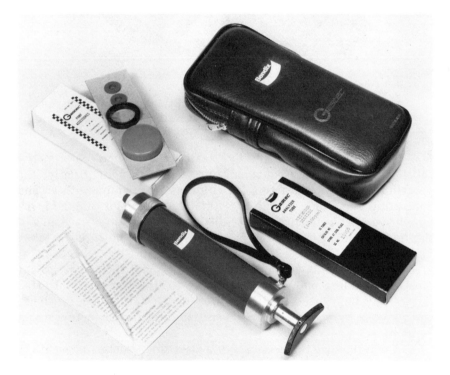

FIGURE 3.1   Syringe type air-sampling unit. (Photo courtesy Bendix Corporation Environmental Science Division.)

Direct Method Portable Air Analyzers. These analyzers can be divided into two main categories: those with which the measurement is based on degree of color change and those which involve physical instrumentation.

FIGURE 3.2   Battery-operated pump type air-
sampling unit. (Photo courtesy Bendix
Corporation Environmental Science Division.)

Liquid reagents, chemically treated paper, and colorimetric
tubes have all been used to measure color changes. Briefly, the
procedure involves contacting the colorimetric medium with air and
comparing the color obtained with a known color standard. To
obtain reasonably quantitative results, the volume or surface area of
the medium, the air-flow rate, and the flow time must all be taken
into account.

Liquid reagents are somewhat inconvenient, since they usually
must be prepared and mixed just prior to use. Chemically treated
papers may be utilized in either a wet or dry state. Chemical chalks
and crayons have been used to sensitize ordinary paper for certain
contaminants.

A colorimetric tube consists of a glass tube, usually one-eighth to
one-fourth inch in diameter and three to five inches long, filled with
a supporting material such as silica gel, alumina, ground glass,
pumice, or resin. This material is impregnated with an indicating
chemical that contrasts with the unexposed color. Upon reaction
with the test gas, a color may be produced throughout the length of
the tube, or a stain may appear at the air inlet end of the tube. Stain
length is proportional to contaminant concentration in the air.

Detector tubes are fairly simple to use. The two sealed ends are
broken open and the tube placed in the manufacturer's holder, which
is fitted with a calibrated squeeze bulb or pump. The recommended

air is drawn through the tube. Even if a squeeze bulb is fully expanded it may still be under partial vacuum and may not have drawn its full volume of air, so adequate time must be allowed for each stroke. The observer evaluates the concentration in the air by comparing the stained portion of the exposed tube with the calibration curve or chart. The manufacturer's sampling instructions must be followed.

Although detector tubes provide a rapid reading, their accuracy, reproducibility, and dependability as detection devices are still questionable, and they are therefore not acceptable for determining exposures in compliance with OSHA standards. The best degree of accuracy that can be expected from even the better-performing tubes is on the order of ±25 percent. Interferences are probably the greatest source of error in making readings. In some cases, a second contaminant may bleach out the color developed by the contaminant being tested for, thus negating the results. In other cases, a second contaminant may reinforce the reaction of the one being tested for, resulting in an abnormally high reading. Persons using detection tubes definitely should be aware of the performance limitations and should have facilities for checking each batch of tubes prior to use.

With analyzers involving physical instrumentation, measurement of the air contaminant is carried out in the instrument and the results indicated by means of a dial or other device. For example, one such device utilizes an exposure chamber with an ultraviolet source at one end and two phototubes at the other end of a sampling chamber. Vapors drawn into the sampling chamber absorb the ultraviolet light. The absorption is proportional to the concentration of the vapor in the air and is indicated with a microammeter.

Indirect Method Portable Air Samplers. Portable air samplers are simple devices to collect and retain the contaminated air. Once the sample has been obtained, it is necessary to analyze it for the collected contaminants by some analytical method such as gas chromatography, infrared spectroscopy, or wet chemical techniques. These sampling methods are particularly useful when absolute identification of the contaminant is necessary and quantitative determinations of several components in air-vapor mixtures are desired.

Air samplers consist of the air mover, collecting apparatus, and, in some cases, a collecting agent. Sampling may be by air displacement, freeze-out, or adsorption. Air displacement is the simplest of all the methods. The contaminated air is simply collected in a plastic bag or glass container designed to insure against any leakage or reaction of the contaminant with the collecting apparatus.

The freeze-out procedure depends on drawing the contaminated air through condensers or traps that have been cooled below the boiling point of the contaminants in order to collect the vapor or gas as a liquid or solid. The collector is immersed in a container of dry ice or liquid nitrogen and the air contaminant drawn through at a measured rate. This procedure is fairly complex and is generally used only after all other air sampling techniques have failed.

Probably the most commonly used solid adsorbents are silica gel and charcoal. Both will adsorb a number of different organic compounds efficiently. The contaminant is usually stripped from the adsorbent by leaching with a suitable solvent such as alcohol or carbon disulfide. The apparatus used to hold the gel or charcoal can be a simple glass tube, U-tube, absorber, or plastic bag.

Portable Air Samplers for Dusts and Mists. Unfortunately, there are no simple methods, techniques, or instruments for the efficient analysis of dust, fumes, and mists. For that matter, there are few complex or sophisticated methods available. Of those devices on the market, most are based on electrostatic precipitation or filtration of the material being detected.

A typical electrostatic precipitator utilizes a grounded metal cylinder as the collecting electrode. A straight wire electrode is located concentrically within the tube. A high voltage is conducted through this wire electrode, causing a corona discharge through the wire and collecting cylinder. Particles drawn through the cylinder become highly charged and are attracted to, and collected on, the oppositely charged collecting cylinder. The sample can be weighed or analyzed by chemical or physical means. Figure 3.3 illustrates such a precipitator.

Dust filters consist essentially of a filter holder, filter, and air mover. The filter may be of glass fiber, paper, or membrane. The selection of a particular filter is a compromise among such factors as cost, collection efficiency, and reliability. Open filter holders are used for routine air sampling and closed holders for sampling from ducts or other systems.

Combustible Gas Indicators. These instruments are used to detect combustible gas or vapor-air mixtures with the object of preventing fires and explosives. For the most part, flammable ranges and explosive limits are 10 to 1,000 times the threshold limit value for a flammable material. Therefore, most industrial hygiene air analyzers may be too sensitive to determine explosive concentrations and, conversely, combustible gas indicators may not be sensitive enough to determine concentrations in the health hazard range.

FIGURE 3.3   Electrostatic
precipitator.
(Photo courtesy Mine Safety
Appliances Company.)

The circuitry in a combustible gas indicator commonly consists of two platinum wires in a Wheatstone bridge circuit. The air sample is drawn through the instrument by squeezing an aspirator bulb. Combustible gas in the air sample is oxidized or burned catalytically on one of the filaments, and the corresponding increase in temperature increases the electrical resistance of the wire. This increased resistance is proportional to the amount of flammable gas or vapor in the sample and is measured by a meter calibrated to give direct readings in percentage of lower explosive limits. Because of the calibration scheme employed, the meters indicate relative hazard rather than concentration.

The meter indication can be interpreted literally only for mixtures of air and the particular gas or vapor that was used to calibrate the instrument. Since any flammable vapor will cause a response, and since it is not necessary to analyze the air, this is not a drawback for evaluating the hazard from flammables. Any positive response may be regarded as indicating the presence of some flammable gas or vapor in the air, but it will neither identify nor quantitate the material. Readings tend to be on the "safe" side, not only because the catalytic combustion allows determination of mixtures that would otherwise be too lean to burn, but also because combustible gas indicators are usually calibrated on hexane (other materials may be specified for special purposes).

Since the response is proportional to the concentration of flammable material in air, the sample rate is not critical. Specificity is poor, since any flammable gas or vapor present in the environment can interfere. Accuracy of readings can be assured only by calibrating the instrument using known concentrations of the specific material of interest; the calibration will change with time as the filaments age. A combustible gas indicator is shown in Figure 3.4.

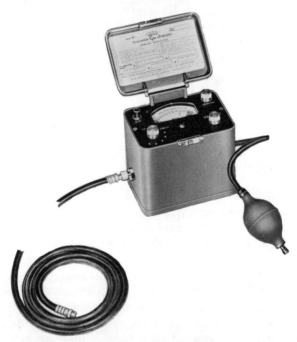

FIGURE 3.4   Combustible gas indicator. (Photo
courtesy Mine Safety Appliances
Company.)

Combustible gas indicators can be made to function as continuous sampler analyzers by replacing the rubber squeeze bulb with a small air pump. They are rugged and reliable if checked before use with standard gas samplers (supplied in convenient form in flammability range by several manufacturers). Malfunctions are signalled by loss of the ability to make zero adjustment.

Portable Oxygen Analyzers. There are two major reasons for oxygen analyses: one to determine whether the atmosphere inside a vessel or closed space is normal with respect to oxygen content before allowing people to enter; the other to confirm the fact that oxygen has been sufficiently excluded to prevent formation of a flammable mixture when a flammable liquid is to be added to a system. A variety of available portable instruments measure the concentration of oxygen in air. The measuring instruments are of three general types: (1) those using a liquid that scavenges oxygen from a sample and displays the decrease in gas volume with a rising column of fluid; (2) those that utilize an electro-chemical cell whose output is proportional to the oxygen concentration in an air sample which diffuses into the cell; and (3) detector tubes that form a stain when gas passed through them contains oxygen. The length of the stain is proportional to the oxygen concentration.

These instruments are typically supplied with a range from 0 to 25 percent oxygen, which is ideal for both applications. The cell varieties can be made to monitor continuously by attaching a small air pump; the tube indicators average the oxygen concentration at a single point over a period of about one minute. Acidic or oxidizing gases when present in more than trace amounts can interfere with the determination by any of these methods. A single reading will require about one minute to complete; response time increases with age of the electrolyte in the cell-type analyzers. Periodic maintenance is therefore necessary; the electrolyte solution should be changed about every two months, and the electrodes require periodic cleaning and replacement. The penalty for extending the maintenance period is to lengthen response time. These instruments can be calibrated by setting the zero on nitrogen (for example) and setting the response to 20.8 percent on air. Readings near 20.8 percent (as for vessel entry) are accurate using the recommended zero setting instead of testing with nitrogen; but for readings below 15 percent (as for confirming exclusion of oxygen) an inert zero gas must be used. The device shown in Figure 3.5 typifies portable oxygen analyzers.

Portable Air Analyzers for Leak Detection. In searching for sources of leaks, leak detectors simply provide a "yes" or "no" answer and cannot be used for quantitative determination of vapor concentrations in air. The mode of operation depends upon the nature of the material being detected. One commercially available device for detecting noncombustible halogenated hydrocarbons consists essentially of a propane gas cylinder, a mixer head, a sampling hose, and a detector head. When the device is lighted at the nozzle, the venturi effect produced by the flow of gas through the

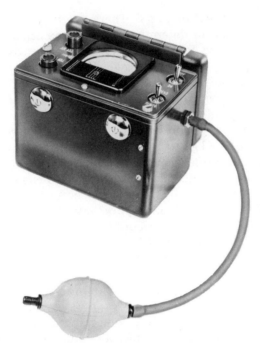

FIGURE 3.5   Portable oxygen analyzer.
(Photo courtesy Mine Safety
Appliances Company.)

mixer head of the detector creates a vacuum in the sampling hose. Any halogenated hydrocarbon drawn through the hose comes in contact with a copper reaction plate heated cherry red and is decomposed to form a copper halide. The copper halide produces a color change in the flame. The difference in color is proportional to the vapor concentration of the material in the air.

A second detector, adaptable to a wide range of gases and vapors, utilizes a sampling pump that draws the air into a probe assembly and through four thermal conductivity cells represented as fixed resistances, which operate at a constant temperature when the detector is balanced. Most gases and vapors have a thermal conductivity value either greater or less than air at the filament operating temperature. The introduction of trace gases with negative thermal conductivities will cause an increase in the filament temperature. As the temperature rises, the increased electrical resistance unbalances the bridge, first in one direction, then another. The unbalance produces both an audible and visual signal. Gases with positive thermal conductivities cool the filament. This decreases the resistance, which again unbalances the bridge.

Sonic detectors can detect leaks of any type of gas or vapor, but only in pressure or vacuum systems. These devices convert inaudible sounds in the 35,000 to 40,000 cycles per second range to a signal that is audible to a user wearing a set of headphones.

Continuous Air Monitoring

Automatic air sampling and analysis provide a relatively new and useful aid for monitoring the concentration of an air contaminant in the work environment. A continuous air monitor may be defined as a stationary or semistationary analyzer, which together with one or more recording devices and a sampling unit automatically samples, measures, and records the concentration of an air-borne contaminant. The analyzer can work on any one of several principles, two of the most commonly used being electroconductivity and infrared spectroscopy.

Continuous monitoring is the best method for measuring environmental conditions to produce a truly quantitative study of worker exposures to airborne chemicals. The most important advantages of a continuous monitoring system are:

1. Accuracy in measurement and validity in conclusions.
2. Documentation of exposures in an efficient manner.
3. Signaled need for selective improvement when concentrations at certain points exceed acceptable levels.
4. Reduced exposures when used properly.
5. Reduced loss of process material when used properly.
6. Reduced fire hazards in those applications where highly flammable or explosive air contaminants can accumulate.

Properly used, continuous air monitors sample work areas that are frequented by workers in particular job classifications. The monitors may have up to 12 sampling probes. The number of probes in a multipoint sampling scheme is reduced by one for infrared applications, since a reference air stream (base line) must be supplied by one of the sample lines. In practice, several sampling lines are strung outward from a centrally located analyzer, and air is drawn through these lines back to the analyzer by means of a gas pump. An automatic sequential sampling system allows a continuous measurement of the contaminant throughout the work area. This information is recorded in one of several ways at the centrally located analyzer. In the case of conductivity, concentrations can be registered directly if the recorder chart is set properly and if the instrument is calibrated properly. In the case of infrared analysis, the concentration is not shown directly but an indirect indication (transmittance) is visually available from the chart attached to the recorder.

The essence of a multipoint continuous system used to monitor an operational classification is the reliability of the time-weighted exposure, which is the factor of most importance to the industrial hygienist. This best estimate of exposure is directly comparable to the industrial hygiene standard for judgment of satisfactory control. If the time-weighted mean exposure exceeds the standard by a substantial margin, corrective action is in order, either generally or at specific trouble spots.

Figure 3.6 shows the schematic diagram of a typical system, in this case an infrared monitoring system, together with suitable instrumentation for computer analysis of the recorded data.

The great volume of data generated by a continuous monitoring device can be reduced to manageable proportions by a digitizer. This computer device translates analog air analysis data from a recorder to punched paper tape, where it is recorded in digital form. This, in turn, allows the data to be processed by the computer to yield a much more complete description of inhalation exposures than is otherwise possible.

Since digital computers are much more versatile than any conductivity, infrared, or other kind of monitoring instrument, they can be adapted to the general situation and can be used "across the board" no matter what the specific application. The computer program can be used to summarize the measured concentrations and time-weighted exposures for certain units of time, such as an eight-hour shift and daily, weekly, and monthly periods. For each summary, the computer will calculate the mean concentration at each sampling location, the standard deviation of that data, and the percentage of time that the concentration exceeds certain prechosen levels. By using job descriptions, the computer extracts the time-weighted mean exposures for the job classifications under consideration.

COMBATING HAZARDS

OSHA stipulates that in controlling occupational diseases caused by contaminated air, the primary objective is to prevent atmospheric contamination. This is accomplished by the use of accepted engineering control methods such as ventilation or enclosure and confinement of the operation. When effective engineering controls are not feasible or are being installed, appropriate respirators must be furnished by the employer.

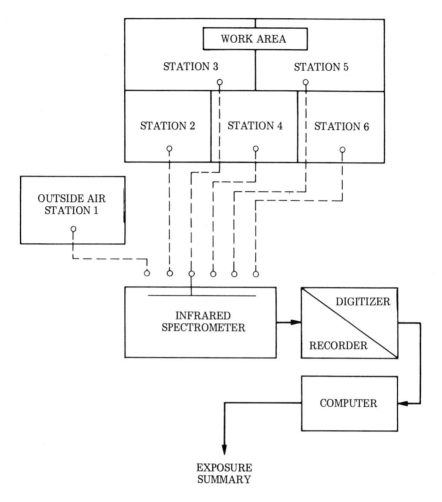

$$\text{Time-weighted mean exposure} = 0.01 \sum_{i=2}^{6} \overline{C}_1 P_1$$

where $\overline{C}_1$ = mean concentration at station 1

P_1 = percent of time spent at location 1 during normal work activity

FIGURE 3.6. Schematic diagram of the infrared monitoring system.

Respiratory Protective Devices

These may be divided into three basic categories: air supplied devices, self-contained breathing apparatus, and air-purifying devices.

Air-Supplied Devices

They include air-supplied suits, air-supplied hoods, and airline masks. The air is supplied to the devices from an outside source and through a hose. Air-supplied suits may be necessary when exposure to substances may result in absorption through the skin, and they are often used in rescue or emergency repair work. Special suit-cooling devices have been helpful in extending the wearing time of the suits.

Air-supplied hoods are used in situations similar to those in which air-supplied suits are employed. Ordinarily, although not always, the hoods are worn in conjunction with complete body protection; for example, rubber suits, rubber boots, and gloves. The hood should be tied off at the bottom and supplied with a minimum of six to seven cubic feet of air per minute. It should be used within 25 feet of a source of good air, so that should the air supply fail, the worker can reach the source by holding his or her breath. Figure 3.7 shows an air-supplied hood.

FIGURE 3.7   General purpose air-supplied hood.
(Photo courtesy 3M Company.)

Airline masks, which are used when full protective clothing is not needed, may be full-face or half-face. There are three classes of airline masks: continuous flow-type, demand-type, and pressure demand-type. It is important that the face piece of such a mask fit tightly. Sometimes a blower is utilized to furnish the air to the mask, in which case the blower must be in an area where clean, fresh air is assured. Figures 3.8 and 3.9 illustrate a continuous flow and a demand-type airline mask respectively.

FIGURE 3.8   Continuous flow airline mask. (Photo courtesy Mine Safety Appliances Company.)

Self-Contained Breathing Apparatus

These may be used in atmospheres that are (1) low in oxygen, (2) of unknown composition, (3) known to be contaminated above levels suitably controlled with canister-type gas masks, and (4) contaminated with substances having few or no warning properties. There are three basic types of self-contained breathing apparatus: the cylinder, the recirculating, and the regenerating types.

The cylinder apparatus obtains its supply of air from a tank carried by the wearer. It generally operates with a regulator valve

FIGURE 3.9   Demand type airline
mask. (Photo courtesy Mine
Safety Appliances Company.)

(demand flow) to conserve the supply of air. If operated continuously (full flow), it has a very limited duration of effectiveness.
These devices have an alarm that warns the wearer when the air
supply is nearly exhausted. Figure 3.10 shows a cylinder device.

The recirculating apparatus passes exhaled air (which still
contains about two-thirds of the original oxygen) through an
absorber that removes carbon dioxide, and then into a breathing bag.
In the bag, the air is mixed with fresh oxygen from the tank and is
passed back to the face piece for rebreathing.

The regenerating apparatus has a canister containing material
that releases oxygen chemically when carbon dioxide and water are
absorbed from the exhaled air. Such units are quite bulky and must
be put on in a clean, uncontaminated atmosphere. Figure 3.11 shows
a regenerating device.

FIGURE 3.10   Cylinder type self-
contained breathing apparatus.
(Photo courtesy Mine Safety
Appliances Company.)

Air-Purifying Devices

These must be used only in atmospheres where: (1) there is adequate oxygen, (2) the contaminant possesses adequate warning properties, (3) the concentration of contaminant will not overwhelm the capacity of the respirator canister, cartridge, or filter, and (4) the canister is known to be effective for the contaminant or contaminants. Air-purifying devices generally consist of two types: the canister and cartridge-type and filters for particulate matter and dust.

A canister respirator consists essentially of a full facepiece with a canister attached directly to it or carried in a harness worn on the body of the user (see Figure 3.12). Chemical cartridge respirators employ either a half-mask facepiece that covers the nose and mouth or a full facepiece. The cartridge respirator is most useful for very short periods against contaminant concentrations that are too high to

FIGURE 3.11   Regenerating type self-contained
breathing apparatus. (Photo courtesy Mine
Safety Appliances Company.)

be breathed without protection but quite low compared with those
that can be handled by canister respirators. The contaminants against
which a canister mask is effective are indicated by labeling and color
coding the canister. Table 4 shows the color code for various
contaminants.

Particulate-removing respirators usually are equipped with
half-mask facepieces that cover the nose and mouth, although some
have full facepieces. Some dust or fume cartridges may be used on
the same facepiece as a chemical cartridge. The removal of the
particulate material is usually by mechanical screening and impingement. The Atomic Energy Commission has developed masks
equipped with ultrafilters that offer better protection against finer
particles than do other types of masks. These masks, which are
recommended for situations where radioactive particles may be
encountered, offer more resistance to air flow than do masks that

FIGURE 3.12   Canister respirator.
(Photo courtesy Mine Safety
Appliances Company.)

bear the U.S. Bureau of Mines approval. Between the toxic dust respirator and the AEC's respirator is a metal fume respirator, which is especially designed to handle the small particles that result when metals are heated.

There are a number of so-called nuisance dust respirators on the market. They are advertised for use against "nuisance, nontoxic dust" and "nontoxic paint sprays." However, these respirators generally do not fit well, are inefficient for removal of particulates of respirable size and, therefore, should not be used.

Table 5 can serve as a guideline in selecting the respirator device best suited for a particular situation.

Fitting of Devices

Federal regulations require that respirator wearers must receive fitting instructions, including demonstrations and practice in how the respirator should be worn and how to adjust it for proper fit and to assure a positive seal. A good face seal is an absolute necessity for safe use. Such conditions as a growth of beard, sideburns, or temple pieces on glasses can seriously affect the seal.

TABLE 4
Color Code for Gas-Mask Canisters[1]

Atmospheric Contaminants to be Protected Against	Colors Assigned
Acid	White
Hydrocyanic acid gas	White with 1/2-inch green stripe completely around the canister near the bottom
Chlorine gas	White with 1/2-inch yellow stripe completely around the canister near the bottom
Organic vapors	Black
Ammonia gas	Green
Acid gases and ammonia gas	Green with 1/2-inch white stripe completely around the canister near the bottom
Carbon monoxide	Blue
Acid gases and organic vapors	Yellow
Hydrocyanic acid gas and chloropicrin vapor	Yellow with 1/2-inch blue stripe completely around the canister near the bottom
Acid gases, organic vapors, and ammonia gases	Brown
Radioactive materials, excepting tritium and noble gases	Purple (Magenta)
Particulates (dusts, fumes, mists, fogs, or smokes) in combination with any of the above gases or vapors	Canister color for contaminant, as designated above, with 1/2-inch gray stripe completely around the canister near the top
All of the above atmospheric contaminants	Red with 1/2-inch gray stripe completely around the canister near the top

[1] Complies with OSHA standards (see *Federal Register*, May 29, 1971, p. 10592).

To assure proper protection, the facepiece fit must be checked by the wearer each time he or she puts on the respirator. This may be done by following the manufacturer's facepiece-fitting instructions, such as these two simple field tests:

1. Positive Pressure Test. Close the exhalation valve and exhale gently into the facepiece. The face fit is considered satisfactory if a slight positive pressure can be built up inside the facepiece without any evidence of outward leakage of air at the seal. For most respirators, this method of leak testing requires that the wearer first remove the exhalation valve cover and then carefully replace it after the test.

2. Negative Pressure Test. Close off the inlet opening of the canister or cartridge(s) by covering with the palm of the hand(s) or by replacing the seal(s), inhale gently so that the facepiece collapses slightly, and hold the breath for ten seconds. If the facepiece remains in its slightly collapsed condition and no inward leakage of air is detected, the tightness of the respirator is probably satisfactory.

Maintenance of Devices

To comply with OSHA standards, the maintenance program must include inspection, cleaning, repair, and storage services.

Inspection. All respirators must be inspected before and after each use and at least monthly to assure that they are in satisfactory working condition. Self-contained breathing apparatus must be inspected monthly. Air cylinders must be fully charged according to the manufacturer's instructions. It must be determined that the regulator and warning devices function properly. These and other recommendations made by the manufacturer should be followed.

Respirator inspection must include a check of the tightness of connections and the condition of the facepiece, headbands, valves, connecting tube, and canisters. A record must be kept of inspection dates and findings for all respirators.

Cleaning. Respirators must be cleaned and disinfected after each use. The following procedure is recommended for cleaning and disinfecting respirators:

1. Remove any filters, cartridges, or canisters. Carefully discard so that reuse is not possible.
2. Wash facepiece and breathing tube in cleaner-disinfectant or solution (see following paragraphs). Use a hand brush to facilitate removal of dirt. It may be desirable to use an analytical procedure to determine if contamination is removed.
3. Rinse completely in clean, warm water.
4. Air dry in a clean area.
5. Clean other respirator parts as recommended by manufacturer.
6. Inspect valves, timing devices, headstraps, and other parts; replace with new parts if defective.
7. Insert new filters, cartridges, or canisters; make sure seal is tight.
8. Place in an easily removable plastic bag or container for storage.

Cleaner-disinfectant solutions are available that effectively decontaminate toxic and radioactive materials. Specific information on the type and use of these solutions should be obtained from safety and industrial hygiene specialists. Detailed information may also be found in the A.N.S.I. Z88.2-1969, Standard Practices for Respiratory Protection. Organic solvents must not be used to clean or disinfect any part of oxygen-generating equipment.

Repair. Replacement or repairs must be made with parts designed for the respirator and only by experienced persons. No attempt must be made to replace components or to make adjustments or repairs beyond the manufacturer's recommendations. Reducing or admission valves on regulators must be returned to the manufacturer or to a trained technician for adjustment or repair.

TABLE 5

Questions to be Answered in Selecting Suitable Respiratory Protection

Does or Will the Device . . .

Type of Device	function in low oxygen concentration?[1]	function in vapor concentration over 2%?	control eye irritation?	control skin irritation or absorption?	reduce the need for tight fit of the facepiece?	maintain the mobility of the wearer?	have a long duration of effectiveness?	provide protection for unknown gas and vapor?	reduce the prob. of mixture of gases?	have other major limitations?
Suit	Yes	Yes	Yes	Yes	Yes	No	Yes	Yes	Yes	High cost. Hot
Hood	Yes	Yes	Yes	No	Yes	No	Yes	Yes	Yes	
Full facepiece; Constant flow	Yes	Yes	Yes	No	Yes	No	Yes	Yes	Yes	
Full facepiece; Demand flow	Yes	Yes	Yes	No	No	No	Yes	Yes	Yes	
Mouth and nose mask; Constant flow	Yes	Yes	No	No	Yes	No	Yes	Yes	Yes	
Mouth and nose mask; Demand flow	Yes	Yes	No	No	No	No	Yes	Yes	Yes	
Hose mask or facepiece; no pump	Yes	Yes	Yes	No	No	No	Yes	Yes	Yes	Hose length must be short.
Hose mask, full face; with pump	Yes	Yes	Yes	No	Yes, unless pump fails	No	Yes	Yes	Yes	

Left-side row groupings: ATMOSPHERE SUPPLYING DEVICES; HOSE OR AIRLINE DEVICE

TABLE 5 (cont.)

Questions to be Answered in Selecting Suitable Respiratory Protection

Does or Will the Device . . .

	Type of Device	function in low oxygen concentration?[1]	function in vapor concentration over 2%?	control eye irritation?	control skin irritation or absorption?	reduce the need for tight fit of the facepiece?	maintain the mobility of the wearer?	have a long duration of effectiveness?	provide protection for unknown gas and vapor?	reduce the prob. of mixture of gases?	have other major limitations?
SELF-CONTAINED	Suit with tank mask inside	Yes	Yes	Yes	Yes	Yes	Yes; 15-30 min.	No	Yes	Yes	Very hot and heavy
	Full facepiece; constant flow	Yes	Yes	Yes	No	Yes	Very short time.	Very short time.	Yes	Yes	Heavy to wear.
	Full face breathing apparatus; demand flow	Yes	Yes	Yes	No	No	Yes; 5-30 min.	No	Yes	Yes	Heavy to wear.
	Full face regenerating apparatus; Chemox®	Yes	Yes	Yes	No	No	Yes	No; 1/2 hour	Yes	Yes	Slow starting. Training for use required.
AIR PURIFYING	Full face canister	Not below 16%	No	Yes	No	No	Yes	No	No	No	
	Half mask canister	Not below 16%	No	No	No	No	Yes	No	No	No	
	Mouth and nose cartridge respirator	Not below 16%	No	No	No	No	Yes	No	No	No	
	Mouth and nose dust respirator	Not below 16%	No	No	No	No	Yes	No	No	No prot. against vapors	
	Mouthpiece	Not below 16%	No	No	No	Yes	Yes	No	No	No	Dirty

[1] Be sure to understand the restrictions described in the text.

SELF-CONTAINED — REQUIRING NO AIRLINE OR HOSE — ATM. SUPPLYING DEVICES
AIR PURIFYING — PURIFYING DEVICES

83

Oil or grease must not be used on any parts of oxygen-generating devices. Expended or damaged canisters of this type of device should be punctured in several places, then dropped into a drum or pail of "oil-free" water. When the bubbling stops the canister contents will be expended. The canister and water may then be discarded. The water will be highly caustic and must be disposed of carefully. Special handling of the expended canister by placing it in a plastic bag for transport to disposal is also recommended.

Storage. After inspection, cleaning, and necessary repair, respirators must be stored to protect against dust, sunlight, heat, extreme cold, excessive moisture, or damaging chemicals. Respirators placed at stations and work areas for emergency use must be readily accessible at all times and clearly marked. Generally, they should be stored near exits outside of immediate contaminant release areas to make them available for emergency procedures. In hazardous areas, where the device is needed for emergency escape, they should be in an accessible location near the work station.

Routinely used respirators, such as dust respirators, may be placed in plastic bags or boxes. Respirators should not be stored in such places as lockers or tool boxes unless they are in carrying cases or boxes. Respirators should be packed or stored so that the facepiece and exhalation valve will rest in a normal position, and function will not be impaired by the elastomer sitting in an abnormal position.

Engineering Control Devices

Engineering control devices are the best means of combating air contaminants, since they actually reduce or eliminate the hazard. Personal protective devices, on the other hand, merely set up a defense against the hazard, and any failure of the defense means immediate exposure to the hazard.

The approaches commonly employed to reduce or eliminate air contaminants include dilution ventilation, local exhaust ventilation, and isolation or enclosure. The first two are discussed in detail in Chapter 10; the last is discussed below.

Isolation is utilized with hazardous processes that may endanger personnel other than those directly concerned with the process. Such processes include the manufacture and use of prussic acid, tetraethyl lead, and other highly toxic substances. In these cases, the entire operation is isolated in a separate building or in a carefully sealed-off area. Operations are made automatic so far as possible and all personnel involved are equipped with suitable personal protective equipment and are trained thoroughly in safe practices not only for day-by-day operations, but for emergencies as well. The very thorough application of this approach in the manufacture of explosives has helped to bring the overall injury rate for this industry down to among the lowest. The isolation of hazardous processes is a widespread practice in the chemical industry.

Many processes can be completely enclosed during all or part of the operation and, when maintained under a slight vacuum, will show no escape of contaminants into the workroom atmosphere. Loading and removing the contents should be mechanical or automatic whenever possible. The means chosen should be as effective as possible, thereby eliminating or reducing the hazard itself rather than depending upon personal protective devices. Examples are the modern bag- and drum-filling equipment, the fully enclosed and exhausted tumblers and shot-blast machines for cleaning castings, glove booths, and the new types of garment dry cleaning machines.

Enclosure is also a method of controlling exposures from such operations as sandblasting, heat treating, mixing, grinding, and screening. The test in each instance is: Does the process at any stage give off any substance that may contaminate the air? If so, can it be completely enclosed? If not, can a partial enclosure combined with exhaust be applied?

EXERCISES: CHAPTER 3

1. Contrast the following types of air contaminants:
   a. dusts and fumes
   b. vapors and mists

2. Contrast:
   a. toxicity and toxic hazard
   b. acute and chronic
   c. LC_{50} and LD_{50}

3. Define time-weighted average (T.W.A.) and tell what it is used for.

4. A degreaser operator working an 8-hour shift spends six 10-minute periods loading and removing work from a degreaser where the trichloroethylene concentration is 300 ppm. He also spends four-and-one-half hours adjacent to this degreaser where the trichloroethylene concentration is 200 ppm. His remaining time is spent in another area of the plant where annealing operations are being performed, and the average carbon

monoxide concentration is 100 ppm. Calculate the worker's time-weighted average exposure to:
   a. trichloroethylene
   b. carbon monoxide

  5. Discuss the four sources of toxicological information and the limitations of each.

  6. Identify the three factors which the toxicologist must take into account in evaluating inhalation exposures.

  7. Discuss the purpose of each of the following types of inhalation studies:
   a. range-finding studies
   b. single-exposure tests
   c. repeated exposure tests

  8. Indicate the parts of the eye of concern to the toxicologist conducting eye-contact studies, and the effects he or she looks for.

  9. Name the factors that must be taken into account in any evaluation of skin-contact toxicity.

 10. Contrast the degree of danger posed by ingestion, inhalation, skin contact, and eye contact, and indicate why ingestion studies are usually included in toxicological evaluations.

 11. Give three reasons why monitoring of air contaminants is carried out.

 12. Distinguish between grab samples and long-term samples.

 13. Discuss the advantages and disadvantages of colorimetric tubes for the analysis of air samples.

 14. Indicate the mode of operation of:
   a. portable air samples for dusts and mists
   b. combustible gas indicators

 15. Name the advantages of continuous monitoring for air contaminants.

 16. Indicate which category of breathing device is represented by each of the following:
   a. canister respirator
   b. hose mask
   c. regenerating-type device

 17. Summarize the proper procedure for maintaining breathing devices.

4

Nonionizing Radiation

Ultraviolet (UV) radiation, microwaves, and laser beams are called *nonionizing radiations* because they are not energetic enough to ionize atoms or molecules. The positions of the first two of these types of radiation are shown in Figure 4.1. Laser beams are comprised of light from the visible portion of the electromagnetic spectrum, which has been concentrated into a comparatively narrow area. Similar devices for concentrating microwaves, ultraviolet radiation, and infrared radiation are also available.

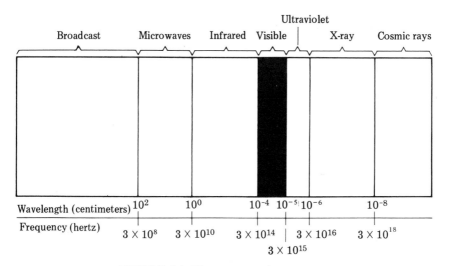

FIGURE 4.1. Electromagnetic spectrum.

No sharp lines of demarcation can be drawn between the different radiation bands in the electromagnetic spectrum. The radiation in each segment differs from the radiation in the others only in frequency, wavelength, and energy level, and each band blends gradually into the next.

DEFINITIONS

Electromagnetic waves may be characterized by wavelength or frequency of vibration. The units used to describe wavelength and frequency are listed below:

ELECTROMAGNETIC WAVE UNITS

Unit	*Symbol and Equivalent*	
Wavelength		
Angstrom	Å	10^{-8} cm
Centimeter	cm	1 cm
Micron	mu	10^{-4} cm
Millimicron	Mμ	10^{-7} cm
Nanometer	nm	10^{-7} cm
Frequency		
Cycles per second	cps	1 cps
Hertz	Hz	1 cps
Kilocycle	kc	1,000 cps
Megacycle	mc	10^6 cps
Gigacycle	gc	10^9 cps

Other units commonly employed in connection with nonionizing radiation include joules (j) and watts (w). The first is a unit of energy equal to 0.239 calorie or 1×10^7 ergs. The econd is a unit of power equal to one joule per second.

ULTRAVIOLET RADIATION

Ultraviolet radiation falls between the visible light and x-ray bands of the electromagnetic spectrum.

Ultraviolet radiation is divided into long-wave radiation, which extends from about 3,900 to 2,900 Å, and short-wave, which extends from about 3,900 to 2,900 Å, and short-wave, which ranges from 3,100 to 4,100 Å is known as the black-light portion of the ultraviolet region, although this is not a very specific term. The peak transmission of black light occurs around the 3,650 Å line. Very long ultraviolet wavelengths from 3,200 to 3,100 Å penetrate deeper into the skin but exert little biological effect. Short rays below 2,300 Å exert very little biological effect and have practically no penetration. Most biological action occurs in the intervening zone between the long wavelengths and the short rays.

Effects of Ultraviolet Radiation

Ultraviolet radiation can have undesirable effects on the eyes and skin. These specific effects are discussed below.

Eye Damage

When eyes are exposed to ultraviolet radiation having a wavelength less than 2,800 Å, inflamation of the eye membrane and corneal ulcers can result. Secondary infection may develop if the individual attempts to allay the painful sensation by rubbing the eyes. Since ultraviolet radiation is absorbed by the cornea, it has no harmful effects on retinal tissues or the lens of the eye.

Skin Damage

Skin that is exposed to ultraviolet radiation develops erythema, commonly known as sunburn. Wavelengths of about 2,500 Å and 2,967 Å are most likely to result in this condition. If exposure is especially severe, blistering can result. The type of erythema depends to some extent on the wavelength of the radiation. Shorter wavelengths, which have less ability to penetrate the skin, produce a more superficial erythema than do longer wavelengths.

Prolonged exposure to ultraviolet energy from 3,400 Å to 2,800 Å, with the greatest potency from 3,000 Å to 2,800 Å, is capable of causing skin cancer.

Welding operations are one of the most frequent causes of overexposure to ultraviolet radiation in industrial operations. In particular, inert-shielded gas welding produces much higher levels of ultraviolet than do other types of welding. Other industrial sources include the plasma jet torch and areas where ultraviolet lamps are used for their germicidal action, such as in hospitals, pharmaceutical plants, and in the sterilization of food.

Maximum Safe Exposure Limits

The Council on Physical Medicine of the American Medical Association has recommended the following limits for exposure to ultraviolet lights: 0.5 milliwatts per centimeter (mw/cm^2) for a 7-hour exposure, and 01. mw/cm^2 for a continuous 24-hour exposure. These figures are based upon the total intensity of ultraviolet radiation at 2,537 Å incident upon the occupants, including that emanating directly from the lights and that reflected from walls and fixtures.

Evaluating Hazards

Devices for detecting ultraviolet radiation include thermal detectors, photoelectric devices, and photographic plates.

Thermal Detectors

These devices act on the principle of producing a detectable physical change (temperature) in a thermally sensitive element. This change is then measured, usually by electrical means. Types of thermal detectors used with ultraviolet radiation include thermopiles, radiometers, and bolometers.

A thermopile consists of a series of thermocouples connected in series, which operates upon the electromotive force generated when the junction between two dissimilar metals is heated. The radiometer comprises a light vane, suspended from a fine quartz fiber, which is deflected when light falls on it. The bolometer is essentially a Wheatstone bridge, one arm of which is heated by the radiation being measured. Bolometers are very sensitive, but also somewhat delicate, being susceptible to damage at high radiation densities. A somewhat more detailed discussion of bolometers can be found in the section of microwaves in this chapter.

Photoelectric Devices

Phototubes and photovoltaic cells are the chief photoelectric devices used for measuring ultraviolet radiation.

The phototube consists of two electrodes in an evacuated quartz or ultraviolet-transmitting glass. When radiation falls on the cathode, it emits electrons, which flow to the anode. The flow is directly proportional to the intensity of the radiation. The device may be used over a very wide range of radiation intensities.

The photovoltaic cell has a thin layer of selenium, which acts as a semiconductor, deposited on iron and covered with a very thin transparent layer of another metal. Light falling on the surface of the cell causes a flow of electrons from the semiconductor to the metal. If the circuit between the two metal electrodes is conducted through a meter, a current flows in the external circuit from the iron to the selenium. The current is directly proportional to the intensity of the ultraviolet radiation.

The photovoltaic cell does not require an external source of current and can be made very compact. It cannot, however, be used for measuring low-intensity radiation. Phototubes can measure very low levels of radiation and can be used with a greater variety of ultraviolet wavelengths.

Photographic Plates

The degree to which ultraviolet radiation blackens a photographic plate is proportional to the intensity of the radiation, hence photographic plates can be utilized as a detection device. The procedure involves measuring the amount of blackening with a densitometer. Very accurate results are possible provided exposure and development conditions are properly controlled.

Combating Hazards

Workers who are exposed to direct or reflected radiation from devices generating ultraviolet waves require eye protection. Such protection is especially vital with ultraviolet radiations of shorter wavelengths. For limited exposures to ordinary ultraviolet lamps, window glass or plate glass is usually sufficient. If, however, radiation is intense, as in welding, or exposure is for long periods of time, special goggles are required. Goggles designed to absorb strongly in particular segments of the ultraviolet spectrum are available commercially.

MICROWAVES

Microwave or radio-frequency radiation comprises that portion of the electromagnetic spectrum having a wavelength of 1 centimeter to 1 meter and a frequency of 300 to 30,000 megacycles. This wide range of frequencies includes commercial, television, and broadcast bands; the X, S, and L bands of radar; and the bands of diathermy and microthermy units. The latter devices are of most concern in industrial hygiene considerations.

Effects of Microwaves

The principal hazard from microwave devices is thermal damage to the skin and underlying tissues. Eye damage is also possible, as are undesirable electrical effects and certain specific biological effects.

Skin Damage

This is produced by the conversion of the electrical energy of the radiations into heat. The amount of microwave radiation that is absorbed and transformed into heat depends upon several factors, one of which is the frequency of the radiation. Longer wavelengths will penetrate the body to a greater depth than shorter wavelengths or higher frequencies. Differences in the thermal conductance of tissues also influence the depth of penetration. As a result, such tissue as fat, bone, and muscle absorb different amounts of energy and become heated to different degrees.

With shorter wavelengths, which do not penetrate body tissues deeply, the effect of overexposure is a general rise in body temperature, similar to a fever. The resulting feeling of warmth may give warning of overexposure, although if the frequencies are such as to bring about a general rise in temperature, a feeling of discomfort may not occur quickly enough to serve as a warning. The human body can compensate to a certain extent for surface heating through perspiring, provided the rise in temperature is gradual.

Longer wavelengths, which penetrate body tissues comparatively deeply, are especially dangerous because of their ability to deep-heat tissue with little or no heating of the surface. A person exposed to damaging levels of longer wavelength radiation does not recognize the danger because there is no sensory warning or sensation of heat. Deep heating is least serious in muscle tissue, which is well-supplied with blood that acts as a coolant and dissipates the heat. The chief danger of overheating is to the internal organs, which lack an adequate supply of blood for cooling.

Eye Damage

When eyes are overexposed to microwave radiation, the viscous material becomes opaque white, in a manner similar to the white of an egg upon cooking. Once whitening has occurred, it cannot be reversed. Experiments carried out with rabbits* showed that opacity developed when eyes were exposed once for periods of 10 to 60 minutes to microwaves having frequencies of 2,450, 8,236, 9,375, and 10,050 mc. Cumulative exposures to subthreshold levels of radiation were likewise found to cause whitening.

Undesirable Electrical Effects

The chief undesirable electrical effect of microwaves is inter-

---

*See R. L. Carpenter, "An Experimental Study of the Biological Effects of Microwave Radiation in Relation to the Eye," *Rome Air Development Center Report* RADC-TDR-62-131, February 28, 1962.

ference with the operation of cardiac pacemakers. Disruption of function has been noted near hospital diothermy units, during operations in which electrocautery was used, in the vicinity of radio stations and television receivers, and near microwave ovens. Recent tests showed that some pacemaker models were stopped at distances of 8 to 100 feet by ovens emitting microwaves at levels 25 to 50 times below current and proposed standards. However, recent modifications have alleviated the problem.

Microwaves can also cause an enforced distribution of the electrical charge in the molecules of tissue undergoing irradiation. This seems to be particularly possible in molecules having loose chemical bands and relatively free electrons; for example, hydrogen and unsaturated bands.

Other Biological Effects

Exposures to microwave radiation have also been reported to cause changes in electrophoretic patterns and to increase antigenic activity.

Maximum Permissible Exposure Limits

The Occupational Safety and Health Act of 1970 has established as an exposure limit a maximum power density of 10 mw/cm^2 for periods of 0.1 hour or more. This value is equivalent to 1 mw-hr/cm^2 during any 0.1 hour period. The guide applies whether the radiation is continuous or intermittent, and pertains to both whole-body irradiation and partial-body irradiation.

The present standard for microwave ovens, as established by the Department of Health, Education, and Welfare, specifies that ovens must not emit radiation at a power density exceeding 0.001 w/cm^2 at any point 5 centimeters from the oven surface. To allow for wear, the standard provides that the power density may increase to 0.005 w/cm^2 after sale.

Evaluating Hazards

Microwave detectors can be divided into two categories: thermal detectors and electrical detectors. These are discussed below.

Thermal Detectors

Thermal detectors act on the principle of producing a detectable physical change (temperature) in a thermally sensitive element. This change is then measured, usually by electrical means. There are three general types of thermally sensitive elements in use today: bolometers, thermocouples, and air-pressure devices.

Bolometers. Thermistors and barreters are types of bolometers. The thermistor consists of a small bead of semiconducting material between two fine parallel wires. The primary resistance is in the bead, which has a negative coefficient of resistance, and the resistance decreases when the temperature of the bead rises. The microwave current causes the heating of the thermistor, but detection of the heating is done by measuring the direct-current resistance. A barreter consists of a fine platinum wire with a positive coefficient of expansion, whose resistance can be impedence-matched at the end of the transmission line. Thermistors are usually more sensitive and less delicate than barreters. In addition, they are less susceptible to damage by high power-density levels, and thus are usually superior as field instruments.

Thermocouples. A thermocouple produces a voltage when heated. Therefore, all that is necessary to make a detection device from a thermocouple is an antenna and a current-measuring or voltage-measuring device. Since the voltage and current levels encountered when using these detectors are very small, a sensitive meter must be used, generally in conjunction with some type of direct-current amplifier. Figure 4.2 shows this type of device.

Air-Pressure Systems. Such a system involves the measurement of a small pressure change in a confined gas when its container is heated slightly by absorbing some microwave energy. The container for the gas is usually some sort of electrical insulator covered by a carbon compound that absorbs the energy quite well.

Electrical Detectors

These instruments utilize a semiconductor diode or rectifier. The alternating microwave signal is rectified into direct current, which may be applied to a meter calibrated to indicate power. From the reading and the radiating area, the power density may be calculated. The diode detector type of indicator may be made extremely sensitive. In fact, this is one of the drawbacks of high-frequency diodes—even moderate signal strengths must be attenuated before reaching the detector in order not to overload it. This is sometimes rather complicated and may lead to inaccuracies of measurement. For this reason, thermal detectors are generally preferred over diode detectors.

Except for the air-pressure systems, all detectors or sensors require an antenna to convert radiation into wire-conducted currents. The chief antenna factor of concern to users is directionality—the extent to which its response depends on the direction from which the wave comes. An antenna should be somewhat directional so that the effect of reflections will be minimal; but at the same time, it

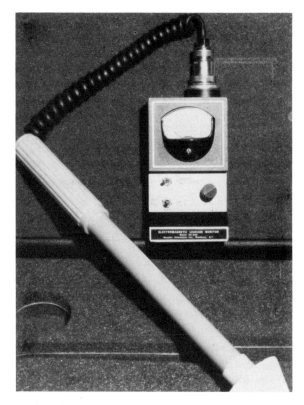

FIGURE 4.2.   Microwave leakage detector (Photo
by Michael D. Ells).

should not be so directional that orientation toward the source
becomes critical.

Most organizations do not have a microwave source of the
appropriate frequency and intensity for calibration of meters; thus, it
is necessary to return them to the manufacturer for periodic
calibration.

Combating Hazards

Hazards can be combated by the use of personal protective
devices, proper engineering design, proper work environment and
practices, and proper maintenance procedures.

Personal Protective Devices

Proper protective clothing should be used by operational and
maintenance workers in fields of high-power densities. Protective
clothing consists of a suit made of a special metalized fabric coated

with neoprene to protect against the discharge of electrical currents. Special closures, such as velcro, are provided because metalized zippers cannot be used. Mittens and boots are fitted to the suit to make it a completely contained unit. The eyes are protected by closely woven copper screening. It is difficult to use a suit of this type, particularly in warm or high temperatures, unless a ventilating or air-conditioning system is provided. Protective clothing as a means of control for microwave radiation is somewhat open to question, and its use as a substitute for other engineering controls and safe operational procedures should never be condoned.

Proper Engineering Design

High-power microwave generating equipment should be connected to a good ground so that no high-frequency potential exists on the enclosure surrounding the device. A low-reactance connection must be used. In buildings with steel frameworks or steel floors, equipment may be grounded directly to the framework or floor. Where there is no steel framework or floor, and the equipment is on the ground floor, the connection may be made directly to a large water pipe.

If a pipe cannot be located, it may be necessary to run a connection to a ground. The resistance value of any ground utilized should meet the requirements of the National Electrical Code. Wide copper strap is recommended as a ground connector. Information concerning the proper size strap for a satisfactory ground connection can be obtained from manufacturers of microwave equipment. The strap should be connected to the base of the enclosure at one end and to the ground at the other end.

Any remote accessory equipment utilized in connection with microwave generators should be shielded and interconnected so that no high-frequency potential exists between or on the exterior surfaces of the generator and the accessories.

With microwave equipment that is installed by simply plugging in a power cord, a separate groundwire, solidly grounded, should be utilized along with the cord. The ground and power cord should be positioned where they cannot be accidentally severed or tripped over. The ground connection should be inspected periodically.

Shielding of high-frequency work areas is necessary in order to protect workers against possible injury. Shielding by the use of wire mesh has proved satisfactory. A nomogram has been developed* for estimating the attenuation of microwave radiation by wire mesh

*See W. H. Kahler, *Industrial Medicine and Surgery*, 27(1958), 556.

under various conditions. Shielding of this type should be used primarily as a means of control in the far field and preferably in areas of occasional occupancy. Another form of shielding is to blank off certain sectors of the microwave field to assure that personnel will not be exposed while in the sector. This type of shielding may reduce the effective operational capability of microwave generators since the area traversed has been reduced. It can sometimes be eliminated by the proper positioning of the transmitters.

Before being placed in operation, any piece of microwave equipment should receive a thorough inspection. Electrical connections, time delay relays, holding devices of fuses, and the like should be checked to insure that they have not become loose during shipment. Fuse capacitances should be checked and meters inspected to insure that the indicating needles are properly set on the scale.

Interlocks on doors, water-supply systems, and temperature control systems should be inspected to see if they operate freely. If any protective device is stuck or inoperative, the situation should be remedied. Tank coils and tubes should be checked to make sure no installation tools have been accidentally left there.

Safe Work Environment and Practices

Safe Environment. Microwave equipment should be located in a clean, level area free from exposure to dust, lint, or oily vapors. Positioning of equipment should be such that access doors can be opened completely. To insure proper ventilation, the rear of the equipment should be at least 3 feet from any wall.

Occupational Safety and Health Act regulations stipulate that the sign in Figure 4.3 be used to denote a radiation hazard.

Safe Work Practices. Only properly trained personnel who are completely familiar with the correct operating procedures should be permitted to operate microwave equipment. Operators should be supervised carefully to insure that they are conducting operations in a proper manner. Where operations are repetitive, consideration should be given to providing relief periods or changing operators periodically.

Operators should never reach into the radiation areas, make contact with electrically hot coils, or reach into hoppers or conveyors to remove or adjust faulty or dislodged work pieces. All such pieces should be removed by a checker or inspector at a point away from the microwave equipment. Any machine that is performing in a defective manner should be shut down by the operator, who should also report the defect to his or her supervisor.

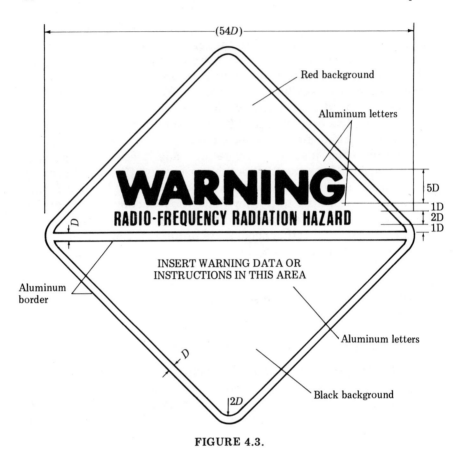

FIGURE 4.3.

Proper Maintenance Procedures

The maintenance schedule recommended by the microwave equipment manufacturer should be followed faithfully. At least two electrical maintenance workers, one a qualified electrician, should be on hand whenever maintenance of high-frequency equipment is carried out.

Before work is started, the source of power to the rectifier or the direct current from the rectifier should be disconnected. All switches should be moved to the "off" position and locked so they cannot be moved to "on" accidentally. Fuses should be removed from branch circuits and carried in the pocket of the electrician carrying out the maintenance. High-voltage leads or terminals to the ground should be shorted and condensers discharged to prevent stored charges from accidentally shocking workers. Only precisely fitting tools and well-insulated equipment should be used by maintenance workers. Tools should be kept clean and never laid across wires or in the tank coils. When maintenance is completed, the worker should check to make sure no tools have been accidentally left in electrical equipment. A final check should be made before the equipment is turned on to insure that all connections have been properly tightened and the equipment is otherwise in good operating condition.

LASERS

A laser is a device for concentrating visible light into a narrow band. Conventional light is comprised of waves of differing length that radiate in all directions and reinforce or cancel one another. Light from a laser travels in just one direction, is of the same length, and is a very narrow beam. Such light is known as coherent light, as contrasted to incoherent, or regular, light. The term "laser" is an acronym, developed from "*l*ight *a*mplification by *s*timulated *e*mission of *r*adiation." A similar device, the maser (from *m*icrowave *a*mplification by *s*timulated *e*mission of *r*adiation) is used for concentrating energy in the microwave segment of the electromagnetic spectrum. Other devices utilize energy in the infrared and ultraviolet frequencies.

How Lasers Work

The list of substances that can produce laser emission is comprised of an impressive number of different solids, liquids, gases, and junction diodes. There are, however, several features (listed below) common to the configuration of all types of lasers.

1. **The Laser Media:** This is the substance, either solid, liquid, gas, of junction between two dissimialr metals, that is capable (because of its atomic and/or molecular makeup) of sustaining stimulated emission.
2. **The Source of Excitation Energy:** To generate a laser beam a redistribution is required in the number of atoms that normally exist in certain atomic energy levels of the laser media. This requires an external source of excitation energy, often called the "pump" energy.
3. **Fabry-Perot Interferometer:** This device is a pair of mirrors aligned plane-parallel to one another. In the case of the laser,

one mirror is placed at each end of the laser media. Usually one mirror is a total reflector, the other a partial reflector (that is, it allows part of the laser beam generated within the active media to pass outside the interferometer).

In the most general manner, the following describes how a laser operates: Excitation energy is vigorously supplied to the active media to produce the specific condition called a population inversion. In this condition, more atoms of the laser media exist in a specific excited-state energy level than are found in the lowest ground-state level. This condition is contrary to the normal condition of a system in thermal equilibrium. One manner for an atom in an excited state to release excess energy is by the spontaneous emission of light in discrete units called photons. The unique feature of a laser device lies in the fact that because of the population inversion the energy release may be accomplished by the process known as *stimulated emission.* In this case a photon released by one excited atom will cause (stimulate) an excited atom it may encounter in its path also to release a photon of excess energy. The result of this interaction is the combination of two photons with identical coherence properties (phase relationships) so that they add completely together to produce a beam of twice the intensity. As the beam progresses through the excited laser media, its amplitude will be rapidly increased while its coherence properties remain unaltered. Upon reaching the total reflection mirror the beam direction is completely reversed, thus allowing another pass through the excited laser media so that the beam may be further amplified. Upon reaching the partial reflecting mirror, a portion of the beam escapes. This escaping portion is the active emission from the laser. The process will continue for as long as sufficient pump energy is supplied to the laser media.

There are three basic types of lasers: (a) continuous wave, which emit laser radiation in a continuous fashion, (b) pulsed lasers, which emit laser radiation in pulses of many nanoseconds or milliseconds duration, and (c) Q-switched lasers, which pack a whole bunch of energy into a very short pulse—nanosecond duration.

Table 6 summarizes the characteristics of lasers utilizing the major types of laser media.

Effects of Lasers

The hazards associated with lasers stem principally from the highly concentrated nature of the beam. As an illustration, an argon gas laser generating 18 watts of continuous wave energy produces

TABLE 6
Laser Characteristics

Laser Media	Predominant Wavelengths (nanometers)	Active Media	Method of Operation	Continuous Power (w)	Peak Power (megawatts)	Beam Divergence (milliradians)
Ruby	694.3	Solid	Pulsed	1.0	1-1000	0.5-20
Neodymium-glass	1,060	Solid	Pulsed		1-500	0.5-20
Neodymium-YAG	1,060	Solid	CW	1-100	1-10	0.5-10
Helium-neon	632.8 1,150 3,390	Gas	CW	0.001-0.100		0.1-1.0
Argon ion	476.5 488.0 514.5	Gas	CW	1-20	10^{-4}	1.0-3.0
Krypton ion	647.1 568.2 520.8 476.2	Gas	CW	0.5-2.0		1.0-3.0
Xenon ion	627.0 597.1 541.9 526.2 504.5 460.3	Gas	CW	0.1-0.5		1.0-3.0
Neon ion	332.4	Gas	CW	0.250		1.0-3.0
Carbon dioxide	10,600	Gas	CW	10-5000	10^{-2}	1.0-5.0
Nitrogen	337.1	Gas	Quasi-CW	0.250	10^{-2}-10^{-1}	Rectangular beam 1-10
Gallium-arsenide	840.0	Semiconductor	Quasi-CW	1-20		Rectangular beam 1-10

power 2,000 times greater than that of the sun at a point a few feet from the emission end.

Although the biological effects of exposure to laser beams are still under observation, some preliminary observations can be made. Of the possible effects, eye damage appears to be the most serious, although skin damage can also occur. These two types of damage are discussed below.

Eye Damage

Any or all portions of the ocular structure may be damaged by a laser beam. Some, however, are more easily damaged than others.

Retinal Effects. Thermal damage to the retina is probably the most serious effect of laser beams. The intensity of the energy traveling to the retina is affected by the action of the iris and by optical absorption of the various media of the eye. The aperture formed by the iris changes with stimulation from about 2 to 8 millimeters in diameter. Thus, contraction of the pupil may reduce the intensity of the energy striking the retina by as much as 16 to 1. This degree of reduction, however, is not very great in view of the tremendous energy that may be present in the beam. Unlike the retina, which does not feel pain, the iris contains a network of pain nerves that provide a warning signal when energy levels are too high. If, however, an energy beam is narrow enough to enter the eye without impinging on the iris, or of such a short duration that the pulse goes unnoticed, the retina may receive excessive exposure and no pain will be felt.

A concentration of energy at the retina sufficient to cause a burn may produce a temporary lesion or a permanent blind spot. Should the lesion produce a permanent retinal scar without damage to the rods and cones, some impairment of vision may result. However, should the lesion be of sufficient extent to cause necrosis of these structures, the victim will forever carry an additional blind spot.

Threshold Values. The threshold values for retinal damage may be defined as the energy density required to produce, within a short interval of time (approximately 5 minutes) after exposure, a clinically observable retinal lesion. Several agencies have established safety standards for laser operations. Thresholds for retinal damage used in determining these standards are given in Table 7.

Because of the understandable scarcity of information on laser effects on the human eye, threshold values are based upon data obtained from experimental animals, mostly rabbits, which has been

TABLE 7

Eye Exposure Guidelines for Laser Radiation as Recommended by Various Organizations

Wavelength and Pulse Duration	Air Force[1] Total Energy or Power Entering Eye	Army/Navy[2] Total Energy or Power Over a 7 mm Aperture (Pupil)
Visible (0.4-0.7 μm) Q switched	0.5×10^{-4} J	1×10^{-6} W/cm^2
Long Pulse	1×10^{-6} J	1×10^{-7} J/cm^2
Continuous Wave	1×10^{-3} W(10-500ms) 2×10^{-3} W(2-10ms)	1×10^{-6} J/cm^2
Q switched (10-100ms)	2.5×10^{-6} J	
Long Pulse (0.2-2ms)	2.0×10^{-5} J	
CW-YAG (2-10ms)	1×10^{-2} W	
CW-YAG (10-500ms)	5×10^{-3} W	

[1] "Laser Health Hazards Control," Department of the Air Force, AFM, Washington, D.C., 1971.
[2] "Control of Hazards to Health from Laser Radiation," TB med. 279/NAVMED P-5052-35, Departments of the Army and Navy, Washington, D.C., February 26, 1969.

extrapolated to the human eye. The values presented in Table 8 should not be taken as the absolute minimum for retinal damage, since injury has been demonstrated at lower levels of irradiation by more refined techniques. Thus, impairment of enzyme activity has been demonstrated at levels 10 to 15 percent below those required to produce visible lesions. Also, retinal lesions may not become clinically visible immediately after exposure at low levels of radiation.

Maximum Permissible Exposure Levels. Table 8 presents maximum permissible limits to which the eye can be exposed without damage to the retina.

Skin Damage

Exposure of the skin to laser radiation produces lesions resembling thermal burns. The amount of damage varies with the intensity of the radiation and the degree of pigmentation. To a limited extent, the presence of hair and its color will also affect the amount of damage.

Effect of Pigmentation. Skin with high pigmentation shows significant changes after impact of a low-energy (5 to 10 joules) laser beam, including superficial scaling, superficial charring, crusting, and even ulceration. No evidence of changes has been noted in skin of

TABLE 8[1]

Maximum Permissible Exposure Levels for Laser Radiation at the Cornea
for Direct Illumination or Specular Reflection at λ = 6,943 Å

Pupil Size	Q-Switched Pulse (J/cm²)	Non-Q-Switched Pulse (J/cm²)	CW Laser (W/cm²)
Daylight 3-mm pupil	5.0×10^{-8}	5.0×10^{-7}	5.0×10^{-5}
Laboratory 5-mm pupil	2.0×10^{-8}	2.0×10^{-7}	2.0×10^{-5}
Night 7-mm pupil	1.0×10^{-8}	1.0×10^{-7}	1.0×10^{-5}

[1] Source: The Dow Chemical Company, based upon material in Supplement No. 7
to "A Guide for Uniform Industrial Hygiene Codes or Regulations," American
Conference of Governmental Industrial Hygienists, 1968.
In the absence of definitive biological data it is impossible to suggest guidelines for
exposure to repetitive pulsed lasers, where cumulative effects are important.

low pigmentation at this energy level. However, significant altera-
tions may develop in low-pigmented skin after a single impact at
densities above 20 to 25 joules. Thus, in one caucasian volunteer,
irradiation of the skin with an unfocused 100-joule ruby laser
produced an area of instantaneous mild inflammation, which was
insensitive to the touch and persisted for four days.

Information is scarce concerning the effect of chronic exposures
to laser radiation. In one instance, an individual with hypersensitive
skin developed a lesion after exposure over a two-month period to a
0.5-joule pulse. Attempts to reproduce these results in other persons
have not been successful.

Maximum Permissible Exposure Levels. The British Aviation
Ministry Code states that the total laser energy falling upon any part
of the body should not exceed 0.1 j/cm². The American Conference
of Government Industrial Hygienists states that the maximum
permissible exposure level for the skin should not exceed daylight
levels for the eye of a factor of more than 10^6.

Evaluating Hazards

The United States Public Health Service has developed the laser
survey profile for evaluating laser hazards.* This profile encompasses
evaluating the laser equipment and the area in which it is used,
calculating safe viewing distances, and establishing an adequate
company policy.

Evaluating Laser Equipment

The form shown in Figure 4.4 can be utilized for evaluating
laser equipment.

*See M. E. Lanier, *National Safety News*, 100 (November 1969), 88.

FIGURE 4.4

Laser Hazard Evaluation Sheet

Installation: Building No._____ Room No._____

LASER EQUIPMENT

1. Wavelength _____ Other _____
2. Mfg._____ Model No._____
3. Type: Solid State ☐   Gas ☐   Injection ☐
4. Mode of Operations:
    S ☐   NP ☐   Q.S. ☐   PRF ☐   CW ☐

                                                    Operational

5. Power (watts)   _____
6. Energy (joules)   _____
7. Beam diameter (cm)   _____
8. Beam divergence (rad) or (Deg)   _____
9. Power density (w/cm^2)   _____
10. Energy density (j/cm^2)   _____
11. Pulse length (sec)   _____
12. Pulse repetition rate   _____
13. Distance to target area   _____
14. Primary application   _____

                                        Yes   No
                                         ☐   ☐
15. Is reporting procedure in case
    of injury known?

The data required can be obtained from the nameplate of the equipment or from the literature supplied by the manufacturer. Since the actual performance characteristics of a piece of equipment may vary from the specifications of the manufacturer, the operation should be checked to ascertain any differences that may exist.

Evaluating the Work Area

When lasers are operated in areas where other persons work, the form shown in Figure 4.5 can be used to evaluate the relative hazard to these persons. The findings can then be used to make changes that will increase the safety of the area. Thus, if the survey discloses a reflective wall surface, an exposed target area, and a room to which other employees have free access, these conditions can be corrected.

Calculating Safe Viewing Distances

The equation shown below can be used to calculate the *minimum* distance at which the beam from a particular device can be viewed safely. The equation excludes the factor of atmospheric attenuation.

FIGURE 4.5

Laser Work Area

1. Inside ☐	Outside ☐		Both ☐	
2. Free Access ☐	Isolated ☐		Limited ☐	

3. Room Lights:
        On ☐   Off ☐   Outside ☐

4. Wall Color
5. Wall Surface   Dull ☐   Reflective ☐
6. Equipment Surface   Dull ☐   Reflective ☐

	Yes	No
7. Warning Signs	☐	☐
8. Warning Lights	☐	☐
9. Interlocked Doors	☐	☐
10. Adequate Ventilation—Room	☐	☐
11. Adequate Illumination—Room	☐	☐
12. Enclosed Beam	☐	☐
13. Enclosed Target Area	☐	☐
14. External Lens Used	☐	☐
15. External Mirrors Used	☐	☐
16. Adequate Ventilation—Process	☐	☐

17. Electrical Equipment Conforms with
    National Electrical Code
    (Grounding & Bonding)
	☐	☐

18. Storage of Flammables in Approved Containers	☐	☐
19. Adequate Room Exits	☐	☐
20. Automatic Sprinkler System	☐	☐

Area Diagram:

Comments:

$$r = \frac{1.2 \sqrt{\dfrac{E}{E_s}} - a}{\Phi}$$

where E = Level of radiant energy leaving laser (watts or joules),
       a = Diameter of emergent beam (cm),
       Φ = Beam divergence (radians),

E_s = Allowable corneal exposure (watts/cm^2 or joules/cm^2) *Laser Health Hazards Control,* Department of the Air Force, AFM, Washington, D.C., 1971.

r = Safe viewing distance (cm): = ?

The following examples illustrate the use of the above equation:

Example No. 1

E = 1 joule > normal pulsed ruby laser
a = 0.5 cm
Φ = 1 $\times$ 10^{-3} radians
E_s= 1 $\times$ 10^{-6} joules/cm^2

Safe viewing distance: $r = \dfrac{1.2 \sqrt{\dfrac{1 \text{ joule}}{1 \times 10^{-6} \text{ joules/cm}^2}} - 0.5}{1 \times 10^{-3} \text{ radians}}$

$$r = \frac{1.2 \sqrt{10^6 \text{ cm}^2} - 0.5 \text{ cm}}{1 \times 10^{-3}}$$

$$= \frac{1.2 \times 10^3}{1 \times 10^{-3}} \quad (0.5 \text{ cm becomes insignificant})$$

$$= 1.2 \times 10^6 \text{ cm or approximately 7.5 miles}$$

Example No. 2

E = 10 milliwatt, continuous wave laser (0.01 watt)
a = 0.5 cm
Φ = 2 $\times$ 10^{-3} radians
E_s = 1 $\times$ 10^{-3} watts/cm^2

Safe viewing distance: $r = \dfrac{1.2 \sqrt{\dfrac{0.01 \text{ watt}}{1 \times 10^{-3} \text{ watt/cm}^2}} - 0.5}{2 \times 10^{-3} \text{ radians}}$

$$r = \frac{1.2 \, (3.16 \text{ cm}) - 0.5 \text{ cm}}{2 \times 10^{-3}}$$

$$= 1.6 \times 10^3 \text{ cm or slightly less than 55 feet.}$$

Establishing Adequate Company Policy

No assessment of hazards can be considered complete without a review of company policy concerning the use of lasers. The checklist shown in Figure 4.6 includes the major points that must be taken into account in any adequate company policy.

FIGURE 4.6

Company Policy Checklist

1. Number of employees potentially at risk:
   Operators_____   Others_____

		Yes	No
2.	Requirements for Employee Vision	☐	☐
3.	Pre-placement Physical Examination	☐	☐
4.	Periodic Physical Examination	☐	☐
5.	Laser Operator Eye Protection Required	☐	☐
6.	Eye Protection Used	☐	☐
7.	Readily Accessible	☐	☐
8.	Adequate Goggles Used	☐	☐
9.	Procedure for Checking Goggles	☐	☐
10.	First Aid Facilities Immediately Available	☐	☐
11.	Disaster or Emergency Planning	☐	☐
	Telephone Number Posted	☐	☐
	Adequate No. of Fire Extinguishers	☐	☐
	Safety Training Program for Employees (Operators, Maint. Crew, etc.)	☐	☐
	Reporting Procedure for Defective Equipment	☐	☐
	Maintenance Program	☐	☐

Combating Hazards

Laser hazards can be combated by personal protective devices, proper engineering design, safe work environment and practices, and medical monitoring.

Personal Protective Devices

Personnel exposed to laser beams should be furnished with suitable eye shields of an optical density (OD) adequate for the energy involved.

Table 9 lists the maximum power or energy density for which adequate protection is afforded by glass of optical density 1 through 9. This table is based on the maximum permissible exposure levels for dark-adapted eyes.

Commercially sold protective devices for the eyes include spectacles and soft-side goggles, each available with lenses in a wide range of optical densities. In addition, there are eye shields with

TABLE 9

Attenuation of Laser Safety Glass

O.D.	Attenuation (db)	Attenuation Factor	Q-Switched Max. Energy Density (j/cm²)	Non-Q-Switched Max. Energy Density (j/cm²)	c.w. Max. Power Density (W/cm²)
1	10	10	10^{-7}	10^{-6}	10^{-4}
2	20	10^2	10^{-6}	10^{-5}	10^{-3}
3	30	10^3	10^{-5}	10^{-4}	10^{-2}
4	40	10^4	10^{-4}	10^{-3}	10^{-1}
5	50	10^5	10^{-3}	10^{-2}	1
6	60	10^6	10^{-2}	10^{-1}	10
7	70	10^7	10^{-1}	1	100
8	80	10^8	1	10	Not applicable
9	90	10^9	10	100	Not applicable

flexible bodies and plateholders that allow quick change of antilaser plates.

The following equation can be used to calculate the optical density of the lenses required in a particular situation.

$$\text{O.D.} = \log_{10} \frac{4E}{\pi a^2 E_s} \quad @ \; \lambda$$

where   E = Level of radiant energy leaving laser (watts or joules)
  a = Diameter of emergent laser beam (cm)
  E_s = Allowable-corneal exposure (watts/cm² or joules/cm²)
    Laser Health Hazards Control, Department of the
    Air Force, AFM, Washington, D.C., 1971.
  O.D. = Optical density of lenses required
  λ = Wavelength of laser (nanometers)
  π = 3.1416

Use of the formula is shown in the two examples below:

Example No. 1

  E = 10 milliwatt, continuous wave helium-neon laser
  a = 0.5 cm
  E_s = 1×10^{-3} watts/cm²
    (allowable corneal exposure limit)

  Wavelength = 632.8 nanometers (6,328 Angstroms)

$$\text{O.D.} = \log_{10} \frac{4E}{\pi a^2 E_s}$$

$$= \log_{10} \frac{4(0.01 \text{ watts})}{(0.5 \text{ cm}^2)(1 \times 10^{-3} \text{ watts/cm}^2)(3.14)}$$

$$= \log_{10} \; 5.1 \times 10^3$$

O.D. = 1.7 @ 632.8 nm

Example No. 2

E = 1 joule, normal pulsed laser

a = 0.5 cm

E_s = 1×10^{-6} joules/cm^2
(allowable corneal exposure limit)

Wavelength = 694.3 nanometers (6,943 Angstroms)

$$\text{O.D.} = \log_{10} \frac{4E}{\pi a^2 E_s}$$

$$= \log_{10} \frac{4(1 \text{ joule})}{(0.5 \text{ cm})^2 (1 \times 10^{-6} \text{ joules/cm}^2)(3.14)}$$

$$= \log_{10} \; 5.1 \times 10^6$$

O.D. = 6.7 @ 694.3 nm

Safety glasses should be evaluated periodically to insure maintenance of adequate optical density at the desired laser wavelength. There should be assurance that laser goggles designed for protection from specific lasers are not mistakenly used with different wavelength lasers. The specific optical density of the filter plate should be printed on the eye wear. Laser safety glasses exposed to very intense energy or power density levels may lose effectiveness and should be discarded.

Protective gloves, clothing, and shields should be used by personnel to prevent exposure to the skin. Impervious, quick-removal gloves, face shields, and safety glasses should be provided as minimum protection for personnel who handle the liquefied gases used as coolants for high-powered pulsed lasers.

Proper Engineering Design

Proper design requires that both electrical and optical safety criteria be built into lasers. To insure proper electrical safety several things are necessary. First, all capacitors should be supplied with adequate grounding and bleeding resistors capable of discharging the capacitors. The resistors are needed because capacitors retain a charge even when the power is disconnected. Capacitors should be shielded to protect employees against possible explosions. High-voltage circuits should be provided with suitable covers marked to indicate high voltage and fitted with interlocks which will insure that the voltage is shut off and the capacitors discharged before any work can be performed on the circuit. Similarly, interlocks should be provided to insure that laser devices cannot be accidentally energized. To guard against short circuiting, high-current fuses should be installed in the energy bank of the power supply. All devices and components should be properly grounded.

Proper optical protection includes shielding the workpiece with black anodized aluminum or some other material capable of absorbing the wavelength of the laser beam being generated. Whenever a microscope is utilized, it should be equipped with an interlock to prevent operation of the laser while the employee is viewing the work. The pumping light of solid-state lasers should be shielded to protect users against the brilliant flashes and the fragments from lamps that may explode. If possible, work should be carried out in a light-tight interlocked enclosure so eye protection is not needed.

Safe Work Environment and Practices

Safe Environment. A safe working environment can be provided through attention to several factors. Laser operations should be conducted in an area isolated from nonlaser equipment, employee travel routes, and heavily populated areas. The area should be classed as a controlled space with access limited to authorized personnel. This fact should be made known by posting of appropriate signs in the area. The work area should be well lighted to prevent dilation of the pupils of workers' eyes. All surfaces in the area should be nonreflective to reduce the danger of eye damage from beam reflections. The laser should be fixed in one position to prevent the operator from directing the beam at an angle that could endanger him- or herself and other employees. Any combustible material used in connection with the operation of the laser should be stored in an approved container in a suitably shielded area to prevent ignition. Closed circuit television, an optical comparator equipped with an appropriate filter, or a microscope with an interlocked shutter should be utilized for viewing the workpiece while the laser is energized.

An audible and/or visual signal should be used to indicate when a laser is being used. The signal should continue for the duration of the operating period. It should be designed so that the employee will look away from, rather than toward, the laser. Before the laser is actively actuated, an audible countdown should be performed.

A complete radiation survey should be conducted at each installation before turning the equipment over to operating personnel. Periodic surveys should be conducted in work areas to insure the adequacy of all control measures.

Safe Operating Practices. Only properly trained personnel who are aware of the hazards associated with lasers should be permitted to operate laser equipment. These employees should be familiar with the latest first-aid methods, especially those needed for high-voltage accidents. Operators should never work alone, and their performance should be audited frequently.

Personnel should never expose their bodies to, or look directly into, a laser beam. If safety goggles or a light-tight interlocked enclosure is not provided, the laser operator should close his or her eyes or count down and look away before firing, to avoid any reflections.

Medical Monitoring

Any adequate medical monitoring program must include a systematic program of medical examination, beginning when the employee is being considered for a laser job and ending when he or she leaves it. Procedures for dealing with workers who report problems must also be established.

Medical Examinations. Any employee considered for work with lasers should have a pre-employment examination. The examination should include a compilation and review of the job candidate's medical history, a complete blood count and urinalysis, an examination of the skin, a complete eye examination with and without glasses for both near and distant vision, and a detailed fundoscopic examination and tonometry by an opthalmologist with a detailed description of findings. The examining physician should evaluate the medical history and results of the tests performed to determine if the candidate should be permitted to work in a laser-controlled area.

As a precautionary measure, persons with an ocular, hematological, or progressive neurological disease; a chronic emotional or mental illness; a history of cancer, arthritis, chronic disabling pulmonary, cardiovascular. or collagen disease; or a poorly controlled endocrine disease such as hyperthyroidism or diabetes should not be permitted to work in laser-controlled areas. In addition, pregnant women should be excluded. Depending upon circumstances, other specific conditions may exclude candidates.

At least once a year, and preferably oftener, persons working with lasers should be re-examined. The examination should include a complete blood count and urinalysis, a skin examination, a complete eye examination with a fundoscopic exam and tonometry, and an evaluation of the general health of the employee.

If the checkup examination reveals any eye changes that could be the result of laser exposure, the employee should be examined by an ophthalmologist familiar with the biological effects of laser radiation. The company physician should advise the industrial hygienist and request a study of the employee's work environment to determine what factors may have contributed to the exposure and what changes, if any, must be made in the work environment to reduce the laser exposure.

If an employee has significant changes in his or her blood or urine, he or she should be re-examined to verify the laboratory findings. If the laboratory results are duplicated, the employee should be removed from the assignment in the laser-control area and additional tests should be conducted under the direction of the company physician. The physician should also consult with the industrial hygienist as prescribed in the preceding paragraph.

Any employee having a skin burn caused by exposure to laser light should be removed from work in the laser-control area. The physician should contact the industrial hygienist, requesting the prescribed investigation and corrective measures, if indicated.

If an employee experiences persistent after-images of a light source after working in a laser-control area, he or she should be removed from the laser assignment, examined by an ophthalmologist, and placed under the surveillance of a company physician.

All employees working in laser-control areas who switch to jobs not involving lasers or who leave the company should be given a termination physical that includes the same checks as the periodic examination. The examining physician should then prepare a report assessing the general health of the employee.

Workers Reporting Medical Problems. Any worker who reports to the medical department with a laser burn or who experiences persistent after-images of any light source should be examined immediately. The physician should treat the employee and determine what additional action should be taken.

EXERCISES: CHAPTER 4

1. Describe the effects of ultraviolet radiation on the eyes and skin.

2. Compare the mode of operation of the phototube and photovoltoic cell, indicating any special advantages or disadvantages of each as ultraviolet radiation detectors.

3. Describe the effects of microwaves on the eyes and skin, as well as any other undesirable effects.

4. Compare the mode of operation of thermal detectors and electrical detectors, indicating any special advantages or disadvantages of each as microwave detectors.

5. Discuss the types of personal protective devices, engineering controls, and work practices employed to combat microwave hazards.

6. Explain briefly the operation of a laser.

7. Describe the effects of laser beams on the eyes and skin.

8. Summarize the laser survey profile procedure for evaluating laser hazards.

9. Discuss the types of personal protective devices, engineering controls, and work practices employed to combat laser hazards.

10. Describe the medical monitoring program for persons working with lasers, with special attention to:
    a. the scope of the pre-employment medical examination.
    b. medical conditions that should exclude persons from working with lasers.
    c. the scope of follow-up medical examinations.
    d. medical conditions calling for the removal of workers from laser jobs.

5

Ionizing Radiation

Ionizing radiation is so named because it has sufficient energy to ionize atoms and molecules. There are five kinds of ionizing radiation: alpha particles, beta particles, X rays, gamma rays, and neutrons.

Alpha particles are helium atom nuclei that are ejected from the nuclei of atoms during the process of radioactive decay. Although the particles have very little penetrating power, being stopped by paper and skin, they are the most highly ionizing per unit path length. Beta rays are high-speed electrons ejected from radioactive materials and are weakly ionizing. Their penetrating power is much greater than that of alpha rays, as they will penetrate up to one-quarter inch of aluminum or one inch of wood. X rays and gamma rays are short wave length electromagnetic radiation, which occupy that part of the electromagnetic spectrum having a wavelength of 10^{-6} to 10^{-8} centimeters and a frequency of 3×10^{16} to 3×10^{18} cycles per second. Both are highly penetrating, requiring shielding of lead or a similar heavy material. Neutrons are electrically uncharged particles of high mass which are ordinarily found in the nuclei of atoms and are released as a result of radioactive decay. Neutrons may be classed as slow, intermediate, or fast, depending upon their energy of ejection. The range of energies for each group is indefinite, different ranges being selected according to specific needs. Like X rays and gamma rays, neutrons are highly penetrating, but because of their nature require shielding that contains a high percentage of hydrogen atoms rather than heavy material shielding.

The energy level of one particular type of radiation may vary considerably depending upon the source of the radiation. Thus, lead-210 emits many times the amount of energy emitted by silver-105. The first four types of radiation are far more commonly encountered in industry than the last and are therefore of greatest concern to industrial hygienists.

Standards adapted under the Federal Occupational Safety and Health Act (OSHA) of 1970 have set rigorous requirements regarding monitoring of employees for exposure, exposure levels, use of caution signs, labels and signals, instruction of personnel regarding radiation hazards, radioactive material storage and waste disposal, notification of incidents involving potentially dangerous exposures, and the maintenance of exposure records. These standards, based on those issued under the Atomic Energy Act of 1954, cover all radioactive sources except radium and X-ray-producing equipment, which are covered by state laws and regulations.

Pertinent provisions of the OSHA standards are incorporated in the appropriate spots in this chapter, as are pertinent provisions of the 1954 act that were not made part of OSHA regulations. Since the provisions are paraphrased and condensed, it is recommended that readers also consult the original OSHA standards, as published in Part II of the *Federal Register* issue of May 29, 1971 (29 CFR 1910), as well as the original AEC regulations as published in the Code of Federal Regulations, Chapter 10, Part 20 (10 CFR 20). Readers should also consult the appropriate state regulations for information concerning radium and X-ray-producing equipment.

DEFINITIONS

The following definitions must be known by anyone seeking an understanding of ionizing radiation.

Curie: That amount of a radioactive material which has a disintegration rate of 3.7×10^{10} atoms per second.

Dose: The quantity of ionizing radiation absorbed, per unit of mass, by the body or by any portion of the body. It may also mean the quantity absorbed, per unit of mass, by the body or any portion of the body during a particular time interval.

Half-Life: The time in which a radioactive material will lose half its activity by radioactive decay. Depending on the material this may be anywhere from a few seconds or microseconds to many thousands of years.

Isotopes: Forms of the same element that differ from one another in the number of neutrons in the nucleus of the atom.

Maximum Permissible Dose: The amount of radiation that a person may receive during a specific period of time without harmful effects.

Rad: A measure of the dose of any ionizing radiation to body tissues in terms of the energy absorbed per unit of mass of the tissue. One rad is the dose corresponding to the absorption of 100 ergs* per gram of tissue.

Radioactivity: The process in which atoms undergo spontaneous disintegration, liberating energy as various forms of ionizing radiation.

Radioisotope: A radiation-emitting isotope of an element.

Rem: A measure of the dose of any ionizing radiation to body tissues in terms of its estimated biological effect relative to a dose of 1 roentgen of X rays. The relation of the rem to other dose units depends upon the biological effect under consideration and upon the conditions for irradiation. Each of the following is considered to be equivalent to a dose of 1 rem:

1. A dose of 1 rad due to X ray, gamma ray, or beta particle radiation.

2. A dose of 0.1 rad due to fast neutrons with an energy of 10 meV.

3. A dose of 0.05 rad due to particles heavier than neutrons and with sufficient energy to reach the lens of the eye.

One rem of neutron radiation may be assumed to be equivalent to 14 million fast neutrons per square centimeter falling on the body.

Mrad and Mrem: One-thousandth of a rad and one-thousandth of a rem, respectively.

Roentgen: The quantity of X rays or gamma rays that produces, directly or indirectly per 0.001293 gram of air, ions carrying one electrostatic unit of charge of either sign.

EFFECTS OF IONIZING RADIATION

Two types of radiation hazard must be distinguished: external hazards and internal hazards.

*An erg is a unit of work equal to the work done by a force of one dyne acting through a distance of one centimeter.

†Slow neutrons are considered to have a relative biological efficient that is about three times greater than that of X rays, gamma rays, or beta particles. This fact should be taken into account when considering rems in slow neutron fluxes.

Radiation that emanates from a source outside the body and is capable of penetrating the body and causing damage is an external hazard. X-ray machines are potential sources of external radiation hazards, and radioactive materials that give off X rays, gamma rays, beta particles, and neutrons are also external hazards.

An internal hazard exists when the body is irradiated by radioactive materials that have been swallowed, inhaled, or absorbed through the intact or broken skin. Radioactive materials that are external hazards become internal hazards when they enter the body by any of the above routes. Radioactive materials that emit only alpha particles or beta particles of low energy are not external hazards because the particles are stopped by the outer layers of skin. When, however, such materials enter the body they become a potent source of damage.

The exposure of living tissue to ionizing radiation has harmful effects to cells, ranging from inhibition of cell function to impairment of function and even to cell death. The extent of damage is dependent upon the radiation dosage and the organs exposed. Different organs have different radiosensitivities. Thus, radiation levels that are too low to result in visible damage to the skin can cause serious damage to one or more internal organs.

The effects of low dosages may be so delayed that they can be detected only by statistical observations of groups of exposed and unexposed individuals. These effects are tied to changes in the structure of the DNA molecule, the basic building block in living tissue, and can result in changes in the function of somatic cells as well as germ cells. The long-term radiation effects may include genetic effects, leukemia, and other types of malignancies, cataracts of the eye, and nonspecific shortening of the life-span.

Higher doses can cause decreases in white blood cells, bleeding, and anemia; ulceration of the gastrointestinal tract; and necrosis and ulceration of the skin.

Exposure to extremely high radiation dosages can result in three forms of radiation injury—the first affecting the brain, the second affecting the gastrointestinal system, and the third affecting the blood-producing system.

The first type begins with the victim experiencing weakness, listlessness, and sleepiness, followed by intermittent stupor, tremors or convulsions, and death. The victim rarely survives more than two days.

The gastrointestinal form of injury is characterized by nausea, vomiting, diarrhea, weakness, and prostration. These symptoms may disappear for a few days but then return in more acute form, perhaps accompanied by delirium or coma. Death usually occurs within two weeks.

The third type is first characterized by nausea, vomiting, and diarrhea, but these soon clear up and are followed by a 2 to 3 week symptom-free period. At the end of this period, the victim begins to lose his or her hair, experiences nosebleeds and hemorrhages, has a recurrence of diarrhea, and becomes extremely emaciated. Death occurs from 3 to 6 weeks following exposure.

EVALUATING HAZARDS

Radiation surveys and monitoring are undertaken to detect contamination in the air and on surfaces, to measure exposure of personnel, and to evaluate the procedures and techniques involved in handling sources of radiation. An initial survey must be made of each new facility, and periodic monitoring conducted as necessary to evaluate changes.

Adequate records must be kept and made available for inspection by a representative of the Atomic Energy Commission.

Radiation-Measuring and Monitoring Instruments

Commonly used devices include film badges, pocket ionization chambers, dosimeters, Geiger counters, scintillation counters, and portable air samplers.

Film Badges

These comprise a metal or plastic holder containing one or more radiation-sensitive films about the same size as the films used for dental X rays. A portion of the film is covered by a thin cadmium filter that measures gamma ray and X-ray dosage levels. The remainder of the film is unshielded and measures beta and low-energy gamma radiation. Special films are available for detecting neutrons. The films are developed and compared to standards to determine the extent of radiation exposure.

The films may be worn for periods ranging from one week to more than a month. The developed film may be filed as a permanent exposure record. Figure 5.1 illustrates a film badge.

Pocket Ionization Chambers

These are useful for nonroutine or hazardous situations where the radiation level varies widely from day to day. The chamber

FIGURE 5.1   Film badge.
(Photo by Michael D. Ells.)

consists essentially of a high-voltage electrode positioned within a case that serves as a second electrode. Before the day's work is begun, the central electrode is charged with a high-voltage battery. As the device is exposed to ionizing radiation, the air in the case becomes ionized and partly conducting, causing the charge on the central electrode to leak away gradually. At the end of the day, the level of the remaining charge is read with an electrometer; the decrease provides a measure of the radiation dosage. Sometimes electricity will leak from the central electrode and cause the instrument to indicate a falsely high dosage level. Misinterpretations because of this can be avoided by having employees wear two chambers, and accepting the lower reading as true. Figure 5.2 illustrates such a device.

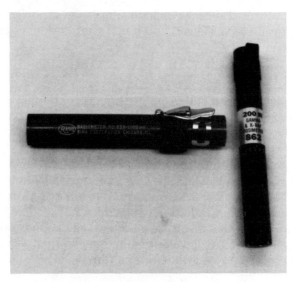

FIGURE 5.2   Pocket ionization chamber.
(Photo by Michael D. Ells.)

Dosimeters

These include units that are carried in the pocket and larger units used for surveys and as permanent monitoring devices. The dosimeter is similar to the ionization chamber but includes a built-in electrometer so that readings can be made at any time, and sometimes also incorporates a charging device. Such a device is particularly useful when employed in especially hazardous situations, since dangerous levels of radiation can be detected quickly and corrective measures taken immediately.

The ordinary dosimeter is sensitive to gamma rays and X rays but not to alpha, beta, and neutron radiation. By use of thin-wall construction, it is possible to make the devices sensitive to beta radiation, but durability its sacrificed. Coating the electrode system with natural boron that has been enriched with boron-10 isotope results in a dosimeter that is sensitive to slow neutrons. Devices for measuring fast neutrons have also been developed, but the sensitivity required can be obtained only in relatively large units. Hence, portable units for fast neutron measurement are not available.

Geiger Counters

This type of device is utilized for conducting radiation surveys and in permanent monitoring installations. The usual Geiger counter consists essentially of a gas-filled chamber with a thin, positively charged wire running longitudinally through the center from one end to the other. The wire is connected to an amplifying device and a recorder or signal, which in turn are connected to an inner shell that serves as the negative pole. Radiation entering the tube ionizes the gas and sets up a tiny current amplified to give a signal that can be seen, heard, or counted. Alpha, beta, and gamma radiation may be detected by Geiger counters. The chief drawback of most models is the long recovery time that results because of secondary electrical discharges from the ionization of the gas. These make more than a few counts a second impossible.

Scintillation Counters

Like Geiger counters, these are used for conducting radiation surveys and in permanent monitoring installations. The heart of the scintillation counter is a screen with a thin layer of crystals or a solution of organic scintillators. Incoming radiation strikes the screen or solution, temporarily disrupting the molecules and producing a tiny flash of light. The energy of the flash is converted to an electrical current, which amplifies to give a signal, as in the Geiger counter. Instruments are available for the detection of alpha, beta, gamma, and slow-neutron radiation.

The meters employed with dosimeters, Geiger counters, and scintillation counters may be calibrated to indicate dose rate.

Portable Air Samplers

These consist essentially of a small vacuum cleaner fitted with a flexible extension head and filter paper holder, as well as with an air-flow measuring device. A filter paper is placed in the holder, the vacuum turned on, and sampling carried out for a measured period. The radiation from the filter paper is determined with a suitable counting device; and from the radiation, air flow, and time, the environmental hazard is calculated.

Similar air samplers may be permanently installed in air ducts, exhaust stacks, and similar locations.

Surveying Procedure

OSHA standards require employers to make whatever surveys are required to insure that they are complying with the stipulations of the act. "Survey" means an evaluation of the radiation hazards incident to the production, use, release, disposal, or presence of radioactive materials or other sources of radiation. When appropriate, the evaluation must include a physical survey of the location of materials and equipment and measurements of the levels of radiation or concentrations of radioactive materials present.

A survey normally includes a study of the techniques and habits of individuals involved, the operating procedures, and the methods used in handling radioactive materials. The survey is conducted under typical operating conditions, and also under conditions of greatest possible hazard. Instruments used for a survey should be checked before the survey is started for proper operation and calibration in the energy range to be encountered. Survey records should include all data and recommendations for corrective action in technique and shielding. Personnel monitoring should be recommended where it is needed, and a recheck made after corrective action has been taken.

Each location or operation must be surveyed individually because of the large number of variables involved. The location and physical form of a radiation source, the type and energy of its radiations, its use, and the habits of the user are important factors. The time and intensity of exposures, the body areas exposed, the biological effectiveness of the particular radiation, and the possibility of internal exposure must also be considered.

The first step in surveying operations involving radioisotopes is to determine what isotopes are present. This is usually known from the history of the operation, but cross-contamination may be a factor if other radioisotopes are being used in nearby areas. Airborne contamination is hazardous and spreads rapidly, so it receives attention first.

Air samples should be taken at the worker's breathing zone if possible, as well as at air ducts and exhaust stacks. Surfaces should be thoroughly checked for contamination. Any possibly contaminated location should be scanned with a Geiger counter, remembering that the response time of the meter limits the speed of scanning unless earphones or flashing lights are used as indicators.

Shielding and containers should be checked for contamination and radiation leaks; all handling techniques should be observed and dose rates for each operation checked, with time limits imposed where necessary; surrounding areas should be checked for contamination and radiation leaks; effluents, air, and sewage should be assayed; and waste disposal practices checked. A program of routine and spot monitoring of radioisotope laboratories usually is desirable.

X-ray installations include radiographic, fluoroscopic, diffraction, diagnostic and therapy units, and electron microscopes. Special consideration is necessary with these units because of their directional beams and the variations in energy of radiations produced. Operating techniques and the operator's habits must be observed, with particular attention to the proper use of protective devices such as interlocks and shielding. Measurements of radiation levels at the operator's position and adjoining rooms and areas should be made to detect shielding leaks and scattered radiation. Normal operating conditions and worst possible hazard conditions should both be checked.

The dose rate or intensity of radiation, adequacy of shielding, and the user's knowledge of the potential hazard are important factors to consider in surveying cobalt-60 sources, radium static eliminator bars, density gauges, hi-vacuum gauges, and beta ray gauges used in the plant. Shielding should be sufficient to permit persons to work near the source without need of time limitations.

The surveyor should measure the dose rate with an ionization chamber instrument, and calculate the safe working distance for a 40-hour week. If workers are required to be closer than the safe working distance, and shielding cannot be added to reduce the intensity, a time limit at the working distance can be calculated to prevent overexposure. All persons concerned should be advised of safe working distances and times.

Sources should be wipe tested for contamination and leakage.

All sources of radiation should be marked with proper signs and registered with the supervisor of the building, so that if they become lost this fact will be quickly known and the harm minimized to persons ignorant of the hazard involved.

Monitoring Procedure

The radiation surveys described above must be supplemented by air, surface, and personnel monitoring to insure an adequate evaluation of ionizing radiation hazards.

Air Monitoring

Laboratory air and effluent should be monitored for radio-activity if there is any possibility of airborne contamination at hazardous levels. Beta and gamma activity may be determined by collecting particulate matter on a filter paper and counting it by appropriate technique. Determination of alpha activity is compli-cated by the presence of naturally occurring decay products of radon and thorium. The samples are collected by means of a portable air sample and the radiation level measured with the appropriate measuring device.

Surface Monitoring

A sample is taken by wiping with a filter paper part of the surface and checking the radioactivity on the paper with a Geiger counter or by liquid scintillation. Wipe samples are not only useful for estimating contamination on surfaces but also for determining whether or not the contamination is likely to rub off.

The radiation types present on a wipe sample can be identified by the use of filters. Alpha activity is completely stopped by a thin absorber such as paper; beta radiation is stopped by a thin metal absorber; and X rays and gamma rays will be only slightly attenuated by the same absorber.

Personnel Monitoring

Every employer must have available for use by employees who work with ionizing radiation sources appropriate personnel monitor-ing devices such as film badges, pocket ionization chambers, and pocket dosimeters.

The employer must require any employee over 18 years of age to use such equipment if he or she is working in an area where he or she will receive, or is likely to receive, a radiation dose that exceeds

25 percent of the value specified in Table 10. Minors are not allowed to work around ionizing radiation sources.

To insure maximum worker protection, it is desirable to issue film badges to persons who work regularly with radiation sources. The badges should be supplemented by pocket ionization chambers whenever an acute exposure of more than 100 mrad per day is possible.

When an unusual or potentially hazardous operation is to be performed, a special monitoring service involving one or more of the instruments used for radiation surveys may be indicated.

The limitations of personnel monitoring devices should be recognized by the user. They measure the exposure only at the spot where they are worn, which should be the part of the body expected to receive the largest exposure. Film badges and pocket chambers are normally worn in a breast pocket or at the waist since this region and the head are the most radiation-sensitive parts of the body. If the hands or feet may receive an unusual exposure, a second-film badge or chamber may be attached to them. In this way, a record is kept of exposure to the extremities as well as to the whole body. A second limitation is that these devices do not accurately measure beta or low energy X-ray exposure unless specially calibrated.

COMBATING HAZARDS

Hazards may be combated by setting safe exposure levels; establishing a safety training program; following safe work practices; having properly designed facilities; utilizing protective equipment and devices; using proper procedures for procuring, transporting, and disposing of radioactive materials; employing suitable signs, labels, and warning devices; and following the proper emergency procedures.

Maximum Permissible Exposure Levels

External Radiation Exposure

OSHA standards stipulate that employees who work in restricted areas may not, during any calendar quarter, receive dosages of radiation which exceed those shown in Table 10.

These radiation dosage levels may be exceeded if the dosage received by the whole body during one calendar quarter does not exceed 3 rems and if this dosage, when added to the total occupational dosage to the whole body, does not exceed 5 × (N-18)

TABLE 10

Radiation Dosage Limits	Rems per Calendar Quarter
Whole body: head and trunk; active blood-forming organs; lens of eyes; or gonads	1.25
Hands and forearms; feet and ankles	18.75
Skin of whole body	7.5

rems. In this formula N = the employee's age in years. When dosages are thus exceeded, the employer must maintain adequate past and present exposure records to show that the dosage is within specified limits.

Internal Radiation Exposure

Control of internal radiation exposure is accomplished by establishing permissible concentrations of radioisotopes in air and water. These permissible limits are calculated to produce over a working lifetime a maximum permissible body burden of the isotope in question. The permissible body burden is based on comparison of the specific isotope with radium for bone-seeking isotopes, or on the amount of isotope that will deliver to the critical organ (the organ in which the greatest isotope concentration accumulates) a maximum of 0.3 rem per week.

OSHA standards forbid exposure of employees to concentrations of airborne material in excess of the limits specified in Part I of Table 11.

These limits are for exposure for 40 hours in any work week of 7 consecutive days. In any 7-day period when the number of hours of exposure is less than 40, the limits specified in the table may be increased proportionately. In any 7-day period where the exposure hours number more than 40, the limits specified must be decreased proportionately.

For individuals less than 18 years old, exposures to airborne contamination must not exceed the levels specified in Part II of the table. Concentrations may be averaged over periods not greater than one week. These levels should also be considered for elderly persons as well as those with impaired health.

Safety Training Program

Several groups must be considered in any training program. First are those who work in the area of a radiation facility but have no responsibility for handling the radioactive materials or the

TABLE 11

Concentrations in Air and Water Above Natural Background

Element (atomic number)	Isotope[1]		Part I Column 1 Air (µc/ml)	Part I Column 2 Water (µc/ml)	Part II Column 1 Air (µc/ml)	Part II Column 2 Water (µc/ml)
nium (80)	Ac 227	S	2×10^{-12}	6×10^{-5}	8×10^{-14}	2×10^{-6}
		I	3×10^{-11}	9×10^{-3}	9×10^{-13}	3×10^{-4}
	Ac 228	S	8×10^{-8}	3×10^{-3}	3×10^{-9}	9×10^{-5}
		I	2×10^{-8}	3×10^{-3}	6×10^{-10}	9×10^{-5}
ericium (95)	Am 241	S	6×10^{-12}	1×10^{-4}	2×10^{-13}	4×10^{-6}
		I	1×10^{-10}	8×10^{-4}	4×10^{-12}	2×10^{-5}
	Am 243	S	6×10^{-12}	1×10^{-4}	2×10^{-13}	4×10^{-6}
		I	1×10^{-10}	8×10^{-4}	4×10^{-12}	3×10^{-5}
imony (51)	Sb 122	S	2×10^{-7}	8×10^{-4}	6×10^{-9}	3×10^{-5}
		I	1×10^{-7}	8×10^{-4}	5×10^{-9}	3×10^{-5}
	Sb 124	S	2×10^{-7}	7×10^{-4}	5×10^{-9}	2×10^{-5}
		I	2×10^{-8}	7×10^{-4}	7×10^{-10}	2×10^{-5}
	Sb 125	S	5×10^{-7}	3×10^{-3}	2×10^{-8}	1×10^{-4}
		I	3×10^{-8}	3×10^{-3}	9×10^{-10}	1×10^{-4}
on (18)	A 37	Sub[2]	6×10^{-3}		1×10^{-4}	
	A 41	Sub	$2 \times 10^{-6-}$		4×10^{-9}	
nic (33)	As 73	S	2×10^{-6}	1×10^{-2}	7×10^{-8}	5×10^{-4}
		I	4×10^{-7}	1×10^{-2}	1×10^{-8}	5×10^{-4}
	As 74	S	3×10^{-7}	2×10^{-3}	1×10^{-8}	5×10^{-5}
		I	1×10^{-7}	2×10^{-3}	4×10^{-9}	5×10^{-5}
	As 76	S	1×10^{-7}	6×10^{-4}	4×10^{-9}	2×10^{-5}
		I	1×10^{-7}	6×10^{-4}	3×10^{-9}	2×10^{-5}
	As 77	S	5×10^{-7}	2×10^{-3}	2×10^{-8}	8×10^{-5}
		I	4×10^{-7}	2×10^{-3}	1×10^{-8}	8×10^{-5}
tine (85)	At 211	S	7×10^{-9}	5×10^{-5}	2×10^{-10}	2×10^{-6}
		I	3×10^{-8}	2×10^{-3}	1×10^{-9}	7×10^{-5}
um (56)	Ba 131	S	1×10^{-6}	5×10^{-3}	4×10^{-8}	2×10^{-4}
		I	4×10^{-7}	5×10^{-3}	1×10^{-8}	2×10^{-4}
	Ba 140	S	1×10^{-7}	8×10^{-4}	4×10^{-9}	3×10^{-5}
		I	4×10^{-8}	7×10^{-4}	1×10^{-9}	2×10^{-5}
elium (97)	Bk 249	S	9×10^{-10}	2×10^{-2}	3×10^{-11}	6×10^{-4}
		I	1×10^{-7}	2×10^{-2}	4×10^{-9}	6×10^{-4}
llium (4)	Be 7	S	6×10^{-6}	5×10^{-2}	2×10^{-7}	2×10^{-3}
		I	1×10^{-6}	5×10^{-2}	4×10^{-8}	2×10^{-3}
uth (83)	Bi 206	S	2×10^{-7}	1×10^{-3}	6×10^{-9}	4×10^{-5}
		I	1×10^{-7}	1×10^{-3}	5×10^{-9}	4×10^{-5}
	Bi 207	S	2×10^{-7}	2×10^{-3}	6×10^{-9}	6×10^{-5}
		I	1×10^{-8}	2×10^{-3}	5×10^{-10}	6×10^{-5}
	Bi 210	S	6×10^{-9}	1×10^{-3}	2×10^{-10}	4×10^{-5}
		I	6×10^{-9}	1×10^{-3}	2×10^{-10}	4×10^{-5}
	Bi 212	S	1×10^{-7}	1×10^{-2}	3×10^{-9}	4×10^{-4}
		I	2×10^{-7}	1×10^{-2}	7×10^{-9}	4×10^{-4}
nine (35)	Br 82	S	1×10^{-6}	8×10^{-3}	4×10^{-8}	3×10^{-4}
		I	2×10^{-7}	1×10^{-3}	6×10^{-9}	4×10^{-5}
mium (48)	Cd 109	S	5×10^{-8}	5×10^{-3}	2×10^{-9}	2×10^{-4}
		I	7×10^{-8}	5×10^{-3}	3×10^{-9}	2×10^{-4}
	Cd 115m	S	4×10^{-8}	7×10^{-4}	1×10^{-9}	3×10^{-5}
		I	4×10^{-8}	7×10^{-4}	1×10^{-9}	3×10^{-5}
	Cd 115	S	2×10^{-7}	1×10^{-3}	8×10^{-9}	3×10^{-5}
		I	2×10^{-7}	1×10^{-3}	6×10^{-9}	4×10^{-5}
ium (20)	Ca 45	S	3×10^{-8}	3×10^{-4}	1×10^{-9}	9×10^{-6}
		I	1×10^{-7}	5×10^{-3}	4×10^{-9}	2×10^{-4}
	Ca 47	S	2×10^{-7}	1×10^{-3}	6×10^{-9}	5×10^{-5}
		I	2×10^{-7}	1×10^{-3}	6×10^{-9}	3×10^{-5}
fornium (98)	Cf 219	S	2×10^{-12}	1×10^{-4}	5×10^{-14}	4×10^{-6}
		I	1×10^{-10}	7×10^{-4}	3×10^{-12}	2×10^{-5}
	Cf 250	S	6×10^{-12}	4×10^{-4}	2×10^{-13}	1×10^{-5}
		I	1×10^{-10}	7×10^{-4}	3×10^{-12}	3×10^{-5}
	Cf 252	S	2×10^{-11}	7×10^{-4}	7×10^{-13}	2×10^{-5}
		I	1×10^{-10}	7×10^{-4}	4×10^{-12}	2×10^{-5}
on (6)	C 14	S	4×10^{-6}	2×10^{-2}	1×10^{-7}	8×10^{-4}
	(CO$_2$)	Sub	5×10^{-5}		1×10^{-6}	
um (58)	Ce 141	S	4×10^{-7}	3×10^{-3}	2×10^{-8}	9×10^{-5}
		I	2×10^{-7}	3×10^{-3}	5×10^{-9}	9×10^{-5}
	Ce 143	S	3×10^{-7}	1×10^{-3}	9×10^{-9}	4×10^{-5}
		I	2×10^{-7}	1×10^{-3}	7×10^{-9}	4×10^{-5}
	Ce 144	S	1×10^{-8}	3×10^{-4}	3×10^{-10}	1×10^{-5}
		I	6×10^{-9}	3×10^{-4}	2×10^{-10}	1×10^{-5}
um (55)	Cs 131	S	1×10^{-5}	7×10^{-2}	4×10^{-7}	2×10^{-3}
		I	3×10^{-6}	3×10^{-2}	1×10^{-7}	9×10^{-4}

Element (atomic number)	Isotope[1]		Part I		Part II	
			Column 1 Air (μc/ml)	Column 2 Water (μc/ml)	Column 1 Air (μc/ml)	Colum Wa (μc/m
	Cs 134m	S	4×10^{-5}	2×10^{-1}	1×10^{-6}	$6 \times$
		I	6×10^{-6}	3×10^{-2}	2×10^{-7}	$1 \times$
	Cs 134	S	4×10^{-8}	3×10^{-4}	1×10^{-9}	$9 \times$
		I	1×10^{-8}	1×10^{-3}	4×10^{-10}	$4 \times$
	Cs 135	S	5×10^{-7}	3×10^{-3}	2×10^{-8}	$1 \times$
		I	9×10^{-8}	7×10^{-3}	3×10^{-9}	$2 \times$
	Cs 136	S	4×10^{-7}	2×10^{-3}	1×10^{-8}	$9 \times$
		I	2×10^{-7}	2×10^{-3}	6×10^{-9}	$6 \times$
	Cs 137	S	6×10^{-8}	4×10^{-4}	2×10^{-9}	$2 \times$
		I	1×10^{-8}	1×10^{-3}	5×10^{-10}	$4 \times$
Chlorine (17)	Cl 36	S	4×10^{-7}	2×10^{-3}	1×10^{-8}	$8 \times$
		I	2×10^{-8}	2×10^{-3}	8×10^{-10}	$6 \times$
	Cl 38	S	3×10^{-6}	1×10^{-2}	9×10^{-8}	$4 \times$
		I	2×10^{-6}	1×10^{-2}	7×10^{-8}	$4 \times$
Chromium (24)	Cr 51	S	1×10^{-5}	5×10^{-2}	4×10^{-7}	$2 \times$
		I	2×10^{-6}	5×10^{-2}	8×10^{-8}	$2 \times$
Cobalt (27)	Co 57	S	3×10^{-6}	2×10^{-2}	1×10^{-7}	$5 \times$
		I	2×10^{-6}	1×10^{-2}	6×10^{-9}	$4 \times$
	Co 58m	S	2×10^{-5}	8×10^{-2}	6×10^{-7}	$3 \times$
		I	9×10^{-6}	6×10^{-2}	3×10^{-7}	$2 \times$
	Co 58	S	8×10^{-7}	4×10^{-3}	3×10^{-8}	$1 \times$
		I	5×10^{-8}	3×10^{-3}	2×10^{-9}	$9 \times$
	Co 60	S	3×10^{-7}	1×10^{-3}	1×10^{-8}	$5 \times$
		I	9×10^{-9}	1×10^{-3}	3×10^{-10}	$3 \times$
Copper (29)	Cu 64	S	2×10^{-6}	1×10^{-2}	7×10^{-8}	$3 \times$
		I	1×10^{-6}	6×10^{-3}	4×10^{-8}	$2 \times$
Curium (96)	Cm 242	S	1×10^{-10}	7×10^{-4}	4×10^{-12}	$2 \times$
		I	2×10^{-10}	7×10^{-4}	6×10^{-12}	$3 \times$
	Cm 243	S	6×10^{-12}	1×10^{-4}	2×10^{-13}	$5 \times$
		I	1×10^{-10}	7×10^{-4}	3×10^{-12}	$2 \times$
	Cm 244a	S	9×10^{-12}	2×10^{-4}	3×10^{-13}	$7 \times$
		I	1×10^{-10}	8×10^{-4}	3×10^{-12}	$3 \times$
	Cm 245	S	5×10^{-12}	1×10^{-4}	2×10^{-13}	$4 \times$
		I	1×10^{-10}	8×10^{-4}	4×10^{-12}	$3 \times$
	Cm 246	S	5×10^{-12}	1×10^{-4}	2×10^{-13}	$4 \times$
		I	1×10^{-10}	8×10^{-4}	4×10^{-12}	$3 \times$
Dysprosium (66)	Dy 165	S	3×10^{-6}	1×10^{-2}	9×10^{-8}	$4 \times$
		I	2×10^{-6}	1×10^{-2}	7×10^{-8}	$4 \times$
	Dy 166	S	2×10^{-7}	1×10^{-3}	8×10^{-9}	$4 \times$
		I	2×10^{-7}	1×10^{-3}	7×10^{-9}	$4 \times$
Erbium (68)	Er 169	S	6×10^{-7}	3×10^{-3}	2×10^{-8}	$9 \times$
		I	4×10^{-7}	3×10^{-3}	1×10^{-8}	$9 \times$
	Er 171	S	7×10^{-7}	3×10^{-3}	2×10^{-8}	$1 \times$
		I	6×10^{-7}	3×10^{-3}	2×10^{-8}	$1 \times$
Europium (63)	Eu 152 (T/2=9.2 hrs)	S I	4×10^{-7} 3×10^{-7}	2×10^{-3} 2×10^{-3}	1×10^{-8} 1×10^{-8}	$6 \times$ $6 \times$
	Eu 152 (T/2=13 yrs)	S I	1×10^{-8} 2×10^{-8}	2×10^{-3} 2×10^{-3}	4×10^{-10} 6×10^{-10}	$8 \times$ $8 \times$
	Eu 154	S	4×10^{-9}	6×10^{-4}	1×10^{-10}	$2 \times$
		I	7×10^{-9}	6×10^{-4}	2×10^{-10}	$2 \times$
	Eu 155	S	9×10^{-8}	6×10^{-3}	3×10^{-9}	$2 \times$
		I	7×10^{-8}	6×10^{-3}	3×10^{-9}	$2 \times$
Fluorine (9)	F 18	S	5×10^{-6}	2×10^{-2}	2×10^{-7}	$8 \times$
		I	3×10^{-6}	1×10^{-2}	9×10^{-8}	$5 \times$
Gadolinium (64)	Gd 153	S	2×10^{-7}	6×10^{-3}	8×10^{-9}	$2 \times$
		I	9×10^{-8}	6×10^{-3}	3×10^{-9}	$2 \times$
	Gd 159	S	5×10^{-7}	2×10^{-3}	2×10^{-8}	$8 \times$
		I	4×10^{-7}	2×10^{-3}	1×10^{-8}	$8 \times$
Gallium (31)	Ga 72	S	2×10^{-7}	1×10^{-3}	8×10^{-9}	$4 \times$
		I	2×10^{-7}	1×10^{-3}	6×10^{-9}	$4 \times$
Germanium (32)	Ge 71	S	1×10^{-5}	5×10^{-2}	4×10^{-7}	$2 \times$
		I	6×10^{-6}	5×10^{-2}	2×10^{-7}	$2 \times$
Gold (79)	Au 196	S	1×10^{-6}	5×10^{-3}	4×10^{-8}	$2 \times$
		I	6×10^{-7}	4×10^{-3}	2×10^{-8}	$1 \times$
	Au 198	S	3×10^{-7}	2×10^{-3}	1×10^{-8}	$5 \times$
		I	2×10^{-7}	1×10^{-3}	8×10^{-9}	$5 \times$
	Au 199	S	1×10^{-5}	5×10^{-3}	4×10^{-8}	$2 \times$
		I	8×10^{-7}	4×10^{-3}	3×10^{-8}	$2 \times$
Hafnium (72)	Hf 181	S	4×10^{-8}	2×10^{-3}	1×10^{-9}	$7 \times$
		I	7×10^{-8}	2×10^{-3}	3×10^{-9}	$7 \times$

		Part I		Part II	
		Column 1 Air ($\mu c/ml$)	Column 2 Water ($\mu c/ml$)	Column 1 Air ($\mu c/ml$)	Column 2 Water ($\mu c/ml$)
ent (atomic number)	Isotope[1]				
ıium (67)	Ho 166 S	2×10^{-7}	9×10^{-4}	7×10^{-9}	3×10^{-5}
	I	2×10^{-7}	9×10^{-4}	6×10^{-9}	3×10^{-5}
rogen (1)	H3 S	5×10^{-6}	1×10^{-1}	2×10^{-7}	3×10^{-3}
	Sub	2×10^{-3}		4×10^{-5}	
ım (49)	In 113m S	8×10^{-6}	4×10^{-2}	3×10^{-7}	1×10^{-3}
	I	7×10^{-6}	4×10^{-2}	2×10^{-7}	1×10^{-3}
	In 114m S	1×10^{-7}	5×10^{-4}	4×10^{-9}	2×10^{-5}
	I	2×10^{-8}	5×10^{-4}	7×10^{-10}	2×10^{-5}
	In 115m S	2×10^{-6}	1×10^{-2}	8×10^{-8}	4×10^{-4}
	I	2×10^{-6}	1×10^{-2}	6×10^{-8}	4×10^{-4}
	In 115 S	2×10^{-7}	3×10^{-3}	9×10^{-9}	9×10^{-5}
	I	3×10^{-8}	3×10^{-3}	1×10^{-9}	9×10^{-5}
ıe (53)	I 126 S	8×10^{-9}	5×10^{-5}	3×10^{-10}	2×10^{-6}
	I	3×10^{-7}	3×10^{-3}	1×10^{-8}	9×10^{-5}
	I 129 S	2×10^{-9}	1×10^{-5}	6×10^{-11}	4×10^{-7}
	I	7×10^{-8}	6×10^{-3}	2×10^{-9}	2×10^{-4}
	I 131 S	9×10^{-9}	6×10^{-5}	3×10^{-10}	2×10^{-6}
	I	3×10^{-7}	2×10^{-3}	1×10^{-8}	6×10^{-5}
	I 132 S	2×10^{-7}	2×10^{-3}	8×10^{-9}	6×10^{-5}
	I	9×10^{-7}	5×10^{-3}	3×10^{-8}	2×10^{-4}
	I 133 S	3×10^{-8}	2×10^{-4}	1×10^{-9}	7×10^{-6}
	I	2×10^{-7}	1×10^{-3}	7×10^{-9}	4×10^{-5}
	I 134 S	5×10^{-7}	4×10^{-3}	2×10^{-8}	1×10^{-4}
	I	3×10^{-6}	2×10^{-2}	1×10^{-7}	6×10^{-4}
	I 135 S	1×10^{-7}	7×10^{-4}	4×10^{-9}	2×10^{-5}
	I	4×10^{-7}	2×10^{-3}	1×10^{-8}	7×10^{-5}
ım (77)	Ir 190 S	1×10^{-6}	6×10^{-3}	4×10^{-8}	2×10^{-4}
	I	4×10^{-7}	5×10^{-3}	1×10^{-8}	2×10^{-4}
	Ir 192 S	1×10^{-7}	1×10^{-3}	4×10^{-9}	4×10^{-5}
	I	3×10^{-8}	1×10^{-3}	9×10^{-10}	4×10^{-5}
	Ir 194 S	2×10^{-7}	1×10^{-3}	8×10^{-9}	3×10^{-5}
	I	2×10^{-7}	9×10^{-4}	5×10^{-9}	3×10^{-5}
(26)	Fe 55 S	9×10^{-7}	2×10^{-2}	3×10^{-8}	8×10^{-4}
	I	1×10^{-6}	7×10^{-2}	3×10^{-8}	2×10^{-3}
	Fe 59 S	1×10^{-7}	2×10^{-3}	5×10^{-9}	6×10^{-5}
	I	5×10^{-8}	2×10^{-3}	2×10^{-9}	5×10^{-5}
ıton[2] (36)	Kr 85m Sub	6×10^{-6}		1×10^{-7}	
	Kr 85 Sub	1×10^{-5}		3×10^{-7}	
	Kr 87 Sub	1×10^{-6}		2×10^{-8}	
hanum (57)	La 140 S	2×10^{-7}	7×10^{-4}	5×10^{-9}	2×10^{-5}
	I	1×10^{-7}	7×10^{-4}	4×10^{-9}	2×10^{-5}
(82)	Pb 203 S	3×10^{-6}	1×10^{-2}	9×10^{-8}	4×10^{-4}
	I	2×10^{-6}	1×10^{-2}	6×10^{-8}	4×10^{-4}
	Pb 210 S	1×10^{-10}	4×10^{-5}	4×10^{-12}	1×10^{-7}
	I	2×10^{-10}	5×10^{-3}	8×10^{-12}	2×10^{-4}
	Pb 212 S	2×10^{-8}	6×10^{-4}	6×10^{-10}	2×10^{-5}
	I	2×10^{-8}	5×10^{-4}	7×10^{-10}	2×10^{-5}
tium (71)	Lu 177 S	6×10^{-7}	3×10^{-3}	2×10^{-8}	1×10^{-4}
ganese (25)	Mn 52 S	5×10^{-7}	3×10^{-3}	2×10^{-8}	1×10^{-4}
	I	2×10^{-7}	1×10^{-3}	7×10^{-9}	3×10^{-5}
		1×10^{-7}	9×10^{-4}	5×10^{-9}	3×10^{-5}
	Mn 54 S	4×10^{-7}	4×10^{-3}	1×10^{-9}	1×10^{-4}
	I	4×10^{-8}	3×10^{-3}	1×10^{-9}	1×10^{-4}
	Mn 56 S	8×10^{-7}	4×10^{-3}	3×10^{-9}	1×10^{-4}
	I	5×10^{-7}	3×10^{-3}	2×10^{-8}	1×10^{-7}
ury (80)	Hg 197m S	7×10^{-7}	6×10^{-3}	3×10^{-8}	2×10^{-4}
	I	8×10^{-7}	5×10^{-3}	3×10^{-8}	2×10^{-4}
	Hg 197 S	1×10^{-6}	9×10^{-3}	4×10^{-8}	3×10^{-4}
	I	3×10^{-6}	1×10^{-2}	9×10^{-8}	5×10^{-4}
	Hg 203 S	7×10^{-8}	5×10^{-4}	2×10^{-9}	2×10^{-5}
	I	1×10^{-7}	3×10^{-3}	4×10^{-9}	1×10^{-4}
bdenum (42)	Mo 99 S	7×10^{-7}	5×10^{-3}	3×10^{-9}	2×10^{-4}
	I	2×10^{-7}	1×10^{-3}	7×10^{-9}	4×10^{-5}
ıymium (60)	Nd 144 S	8×10^{-11}	2×10^{-3}	3×10^{-12}	7×10^{-5}
	I	3×10^{-10}	2×10^{-3}	1×10^{-11}	8×10^{-5}
	Nd 147 S	4×10^{-7}	2×10^{-3}	1×10^{-8}	6×10^{-5}
	I	2×10^{-7}	2×10^{-3}	8×10^{-9}	6×10^{-5}
	Nd 149 S	2×10^{-6}	8×10^{-3}	6×10^{-8}	3×10^{-4}
	I	1×10^{-6}	8×10^{-3}	5×10^{-8}	3×10^{-4}
unium (93)	Np 237 S	4×10^{-12}	9×10^{-5}	1×10^{-13}	3×10^{-6}
	I	1×10^{-10}	9×10^{-4}	4×10^{-12}	3×10^{-5}

Element (atomic number)	Isotope[1]		Part I		Part II	
			Column 1 Air (µc/ml)	Column 2 Water (µc/ml)	Column 1 Air (µc/ml)	Colum Wate (µc/m
	Np 239	S	8×10^{-7}	4×10^{-3}	3×10^{-8}	1×1
		I	7×10^{-7}	4×10^{-3}	2×10^{-9}	1×1
Nickel (28)	Ni 59	S	5×10^{-7}	6×10^{-3}	2×10^{-8}	2×1
		I	8×10^{-7}	6×10^{-2}	3×10^{-8}	2×1
	Ni 63	S	6×10^{-8}	8×10^{-4}	2×10^{-9}	3×1
		I	3×10^{-7}	2×10^{-2}	1×10^{-8}	7×1
	Ni 65	S	9×10^{-7}	4×10^{-3}	3×10^{-9}	1×1
		I	5×10^{-7}	3×10^{-3}	2×10^{-8}	1×1
Niobium (Columbium) (41)	Nb 93m	S	1×10^{-7}	1×10^{-2}	4×10^{-9}	4×1
		I	2×10^{-7}	1×10^{-2}	5×10^{-9}	4×1
	Nb 95	S	5×10^{-7}	3×10^{-3}	2×10^{-9}	1×1
		I	1×10^{-7}	3×10^{-3}	3×10^{-9}	1×1
	Nb 97	S	6×10^{-6}	3×10^{-2}	2×10^{-7}	9×1
		I	5×10^{-6}	3×10^{-2}	2×10^{-7}	9×1
Osmium (76)	Os 185	S	5×10^{-7}	2×10^{-3}	2×10^{-8}	7×1
		I	5×10^{-8}	2×10^{-3}	2×10^{-9}	7×1
	Os 191	S	2×10^{-5}	7×10^{-2}	6×10^{-7}	3×1
		I	9×10^{-6}	7×10^{-2}	3×10^{-7}	2×1
	Os 191m	S	1×10^{-6}	5×10^{-3}	4×10^{-8}	2×1
		I	4×10^{-7}	5×10^{-3}	1×10^{-8}	2×1
	Os 193	S	4×10^{-7}	2×10^{-3}	1×10^{-8}	6×1
		I	3×10^{-7}	2×10^{-3}	9×10^{-9}	6×1
Palladium (46)	Pd 103	S	1×10^{-6}	1×10^{-2}	6×10^{-8}	3×1
		I	7×10^{-7}	8×10^{-3}	3×10^{-8}	3×1
	Pd 109	S	6×10^{-7}	3×10^{-3}	2×10^{-8}	9×1
		I	4×10^{-7}	2×10^{-3}	1×10^{-8}	7×1
Phosphorus (15)	P32	S	7×10^{-8}	5×10^{-4}	2×10^{-9}	2×1
		I	8×10^{-8}	7×10^{-4}	3×10^{-9}	2×1
Platinum (78)	Pt 191	S	8×10^{-7}	4×10^{-3}	3×10^{-8}	1×1
		I	6×10^{-7}	3×10^{-3}	2×10^{-8}	1×1
	Pt 193m	S	7×10^{-6}	3×10^{-2}	2×10^{-7}	1×1
		I	5×10^{-6}	3×10^{-2}	2×10^{-7}	1×1
	Pt 197m	S	6×10^{-6}	3×10^{-2}	2×10^{-7}	1×1
		I	5×10^{-6}	3×10^{-2}	2×10^{-7}	9×1
	Pt 197	S	8×10^{-7}	4×10^{-3}	3×10^{-8}	1×1
		I	6×10^{-7}	3×10^{-3}	2×10^{-8}	1×1
Plutonium (94)	Pu 238	S	2×10^{-12}	1×10^{-4}	7×10^{-14}	5×1
		I	3×10^{-11}	8×10^{-4}	1×10^{-12}	3×1
	Pu 239	S	2×10^{-12}	1×10^{-4}	6×10^{-14}	5×1
		I	4×10^{-11}	8×10^{-4}	1×10^{-12}	3×1
	Pu 240	S	2×10^{-12}	1×10^{-4}	6×10^{-14}	5×1
		I	4×10^{-11}	8×10^{-4}	1×10^{-12}	3×1
	Pu 241	S	9×10^{-11}	7×10^{-3}	3×10^{-12}	2×1
		I	4×10^{-8}	4×10^{-2}	1×10^{-9}	1×1
	Pu 242	S	2×10^{-12}	1×10^{-4}	6×10^{-14}	5×1
		I	4×10^{-11}	9×10^{-4}	1×10^{-12}	3×1
Polonium (84)	Po 210	S	5×10^{-10}	2×10^{-5}	2×10^{-11}	7×1
		I	2×10^{-10}	8×10^{-4}	7×10^{-12}	3×1
Potassium (19)	K 42	S	2×10^{-6}	9×10^{-3}	7×10^{-8}	3×1
		I	1×10^{-7}	6×10^{-4}	4×10^{-9}	2×1
Praseodymium (59)	Pr 142	S	2×10^{-7}	9×10^{-4}	7×10^{-9}	3×1
		I	2×10^{-7}	9×10^{-4}	5×10^{-9}	3×1
	Pr 143	S	3×10^{-7}	1×10^{-3}	1×10^{-8}	5×1
		I	2×10^{-7}	1×10^{-3}	6×10^{-9}	5×1
Promethium (61)	Pm 147	S	6×10^{-8}	6×10^{-3}	2×10^{-9}	2×1
		I	1×10^{-7}	6×10^{-3}	3×10^{-9}	2×1
	Pm 149	S	3×10^{-7}	1×10^{-3}	1×10^{-8}	4×1
		I	2×10^{-7}	1×10^{-3}	8×10^{-9}	4×1
Protoactinium (91)	Pa 230	S	2×10^{-9}	7×10^{-3}	6×10^{-11}	2×1
		I	8×10^{-10}	7×10^{-3}	3×10^{-11}	2×1
	Pa 231	S	1×10^{-12}	3×10^{-5}	4×10^{-14}	9×1
		I	1×10^{-10}	8×10^{-4}	4×10^{-12}	2×1
	Pa 233	S	6×10^{-7}	4×10^{-3}	2×10^{-8}	1×1
		I	2×10^{-7}	3×10^{-3}	6×10^{-9}	1×1
Radium (88)	Ra 223	S	2×10^{-9}	2×10^{-5}	6×10^{-11}	7×1
		I	2×10^{-10}	1×10^{-4}	8×10^{-12}	4×1
	Ra 224	S	5×10^{-9}	7×10^{-5}	2×10^{-10}	2×1
		I	7×10^{-10}	2×10^{-4}	2×10^{-11}	5×1
	Ra 226	S	3×10^{-11}	4×10^{-7}	1×10^{-12}	1×1
		I	5×10^{-11}	9×10^{-4}	2×10^{-12}	3×1
	Ra 228	S	7×10^{-11}	8×10^{-7}	2×10^{-12}	3×1

ment (atomic number)	Isotope¹		Part I — Column 1 Air (µc/ml)	Part I — Column 2 Water (µc/ml)	Part II — Column 1 Air (µc/ml)	Part II — Column 2 Water (µc/ml)
		I	4×10^{-11}	7×10^{-4}	1×10^{-12}	3×10^{-5}
on (86)	Rn 220	S	3×10^{-7}		1×10^{-8}	
		I				
	Rn 222	S	1×10^{-7}		3×10^{-9}	
nium (75)	Re 183	S	3×10^{-6}	2×10^{-2}	9×10^{-8}	6×10^{-4}
		I	2×10^{-7}	8×10^{-3}	5×10^{-9}	3×10^{-4}
	Re 186	S	6×10^{-7}	3×10^{-3}	2×10^{-8}	9×10^{-5}
		I	2×10^{-7}	1×10^{-3}	8×10^{-9}	5×10^{-5}
	Re 187	S	9×10^{-6}	7×10^{-2}	3×10^{-7}	3×10^{-3}
		I	5×10^{-7}	4×10^{-2}	2×10^{-8}	2×10^{-3}
	Re 188	S	4×10^{-7}	2×10^{-3}	1×10^{-8}	6×10^{-5}
		I	2×10^{-7}	9×10^{-4}	6×10^{-9}	3×10^{-5}
dium (45)	Rh 103m	S	8×10^{-5}	4×10^{-1}	3×10^{-6}	1×10^{-2}
		I	6×10^{-5}	3×10^{-1}	2×10^{-6}	1×10^{-2}
	Rh 105	S	8×10^{-7}	4×10^{-3}	3×10^{-8}	1×10^{-4}
		I	5×10^{-7}	3×10^{-3}	2×10^{-8}	1×10^{-4}
idium (37)	Rb 86	S	3×10^{-7}	2×10^{-3}	1×10^{-8}	7×10^{-5}
		I	7×10^{-8}	7×10^{-4}	2×10^{-9}	2×10^{-5}
	Rb 87	S	5×10^{-7}	3×10^{-3}	2×10^{-8}	1×10^{-4}
		I	7×10^{-8}	5×10^{-3}	2×10^{-9}	2×10^{-4}
henium (44)	Ru 97	S	2×10^{-6}	1×10^{-2}	8×10^{-8}	4×10^{-4}
		I	2×10^{-6}	1×10^{-2}	6×10^{-8}	3×10^{-4}
	Ru 103	S	5×10^{-7}	2×10^{-3}	2×10^{-8}	8×10^{-5}
		I	8×10^{-8}	2×10^{-3}	3×10^{-9}	8×10^{-5}
	Ru 105	S	7×10^{-7}	3×10^{-3}	2×10^{-8}	1×10^{-4}
		I	5×10^{-7}	3×10^{-3}	2×10^{-8}	1×10^{-4}
	Ru 106	S	8×10^{-8}	4×10^{-4}	3×10^{-9}	1×10^{-5}
		I	6×10^{-9}	3×10^{-4}	2×10^{-10}	1×10^{-5}
arium (62)	Sm 147	S	7×10^{-11}	2×10^{-3}	2×10^{-12}	6×10^{-5}
		I	3×10^{-10}	2×10^{-3}	9×10^{-12}	7×10^{-5}
	Sm 151	S	6×10^{-8}	1×10^{-2}	2×10^{-9}	4×10^{-4}
		I	1×10^{-7}	1×10^{-2}	5×10^{-9}	4×10^{-4}
	Sm 153	S	5×10^{-7}	2×10^{-3}	2×10^{-8}	8×10^{-5}
		I	4×10^{-7}	2×10^{-3}	1×10^{-8}	8×10^{-5}
ndium (21)	Sc 46	S	2×10^{-7}	1×10^{-3}	8×10^{-9}	4×10^{-5}
		I	2×10^{-8}	1×10^{-3}	8×10^{-10}	4×10^{-5}
	Sc 47	S	6×10^{-7}	3×10^{-3}	2×10^{-8}	9×10^{-5}
		I	5×10^{-7}	3×10^{-3}	2×10^{-8}	9×10^{-5}
	Sc 48	S	2×10^{-7}	8×10^{-4}	6×10^{-9}	3×10^{-5}
		I	1×10^{-7}	8×10^{-4}	5×10^{-9}	3×10^{-5}
nium (34)	Se 75	S	1×10^{-6}	9×10^{-3}	4×10^{-8}	3×10^{-4}
		I	1×10^{-7}	8×10^{-3}	4×10^{-9}	3×10^{-4}
on (14)	Si 31	S	6×10^{-6}	3×10^{-2}	2×10^{-7}	9×10^{-4}
		I	1×10^{-6}	6×10^{-3}	3×10^{-8}	2×10^{-4}
r (47)	Ag 105	S	6×10^{-7}	3×10^{-3}	2×10^{-8}	1×10^{-4}
		I	8×10^{-9}	3×10^{-3}	3×10^{-9}	1×10^{-4}
	Ag 110m	S	2×10^{-7}	9×10^{-4}	7×10^{-9}	3×10^{-5}
		I	1×10^{-5}	9×10^{-4}	3×10^{-10}	3×10^{-5}
	Ag 111	S	3×10^{-7}	1×10^{-3}	1×10^{-8}	4×10^{-5}
		I	2×10^{-7}	1×10^{-3}	8×10^{-9}	4×10^{-5}
ium (11)	Na 22	S	2×10^{-7}	1×10^{-3}	6×10^{-9}	4×10^{-5}
		I	9×10^{-9}	9×10^{-4}	3×10^{-10}	3×10^{-5}
	Na 24	S	1×10^{-6}	6×10^{-3}	4×10^{-8}	2×10^{-4}
		I	1×10^{-7}	8×10^{-4}	5×10^{-9}	3×10^{-5}
ntium (38)	Sr 85m	S	4×10^{-5}	2×10^{-1}	1×10^{-6}	7×10^{-3}
		I	3×10^{-5}	2×10^{-1}	1×10^{-6}	7×10^{-3}
	Sr 85	S	2×10^{-7}	3×10^{-3}	8×10^{-9}	1×10^{-4}
		I	1×10^{-7}	5×10^{-3}	4×10^{-9}	2×10^{-4}
	Sr 89	S	3×10^{-9}	3×10^{-4}	1×10^{-10}	1×10^{-5}
		I	4×10^{-9}	8×10^{-4}	1×10^{-9}	3×10^{-5}
	Sr 90	S	3×10^{-10}	4×10^{-6}	1×10^{-11}	1×10^{-7}
		I	5×10^{-9}	1×10^{-3}	2×10^{-10}	4×10^{-5}
	Sr 91	S	4×10^{-7}	2×10^{-3}	2×10^{-8}	7×10^{-5}
		I	3×10^{-7}	1×10^{-3}	9×10^{-9}	5×10^{-5}
	Sr 92	S	4×10^{-7}	2×10^{-3}	2×10^{-8}	7×10^{-5}
		I	3×10^{-7}	2×10^{-3}	1×10^{-8}	6×10^{-5}
ur (16)	S 35	S	3×10^{-7}	2×10^{-3}	9×10^{-9}	6×10^{-5}
		I	3×10^{-7}	8×10^{-3}	9×10^{-9}	3×10^{-4}
talum (73)	Ta 182	S	4×10^{-8}	1×10^{-3}	1×10^{-9}	4×10^{-5}
		I	2×10^{-8}	1×10^{-3}	7×10^{-10}	4×10^{-5}

			Part I		Part II	
Element (atomic number)	*Isotope*[1]		*Column 1 Air (μc/ml)*	*Column 2 Water (μc/ml)*	*Column 1 Air (μc/ml)*	*Colu Wa (μc/*
Technetium (43)	Tc 96m	S	8×10^{-5}	4×10^{-1}	3×10^{-6}	$1 \times$
		I	3×10^{-5}	3×10^{-1}	1×10^{-6}	$1 \times$
	Tc 96	S	6×10^{-7}	3×10^{-3}	2×10^{-8}	$1 \times$
		I	2×10^{-7}	1×10^{-3}	8×10^{-9}	$5 \times$
	Tc 97m	S	2×10^{-6}	1×10^{-2}	8×10^{-8}	$4 \times$
		I	2×10^{-7}	5×10^{-3}	5×10^{-9}	$2 \times$
	Tc 97	S	1×10^{-5}	5×10^{-2}	4×10^{-7}	$2 \times$
		I	3×10^{-7}	2×10^{-2}	1×10^{-8}	$8 \times$
	Tc 99m	S	4×10^{-5}	2×10^{-1}	1×10^{-6}	$6 \times$
		I	1×10^{-3}	8×10^{-2}	5×10^{-7}	$3 \times$
	Tc 99	S	2×10^{-6}	1×10^{-2}	7×10^{-8}	$3 \times$
		I	6×10^{-8}	5×10^{-3}	2×10^{-9}	$2 \times$
Tellurium (52)	Te 125m	S	4×10^{-7}	5×10^{-3}	1×10^{-8}	$2 \times$
		I	1×10^{-7}	3×10^{-3}	4×10^{-9}	$1 \times$
	Te 127m	S	1×10^{-7}	2×10^{-3}	5×10^{-9}	$6 \times$
		I	4×10^{-8}	2×10^{-3}	1×10^{-9}	$5 \times$
	Te 127	S	2×10^{-6}	8×10^{-3}	6×10^{-9}	$3 \times$
		I	9×10^{-7}	5×10^{-3}	3×10^{-8}	$2 \times$
	Te 129m	S	8×10^{-8}	1×10^{-3}	3×10^{-9}	$3 \times$
		I	3×10^{-8}	6×10^{-4}	1×10^{-9}	$2 \times$
	Te 129	S	5×10^{-6}	2×10^{-2}	2×10^{-7}	$8 \times$
		I	4×10^{-6}	2×10^{-2}	1×10^{-7}	$8 \times$
	Te 131m	S	4×10^{-7}	2×10^{-3}	1×10^{-8}	$6 \times$
		I	2×10^{-7}	1×10^{-3}	6×10^{-9}	$4 \times$
	Te 132	S	2×10^{-7}	9×10^{-4}	7×10^{-9}	$3 \times$
		I	1×10^{-7}	6×10^{-4}	4×10^{-9}	$2 \times$
Terbium (65)	Tb 160	S	1×10^{-7}	1×10^{-3}	3×10^{-9}	$4 \times$
		I	3×10^{-8}	1×10^{-3}	1×10^{-9}	$4 \times$
Thallium (81)	Tl 200	S	3×10^{-6}	1×10^{-2}	9×10^{-8}	$4 \times$
		I	1×10^{-6}	7×10^{-3}	4×10^{-8}	$2 \times$
	Tl 201	S	2×10^{-6}	9×10^{-3}	7×10^{-8}	$3 \times$
		I	9×10^{-7}	5×10^{-3}	3×10^{-8}	$2 \times$
	Tl 202	S	8×10^{-7}	4×10^{-3}	3×10^{-8}	$1 \times$
		I	2×10^{-7}	2×10^{-3}	8×10^{-9}	$7 \times$
	Tl 204	S	6×10^{-7}	3×10^{-3}	2×10^{-8}	$1 \times$
		I	3×10^{-8}	2×10^{-3}	9×10^{-10}	$6 \times$
Thorium (90)	Th 228	S	9×10^{-12}	2×10^{-4}	3×10^{-13}	$7 \times$
		I	6×10^{-12}	4×10^{-4}	2×10^{-13}	$1 \times$
	Th 230	S	2×10^{-12}	5×10^{-5}	8×10^{-14}	$2 \times$
		I	1×10^{-11}	9×10^{-4}	3×10^{-13}	$3 \times$
	Th 232	S	3×10^{-11}	5×10^{-5}	1×10^{-12}	$2 \times$
		I	3×10^{-11}	1×10^{-3}	1×10^{-12}	$4 \times$
	Th natural	S	3×10^{-11}	3×10^{-5}	1×10^{-12}	$1 \times$
		I	3×10^{-11}	3×10^{-4}	1×10^{-12}	$1 \times$
	Th 234	S	6×10^{-8}	5×10^{-4}	2×10^{-9}	$2 \times$
		I	3×10^{-8}	5×10^{-4}	1×10^{-9}	$2 \times$
Thulium (69)	Tm 170	S	4×10^{-8}	1×10^{-3}	1×10^{-9}	$5 \times$
		I	3×10^{-8}	1×10^{-3}	1×10^{-9}	$5 \times$
	Tm 171	S	1×10^{-7}	1×10^{-2}	4×10^{-8}	$5 \times$
		I	2×10^{-7}	1×10^{-2}	8×10^{-9}	$5 \times$
Tin (50)	Sn 113	S	4×10^{-7}	2×10^{-3}	1×10^{-8}	$9 \times$
		I	5×10^{-8}	2×10^{-3}	2×10^{-9}	$8 \times$
	Sn 125	S	1×10^{-7}	5×10^{-4}	4×10^{-9}	$2 \times$
		I	8×10^{-8}	5×10^{-4}	3×10^{-9}	$2 \times$
Tungsten (Welfram) (74)	W 181	S	2×10^{-6}	1×10^{-2}	8×10^{-9}	$4 \times$
		I	1×10^{-7}	1×10^{-2}	4×10^{-9}	$3 \times$
	W 185	S	8×10^{-7}	4×10^{-3}	3×10^{-8}	$1 \times$
		I	1×10^{-7}	3×10^{-3}	4×10^{-9}	$1 \times$
	W 187	S	4×10^{-7}	2×10^{-3}	2×10^{-8}	$7 \times$
		I	3×10^{-7}	2×10^{-3}	1×10^{-8}	$6 \times$
Uranium (92)	U 230	S	3×10^{-10}	1×10^{-4}	1×10^{-11}	$5 \times$
		I	1×10^{-10}	1×10^{-4}	4×10^{-12}	$5 \times$
	U 232	S	1×10^{-10}	8×10^{-4}	3×10^{-12}	$3 \times$
		I	3×10^{-11}	8×10^{-4}	9×10^{-13}	$3 \times$
	U 233	S	5×10^{-10}	9×10^{-4}	2×10^{-11}	$3 \times$
		I	1×10^{-10}	9×10^{-4}	4×10^{-12}	$3 \times$
	U 234	S	6×10^{-10}	9×10^{-4}	2×10^{-11}	$3 \times$
		I	1×10^{-10}	9×10^{-4}	4×10^{-12}	$3 \times$
	U 235	S	5×10^{-10}	8×10^{-4}	2×10^{-11}	$3 \times$
		I	1×10^{-10}	8×10^{-4}	4×10^{-12}	$3 \times$

		Part I		Part II	
		Column 1 Air ($\mu c/ml$)	Column 2 Water ($\mu c/ml$)	Column 1 Air ($\mu c/ml$)	Column 2 Water ($\mu c/ml$)
ent (atomic number)	Isotope[1]				
	U 236 S	6×10^{-10}	1×10^{-3}	2×10^{-11}	3×10^{-5}
	I	1×10^{-10}	1×10^{-3}	4×10^{-12}	3×10^{-5}
	U 238 S	7×10^{-11}	1×10^{-3}	3×10^{-12}	4×10^{-5}
	I	1×10^{-10}	1×10^{-3}	5×10^{-12}	4×10^{-5}
	U-natural S	7×10^{-11}	5×10^{-4}	3×10^{-12}	2×10^{-5}
	I	6×10^{-11}	5×10^{-4}	2×10^{-12}	2×10^{-5}
dium (23)	V 48 S	2×10^{-7}	9×10^{-4}	6×10^{-9}	3×10^{-5}
	I	6×10^{-8}	8×10^{-4}	2×10^{-9}	3×10^{-5}
n (54)	Xe 131m Sub	2×10^{-5}		4×10^{-7}	
	Xe 133 Sub	1×10^{-5}		3×10^{-7}	
	Xe 135 Sub	4×10^{-6}		1×10^{-7}	
bium (70)	Yb 175 S	7×10^{-7}	3×10^{-3}	2×10^{-8}	1×10^{-4}
	I	6×10^{-7}	3×10^{-3}	2×10^{-8}	1×10^{-4}
m (39)	Y 90 S	1×10^{-7}	6×10^{-4}	4×10^{-9}	2×10^{-5}
	I	1×10^{-7}	6×10^{-4}	3×10^{-9}	2×10^{-5}
	Y 91m S	2×10^{-5}	1×10^{-1}	8×10^{-7}	3×10^{-3}
	I	2×10^{-5}	1×10^{-1}	6×10^{-7}	3×10^{-3}
	Y 91 S	4×10^{-8}	8×10^{-4}	1×10^{-9}	3×10^{-5}
	I	3×10^{-8}	8×10^{-4}	1×10^{-9}	3×10^{-5}
	Y 92 S	4×10^{-7}	2×10^{-3}	1×10^{-8}	6×10^{-5}
	I	3×10^{-7}	2×10^{-3}	1×10^{-8}	6×10^{-5}
	Y 93 S	2×10^{-7}	8×10^{-4}	6×10^{-9}	3×10^{-5}
	I	1×10^{-7}	8×10^{-4}	5×10^{-9}	3×10^{-5}
(30)	Zn 65 S	1×10^{-7}	3×10^{-3}	4×10^{-9}	1×10^{-4}
	I	6×10^{-8}	5×10^{-3}	2×10^{-9}	2×10^{-4}
	Zn 69m S	4×10^{-7}	2×10^{-3}	1×10^{-8}	7×10^{-5}
	I	3×10^{-7}	2×10^{-3}	1×10^{-8}	6×10^{-5}
	Zn 69 S	7×10^{-6}	5×10^{-2}	2×10^{-7}	2×10^{-3}
	I	9×10^{-6}	5×10^{-2}	3×10^{-7}	2×10^{-3}
nium (40)	Zr 93 S	1×10^{-7}	2×10^{-2}	4×10^{-9}	8×10^{-4}
	I	3×10^{-7}	2×10^{-2}	1×10^{-8}	8×10^{-4}
	Zr 95 S	1×10^{-7}	2×10^{-3}	4×10^{-9}	6×10^{-5}
	I	3×10^{-8}	2×10^{-3}	1×10^{-9}	6×10^{-5}
	Zr 97 S	1×10^{-7}	5×10^{-4}	4×10^{-9}	2×10^{-5}
	I	9×10^{-8}	5×10^{-4}	3×10^{-9}	2×10^{-5}

uble ( ); Insoluble (I).
b" means that values given are for submersion in an infinite cloud of gaseous material.

radiation-producing machinery. These persons must be informed of the extent of the hazard; without such information they will be apprehensive naturally. Also, these persons might otherwise tamper with radioactive sources, thus increasing the hazard to themselves as well as to others.

The second group includes technical and nontechnical personnel who handle radioactive materials or radiation-producing machinery, either routinely or incidentally in connection with projects that are primarily nonradioactive in nature. The training for this group must be quite different from that of the first group.

The third group is made up of personnel who would be involved in case of an emergency, such as a fire or explosion. These people will include plant protection, safety department, medical department, fire department, and industrial hygiene personnel.

A suitable training program should be designed in units so that people may participate only in those parts that they need. Such a program is described below.

Employees who have completed the training program should be given periodic reviews and kept up to date on new developments through a continuing program of safety meetings. For a review of safety meetings, see Chapter 2.

Preliminary Training Unit

The first unit of the training course should be aimed directly at explaining radiation in an elementary manner and putting it in perspective with the other, better-known hazards that workers have encountered.

This unit is needed by all persons who will have any contact at all with radioactive materials and radiation-producing machinery. (Those who simply work in the area but have no direct contact with the radiation source will need no further training.) It may be conveniently built around the series of explanatory cartoons that has been prepared by the fire and safety branch of the Atomic Energy Commission. Presentation of the cartoons should be followed by a discussion related directly to the radiation problems of specific interest to the group. Pocket-sized books containing the cartoons and any other pertinent matter may also be furnished to the participants.

Training of Persons Handling Radioactive Sources

Two groups of persons are involved, each requiring a different type of training. The first group comprises both technical and nontechnical personnel who routinely handle radioactive materials or radiation-producing machinery. These people must know in detail what is expected of them. The company policy should be outlined in a safety standard for the control of radiation hazards. The standard should spell out procedures for purchasing, receiving, registering, and disposing of radiation-producing equipment and materials. It should also describe the recordkeeping and monitoring responsibilities of persons having jurisdiction over radiation-producing equipment and materials, provide information concerning the availability of outside assistance for monitoring work areas and educating workers, and describe the responsibilities of a radiation hazards committee. Each person who will actually handle or work with radioactive materials or radiation-producing machines should be expected to know in detail the information set forth in the standard.

A radiation protection manual can be prepared to supplement the standard. Such a manual supplies detailed information concerning all major aspects of radiation hazards. It can be used as a basis for a periodic review of hazard control within departments where radioactive materials or radiation-producing machines are used. This normally can be done in regular safety meetings, which are handled by supervision within the department.

The second group is made up of technically trained personnel who have no experience with either radiochemistry or radiation physics but who wish to use these techniques to assist them in other forms of research. Such persons will be directly responsible for the work to be done. Thus, they must be taught to carry out their duties in a safe manner and according to regulations imposed by the Atomic Energy Commission, the state health department, and company policy.

This particular phase of the course conveniently may be built around a series of films available from the AEC. The films have the following titles: "Fundamentals of Radioactivity," "Properties of Radiation," "Principles of Radiological Safety," "Practice of Radiological Safety," "Biological Effects of Radiation," and "Practical Procedures of Measurement."

The whole formal training course for these two groups should culminate in a written examination so that there will be a record of individual participation, completion, and understanding of radiation safety training. No laboratory work is included in the training program; however, "cold runs" may be carried out in order to establish the proper techniques.

A continuing educational program can be built into the record keeping and monitoring systems. Each individual user of radioactive materials must keep records of use and disposal. These records become a part of overall records, which are kept so that the requirements of the state health department and the AEC will be satisfied. This provides for regularly scheduled contact between the record keeper and the user. At that time, the hazards and precautions associated with a particular use can be discussed. Similar discussions of hazards and their control should accompany the periodic survey of all sealed sources.

Training of Special Groups

Special groups include those who may be involved in handling emergencies—the medical, safety, fire, plant protection, and industrial hygiene departments. All of these people will require the preliminary training already described. In addition, fire, safety, and plant protection personnel should have a quarterly review of information important to them. Such information should include a review of the possible extent of hazards that may occur if various sources of radioactivity are involved in fire or explosion. It should also include practice in the use of the survey instruments that the groups employ for checking external radiation. Personnel should be advised not to enter areas where their instruments show excessive radiation until a monitor from the industrial hygiene group arrives and approves entry. These groups should be advised to wear gas masks whenever called upon to fight a fire in an area where loose radioactive contamination may exist. They should be trained to remain in the area with their equipment until it has been monitored following the fire, so that assurance may be given that there will be no spread of contamination. These personnel must be trained to handle injured persons with possible radioactive contamination.

Medical department and selected safety department personnel should receive thorough training in how to proceed in the event of an emergency. The personnel should be impressed with the importance of notifying the industrial hygiene department so that a qualified person with the equipment to handle wounds can be dispatched to the scene. Details of personal protection for medical and safety personnel should also be covered, including removal of wristwatches, jewelry, and similar items before contamination can occur; use of caps, gowns, rubber gloves, and respirators; and the desirability of minimizing contact with victims and contaminated items. Procedures for cleaning up both ambulatory patients and those on stretchers should be discussed thoroughly. The desirability of saving biological specimens such as blood, urine, and tissues for later radioactive analysis should be emphasized. Finally, personnel must be made to understand the importance of monitoring everything—ambulances, equipment, supplies, and personnel—before they leave the area.

Members of the industrial hygiene department and selected safety department personnel should review the radiation protection manual on a quarterly basis. This manual should describe the location and purpose of the various portable monitoring devices. Each individual should also practice using the devices. The various locations where one may find radioactive materials or radiation-producing machinery should be reviewed, along with the type of hazard to be expected at each location and the procedure for monitoring. Stress should be placed on the importance of utilizing proper respiratory protective equipment in emergencies involving loose radioactive material and on the importance of carefully monitoring all persons or equipment before they are allowed to leave the area. Where external radiation is a possibility, the extent of the hazard should be described for each location.

Predetermined total permissible exposures for various conditions should be reviewed and practice provided in converting meter readings to acceptable times of exposure. The radiation manual should contain a "life-saving" table by which the radiation intensity near a lethal radiation source can be determined from a meter reading taken at a relatively low intensity site farther from the source. This table should be reviewed diligently during the training sessions.

Safe Work Practices

The danger of injury from radioactive isotopes can be materially lessened by observing a number of general handling precautions and personal hygiene practices.

Manipulations should be carried out over a suitable drip tray or some form of double container, which will minimize the importance of breakages or spills. The work area should be covered with absorbent material to soak up minor spills. The absorbent material should be changed when unsuitable for further work, and considered as radioactive waste.

Whenever a choice among several isotopes of different toxicity is possible, one of relatively low toxicity should be selected and the smallest quantity necessary for the purpose should be used. Wet operations should be used in preference to dry ones, and frequent transfers of isotopes should be avoided if possible.

Skin contamination should be avoided by wearing gloves whenever radioactive material is suspected of being on the outside of bottles. Likewise, to prevent ingestion, no solution of radioactive isotopes should ever be pipetted by mouth. No person with an open skin wound below the wrist should work with radioactive isotopes unless the wound is firmly protected, preferably with gloves, against contamination.

Used pipettes, stirring rods, beakers, and other equipment should be placed on absorbent material, never directly on the table or bench. Such equipment should receive special handling when it is cleaned. To avoid the spread of contamination, equipment and tools employed in a radiation area should not be used in a nonradiation area.

When not in use, radioactive sources should be kept in a well-designed and suitably shielded area. Volatile compounds should be stored in a well-ventilated location. All sources should be suitably labeled, giving information on the activity and nature of the material. Self-adhesive labels are preferable to those that must be wetted. Their use will prevent the possibility of an employee becoming contaminated by licking a label that has started to fall off a bottle. High alpha radiation sources should not be stored in thin-walled bottles because irradiation may weaken the glass.

Persons working with radioactive materials should wash exposed parts of the body thoroughly whenever leaving the active area, even though no known contact has occurred. Wash hands 2 to 3 minutes by the clock with a mild, pure soap in tepid water, covering the entire surface with a good lather.

Protective clothing and equipment should be monitored routinely each day, or when a particular job with radioactive material has been finished. As an added precaution, exposed portions of street clothes, such as trouser cuffs and shoes, should be monitored at the end of each work day.

To guard against contamination caused by breakage, all bottles containing radioactive materials should be stored within a larger vessel, such as a tray, that is big enough to hold the contents of all of the bottles. This precaution guards against the spread of contamination and greatly lessens the time and effort needed to decontaminate the area.

Suitable waste receptables should be handy. After the waste is placed in the receptable, the amount of activity being disposed of should be documented. Solid wastes should be segregated into high-level and low-level waste. Low-level waste, such as paper tissue, filter paper, glassware, and generally the absorbent pads, should be placed in plastic bags inside step-lid cans or fiber packs. This will provide for easy removal and containment of contamination. (Waste disposal is discussed in detail later in this chapter.)

Food, candy, or beverages should not be consumed in areas where radioactive materials are used. Smoking should not be permitted in such areas.

Design of Facility

To insure a safer working environment and to minimize the spread of radioactive contamination, offices and similar nonactive work areas should be separated from the active areas in which radioactive materials and radiation-producing equipment are used. Separation may be achieved by constructing a doorless wall between the two areas and having separate doors to the two sections. If the installation is sufficiently large, offices may be placed in a separate building.

Within the active area, low-level and high-level radiation rooms should be arranged so that the worker proceeds gradually from one to the other. The entrance to the active areas should open into a room containing monitoring devices for checking workers when they leave the area. A small room for storing coats, rubbers, and such items, and for washing up should be located off the monitor room. A room for changing into laboratory clothing should be located just beyond the door leading from the monitor room into the laboratory area. This room should contain washing facilities as well as hooks for laboratory clothing. The active area should be provided with an emergency exit that is ordinarily kept closed.

The ventilation system should provide 10 to 20 changes of air per minute. The air should be filtered to prevent extensive intake of dust, and the pressure inside the building should be slightly lower than the pressure outside to prevent contaminants from escaping outside. As an added safety measure, exhaust air can also be filtered. The flow of air should be so arranged that contaminants do not pass from areas of high-level to low-level radiation.

Drains for radioactive wastes should be completely separated from other drains and should lead to tanks where the wastes can be stored until disposed of. The system should be designed so that there is no possibility of wastes from high-level areas backing up into low-level areas. In some instances, separate drain systems for high-level and low-level wastes may be desirable.

High-gloss paint is suitable for the walls and ceilings of radiation facilities. Floors are preferably covered with waxed linoleum, and bench tops with waxed linoleum or plastic. Periodic cleaning and rewaxing of the linoleum will remove minor contamination. Badly contaminated linoleum is readily and inexpensively replaced.

Detailed information on designing a laboratory facility is available from the Atomic Energy Commission. Inquiries should be addressed to: USAEC, P.O. Box 62, Oak Ridge, Tennessee 37830.

Protective Equipment and Devices

Personal Protective Devices and Clothing

Respiratory Equipment. Whenever there is danger of inhaling radioactive particles, self-contained breathing apparatus should be worn. This may be either the type in which air is supplied from a tank or the type in which the air is regenerated. Filter-type devices may be utilized for short periods of time with low-level contaminants, but they cannot be generally recommended. Apparatus should be thoroughly decontaminated following use.

Protective Clothing. Laboratory clothing or coveralls should be worn whenever there is a significant chance that radioactive materials will contaminate street clothing. Rubber gloves should be used to guard against hand contamination and canvas shoe covers to protect against shoe contamination. Clothing should be removed at the end of the work day or whenever the wearer leaves the work area. If clothing is suspected of being contaminated, it should be checked immediately and removed if found to be so. The clothing should be handled carefully, washed separately from other garments, and not reused until it is checked and found free of radioactivity.

Face and Eye Protection. Face shields will protect against splashing or spattering onto the face, and safety goggles will guard against eye contamination.

Air-Supplied Suits. Where accidents have caused an area to become grossly contaminated with radiation, or other circumstances make it likely that an individual may become contaminated by a number of routes, air-supplied suits may be worn. These may or may not have integral feet and hands. If separate boots and gloves are used, the sleeves of the suit should be worn outside the gauntlets of the gloves and the trouser legs outside the top of the boots. Contaminated suits should be washed thoroughly with soap and water before they are removed.

Hoods

Any quantity of radioactive material in excess of a few microcuries should be handled in a hood or glove box. The hood should provide a face velocity of 200 feet per minute with the sash raised 18 inches. The sides and bottom should be designed so as to provide a smooth,· nonturbulent influx of air. Hood limitations should be understood and respected: hoods are not designed to control fires, explosions, or other violent disturbances; they are not maintenance-free, corrosion-proof, or fail-safe; and they should not be used as storage cabinets.

Glove Boxes

A glove box may be more desirable than a hood for working with dusty solids or volatile liquids. This device is a sealed box with one or more transparent sides and with a pair of rubber gloves mounted in glove ports and protruding into the interior. To guard against the escape of radioactive materials into the laboratory, the box is equipped with an exhaust system comprising a blower mounted behind a filter. This system maintains a slightly reduced pressure within the box. To decrease the laboratory contamination further, radioactive materials are introduced into and removed from the glove box through a double-doored chamber at one end.

Shielding

Shielding is an important consideration in working with radioactive sources and storing radioactive materials. The type of shielding required will depend upon the type and strength of radiation emitted.

Radioactive materials classified as soft beta emitters—such as carbon-14, sulfur-35, and hydrogen-3—do not require shielding. Hard beta emitters—for example, phosphorus-32—should be shielded by a three-eighths-inch-thick piece of plastic or glass. If the source produces appreciable amounts of secondary X rays, some lead shielding may also be required.

Gamma-ray and X-ray emitters must be stored in closed containers with adequate lead or concrete shielding. The thickness of the shielding will depend upon the strength of the source involved and the degree of attenuation desired. For example, 120 millimeters of lead or 26 inches of concrete are required to attenuate X rays with a peak energy of 2,000 kilovolts by a factor of 1,000.

Water, paraffin, or some other substance containing large amounts of hydrogen is effective for shielding neutron emitters. Thin sheets of cadmium may also be utilized if the amount of shielding is small enough to make its use economically feasible. Under some circumstances, shielding against neutrons results in the production of secondary gamma radiation, which must be taken into account in providing the proper shielding protection.

The tube in X-ray-producing equipment should be completely shielded except for the exit window, and emanations from the window controlled by filters and other appropriate devices. State regulations set detailed requirements for the shielding of various X-ray-producing equipment.

Where the type and strength of radiation are such that the material must be completely enclosed by shielding, remotely operated tongs or similar devices will be required for carrying out handling operations. In such cases, operations are viewed through a lead glass window or a dense transparent liquid.

Procuring, Transporting, and Disposing of Radioactive Materials

Procuring

According to the Atomic Energy Act of 1954 and the regulations established under its authority, organizations wishing to use radioactive materials other than radium must obtain from the Atomic Energy Commission a license for each location where the materials will be employed. Each individual license specifies the kinds and quantities of materials authorized for use.

When purchasing radioactive materials from a commercial supplier for the first time, the supplier must be provided with a copy of the license to establish the purchaser's right to possess the material. For subsequent purchases from the same supplier, only the license number is required.

It is important to keep an accurate record of the materials purchased and where they are being used. To facilitate this, a copy of each purchase order should be sent to the industrial hygiene department of the organization. If the organization has no such department, some other individual or group—for example, the medical department—should be designated to receive the copy. Each group using radioactive isotopes should keep records of the amounts received and dispensed. These records should be sent to the industrial hygiene department or other designated recipient. Periodically, the two sets of records should be checked against each other to insure that they are up to date. Similar records should be kept of the purchase and location of radiation-producing equipment.

Transporting

Before radioactive material is shipped to another location or organization, the shipper must have a copy of the recipient's license.

Each container of material must be packaged in such a manner that damage to the package will not permit the radioactive source and container to be separated from the shipping carton.

The outside of the shipping carton should bear appropriate labels. The Department of Transportation (DOT) requires one of three labels, depending upon the radiation level from the obsolete container. The kinds and amounts of radioactive material must be listed and the radiation level indicated.

If quantities shipped are sufficiently large, a caution label bearing the radiation hazard symbol is required by OSHA regulations (Figure 5.3, page 145), and a special DOT permit also may be required.

Waste Disposal

All radioactive wastes must be handled in accordance with the regulations of the Atomic Energy Commission. The person or group generating the wastes should identify the isotopes present, specify their chemical state if known, and estimate the amount of activity involved. The industrial hygiene group should evaluate the health hazard involved in handling the waste and, in conjunction with waste disposal personnel, should choose the most suitable means of disposal.

Two general techniques are employed for the disposal of radioactive wastes. First, whenever possible, the wastes should be diluted with stable isotopes. Second, materials with short half-lives should be stored for several half-lives before final disposal.

Specific procedures for solid and liquid wastes are discussed below.

Solid Wastes. Solid wastes include such items as the absorbent material used on benches and tables, glass beakers and pipettes, rubber gloves, and filters. Solid wastes should be segregated into high- and low-level wastes. Low-level wastes should be placed in lined step-lid cans or fiber packs. High-level wastes should be placed in lined wide-mouth cans or doubly wrapped in plastic and sealed with tape.

The tops of filled liners should be folded over and sealed with adhesive tape. This should be done carefully to avoid disturbing the contents and thereby releasing radioactive particles into the atmosphere.

Burial and incineration are the two courses open for the disposal of solid wastes. Wastes are prepared for burial by placing them in a concrete container, which is then sealed with wet concrete. AEC regulations specify that the total quantity of buried material at any one location and time cannot exceed 1,000 times the amount specified in Table 12. Burial must be at a depth of at least 4 feet; successive burials must be separated by distances of at least 6 feet; and no more than 12 burials may be made in one year. Special approval is required for the incineration of radioactive wastes.

Liquid Wastes. Such wastes should be retained in polyethylene bottles, lined five-gallon cans, or—in the case of effluents escaping to the drains—in special storage tanks. Disposal is commonly by dilution and discharge into sanitary sewage systems. This can be done, however, only if the stipulations of the Atomic Energy Commission are followed. To begin, the waste must be readily soluble or dispersible in water. Second, certain limitations on the quantity must be observed. This permissible quantity can be the larger of the following:

1. The quantity which, if diluted by the average daily quantity of sewage released into the sewer by the licensee, will give an average concentration equal to the limits in Part I, Column 2, Table 11.
2. Ten times the quantity of material specified in Table 12.

The second value may, however, be used only if (1) the quantity of radioactive material released in any one month, if diluted by the average monthly quantity of water released by the licensee, will not result in an average concentration exceeding the limits specified in Part I, Column 2, of Table 11, and (2) the gross quantity

TABLE 12

Material	Micro-curies	Material	Micro-curies
Ag^{105}	1	Pd^{109}	10
Ag^{111}	10	Pm^{147}	10
As^{76}, As^{77}	10	Po^{210}	0.1
Au^{198}	10	Pr^{143}	10
Au^{199}	10	Pu^{239}	1
$Ba^{140}+La^{140}$	1	Ra^{226}	0.1
Be^{7}	50	Rb^{86}	10
C^{14}	50	Re^{186}	10
Ca^{45}	10	Rh^{105}	10
$Cd^{109}+Ag^{109}$	10	$Ru^{106}+Rh^{106}$	1
$Ce^{144}+Pr^{144}$	1	S^{35}	50
C^{136}	1	Sb^{124}	1
Co^{60}	1	Sc^{46}	1
Cr^{51}	50	Sm^{153}	10
$Cs^{137}+Ba^{137}$	1	Sn^{113}	10
Cu^{64}	50	Sr^{89}	1
Eu^{154}	1	$Sr^{90}+Y^{90}$	0.1
F^{18}	50	Ta^{182}	10
Fe^{55}	50	Tc^{96}	1
Fe^{59}	1	Tc^{99}	1
Ga^{72}	10	Te^{127}	10
Ge^{71}	50	Te^{129}	1
H^{3} (HTO or H^{3},O)	250	Th (natural)	50
I^{131}	10	Tl^{204}	50
In^{114}	1	U (natural)	50
Ir^{192}	10	U^{233}	1
K^{42}	10	$U^{234}-U^{235}$	50
La^{140}	10	V^{48}	1
Mn^{52}	1	W^{185}	10
Mn^{56}	50	Y^{90}	1
Mo^{99}	10	Y^{91}	1
Na^{22}	10	Zn^{65}	10
Na^{24}	10	Unidentified radioactive materials	
Nb^{95}	10	or any of the above in unknown	
Ni^{59}	1	mixtures	0.1
Ni^{63}	1		
P^{32}	10		
$Pd^{103}+Rh^{103}$	50		

of radioactive material released into the sewage system by the licensee does not exceed one curie per year.

Signs, Labels, Warning Devices

OSHA standards set strict requirements for posting and labeling areas in which radioactive sources are utilized and containers in which they are stored and transported. They also stipulate the use of warning devices under certain conditions.

RADIATION SYMBOL
1. Cross-hatched area is to be magenta or purple.
2. Background is to be yellow.

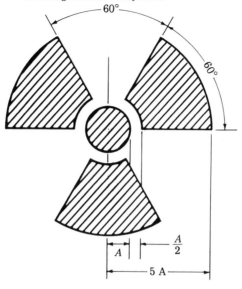

FIGURE 5.3.

All of the specified signs and labels must bear the conventional radiation symbol shown in Figure 5.3 The details of posting, labeling, and the use of warning devices are covered in the remaining portions of this section.

Posting

OSHA regulations distinguish among three types of areas for posting purposes:

A "radiation area" is any area accessible to personnel in which radiation exists at such levels that a major portion of the body could receive in any one hour a dose in excess of 5 millirems, or in any consecutive 5 days a dose in excess of 100 millirems.

A "high radiation area" is any area accessible to personnel in which radiation exists at such levels that a major portion of the body could receive in any one day a dose in excess of 100 millirems.

An "airborne radioactivity area" is a room, enclosure, or operating area in which airborne radioactive materials exist in concentrations in excess of those specified in Part I, Column I, of Table 11. Rooms or other areas in which airborne concentrations exceed 25 percent of these same limits when averaged over the number of hours per week during which individuals are in the area are also designated as airborne radioactivity areas.

These three areas must be conspicuously posted with a sign or signs bearing the radiation symbol and the following appropriate phrasing: "Caution—Radiation Area," "Caution—High Radiation Area," or "Caution—Airborne Radioactive Area."

Posting is, however, not required in a room with a sealed source of radioactivity, provided the radioactivity level 12 inches from the container or housing does not exceed 5 millirems per hour, nor is it required for rooms containing radioactive materials for less than 8 hours.

Labeling

Each container in which is transported, stored, or used a quantity of any radioactive material (other than natural uranium or thorium) greater than the quantity specified in Table 12 must have a label bearing the radiation symbol and the words "Caution—Radioactive Materials." This label also must be placed on any container in which natural uranium or thorium is transported, stored, or used in a quantity 10 times that specified in Table 12.

Notwithstanding this provision, however, labeling is not required if the concentration of the material in the container does not exceed that specified in Part I, Column 2, of Table 11. Also, laboratory containers used transiently in laboratory procedures need not be labeled when the user is present.

When containers are used for storage, the labels must also state the quantities and kinds of materials and the date that the quantities were measured.

Warning Devices

Each high radiation area must be equipped with a control device that will either cause the level of radiation to be reduced below that at which an entering individual might receive a dose of 100 millirems in one hour or energize a conspicuous visible or audible alarm signal. Signals are not required where areas are established for less than 30 days.

In addition to this type of signal, OSHA standards require that audible warning signals be installed at every location where an emergency involving radioactive material might make immediate, rapid, and complete evacuation essential. The signal must be unique in the plant or facility and of sufficient duration to insure that all persons who might be affected by the emergency will hear it. The signal-generating system must be one that will respond automatically to an initiating event without requiring any human action to sound the signal. Detailed stipulations concerning the sound characteristics of the signal and the design characteristics of the system are spelled out in the standards.

Emergency Procedures

These include procedures for decontaminating minor and major spills and those to be followed in case of fire.

All groups using or storing potentially hazardous amounts of radioactive materials should have a written emergency plan and periodically review it in their safety meetings. The plan should include the location of emergency equipment such as respiratory devices and fire extinguishers. It should also provide a plan of action for each type of foreseeable emergency: fire, explosion, minor wounds, major injuries, and release of radioactive materials to other areas by leakage, spills, or other accidents.

Decontaminating Minor Spills

A minor spill is one that does not constitute a radiation hazard to personnel and may be cleaned up by laboratory workers. Should a minor spill occur, all other personnel in the room should be notified at once and only the minimum number needed to deal with it allowed into the area.

Persons carrying out decontamination should don protective gloves and respirators before doing anything with the spill. If the spill is dry, the material should be dampened thoroughly, taking care not to spread the contamination or create a dust hazard. The material should then be removed from the contaminated surface by swabbing repeatedly, taking care not to spread the decontamination. Once decontamination is complete, all persons involved in the spill and cleanup should be monitored and a complete history of the accident and subsequent activity prepared for the laboratory records.

Decontaminating Major Spills

A major spill is one in which personnel and large portions of the work area become contaminated. If a major spill occurs, all persons in the area should be evacuated to a safe location. If there is a chance the contamination might spread to adjacent buildings, their person-nel should be immediately warned of the impending danger. The

industrial hygiene and plant protection departments, or corresponding personnel, should be notified to give aid in isolating the area and
in decontamination.

Handling Personnel. All personnel in the area should be
checked for contamination of skin or clothing. If contamination is
found, the individual should be made to remove all clothing and
shower thoroughly with soap and water for 15 minutes, then be
taken to a medical facility for re-examination and possible further
treatment.

If an injury accompanies the exposure, the ambulance should be
called, clothing removed, and decontamination of the skin begun by
flushing with large amounts of running water. Anyone who helps a
contaminated person may likewise become contaminated, and should
decontaminate him- or herself as soon as possible.

Handling Contaminated Clothing and Personal Equipment. All
contaminated clothing should be isolated until it is decontaminated.
Contaminated clothing must be laundered separately in a special
washing machine with the wash and rinse water monitored or, if the
volume of washing is too low to justify such a procedure, disposed of
as dry radioactive waste. (See earlier section in this chapter on waste
disposal.)

Rubber gloves usually decontaminate readily by washing with
plenty of suds and hot water, but excessive time cannot be justified
in this operation because of the low replacement cost. Leather goods
such as shoes cannot be readily decontaminated; therefore, such
items should be disposed of as dry waste. Respirators and safety
goggles may be decontaminated by washing with soap and water, a
hot 20 percent solution of sodium citrate, or similar agents.

Handling Laboratory Tools and Glassware. Contaminated
equipment should be placed inside a plastic bag and set aside in an
isolated location until proper decontamination can be carried out.
This equipment should not be released from the laboratory for repair
or any other purpose until the activity has been reduced to a safe
level.

When the half-life of the contaminant is short, it may be
desirable to store the equipment for decay of activity rather than
attempt to decontaminate. Also, if the items are cheap or easily
replaced, it may be simpler to dispose of them as solid waste.

Glass and porcelain articles may be cleaned with mineral acids,
ammonium citrate, trisodium phosphate, or chromic acid cleaning
solution. When the glaze on porcelain is broken, decontamination is
very difficult, and it is usually cheaper to replace the item.

Metal objects may be decontaminated by dilute mineral acids such as nitric acid, or with a 10 percent solution of sodium citrate. Brass polish is an excellent decontaminant for brass.

Plastics may be cleaned with ammonium citrate, dilute acids, or organic solvents.

In general, decontamination seldom exceeds 99.9 percent efficiency and usually runs about 98 to 99.5 percent. If the residual contamination indicates that the level of activity still exceeds permissible limits, the equipment should be regarded as radioactive waste.

Handling Floors, Benches, Hoods, and Other Items. Linoleum may be decontaminated by swabbing with absorbent paper moistened with perchloroethylene, kerosene, ammonium citrate solution, or dilute mineral acids. Care should be taken not to dissolve sealing compounds between abutting edges of the linoleum.

Ceramic tile may be decontaminated by use of mineral acids, ammonium citrate, or trisodium phosphate.

Paint is sometimes successfully decontaminated by perchloroethylene or 10 percent hydrochloric acid. However, such treatment may partially dissolve the paint. If so, the paint should be completely removed and a new coating applied.

Concrete has been successfully decontaminated by the use of 32 percent hydrochloric acid. If this is not successful, the contaminated surface may be covered with another layer of fresh concrete.

Contaminated wood surfaces may be painted if the contaminant emits only alpha or weak beta radiation. Otherwise, the wood must be replaced, or else sanded or planed.

Sinks may be decontaminated by flushing thoroughly with a large volume of water, scouring with a rust remover, soaking in a solution of citric acid made by adding one pound of acid to one gallon of water, and flushing once again with a large volume of water.

Fire Fighting Emergencies

In the event of a fire where radioactive materials may be present, measures must be taken to prevent fire fighting personnel and their equipment from contamination. Upon reaching the scene of the fire, apparatus should be parked upwind, out of the smoke area if possible, and tools and equipment should be dispensed to fire fighters by a person who has not been contaminated.

The amount of radiation in the fire area should be determined by fire department or plant protection personnel furnished with radiation-detection instruments. In addition, each person entering the radiation area should wear an emergency badge identifying him or her by name and number.

Each fire fighter entering the emergency area must wear a respirator until such time as it has been deemed unnecessary by personnel with monitoring devices.

Any fire fighter receiving a wound while working in a contaminated area must receive *immediate attention* to prevent a hazardous amount of radioactive material from entering the body.

Everyone leaving the area must be monitored with a suitable meter and, if found to be contaminated, required to shower and put on clean clothing. Film badges must be surrendered to the appropriate person at this time. Tools, hoses, and other materials and equipment that may have been contaminated must be segregated until monitored with a suitable instrument.

Investigating and Reporting of Emergency. Following the emergency, a thorough investigation should be carried out and a report made. For a detailed description of the procedure to be followed, review Chapter 2.

NOTIFICATION OF GOVERNMENTAL AGENCIES

OSHA and AEC standards require that certain types of incidents be reported to the appropriate officials. For employers regulated by OSHA but not by the AEC, this official is the Assistant Secretary of Labor or authorized representative. For employers regulated by the AEC, the official is the director of the appropriate AEC Regional Compliance Office. Both sets of standards require that overexposures and excessive levels and concentrations be reported. These requirements are discussed below.

Reports of Incidents

Incidents are divided into two categories: those requiring immediate notification and those requiring 24-hour notification.

Immediate Notification

Both standards require immediate notification of any incident involving radiation that may have caused or threatened to cause:

1. Exposure of the whole body of any individual to 25 rems or more of radiation; exposure of the skin of the whole body of

any individual to 150 rems or more of radiation; or exposure of the feet, ankles, hands, or forearms of any individual to 375 rems or more of radiation.

2. The release of radioactive materials in concentrations which, if averaged over a period of 24 hours, would exceed 5,000 times the limits specified for these materials in Part II of Table 11.

3. A loss of one working week or more in the operation of any facilities affected.

4. Damage to property in excess of $100,000.

Twenty-Four-Hour Notification

Such notification is required by both standards for any incident that may have caused or threatened to cause:

1. Exposure of the whole body of any individual to 5 rems or more of radiation; exposure of the skin of the whole body of any individual to 30 rems or more of radiation; or exposure of the feet, ankles, hands, or forearms to 75 rems or more of radiation.

2. A loss of one day or more in the operation of any facilities affected.

3. Damage to property in excess of $10,000 (for employers covered by AEC standards, this figure is $1,000).

In addition, only AEC standards require notification of the release of radioactive material in concentrations which, if averaged over a period of 24 hours, would exceed 500 times the limits specified for these materials in Part II of Table 11.

Other Reports

OSHA and AEC standards require employers to report within 30 days each exposure of an individual to radiation or concentrations of radioactive materials in excess of any limit specified by the standards.

The report must describe the extent of the exposure, the levels of radiation and concentration of radioactive materials involved, the cause of the exposure levels, and the steps taken or planned to assure against a recurrence. Each exposed individual must be notified by the employer in writing of the nature and extent of exposure. The notification must advise him or her to preserve the written notice for future reference.

The AEC also requires that licensees report levels of radiation or concentrations of radioactive material not involving excessive exposure of any individual, in any unrestricted area in excess of 10 times the limit set forth in the license.

OSHA standards stipulate that employers maintain records of the radiation exposure of all employees for whom personnel monitoring is required and inform each employee at least once a year of his or her individual exposure.

EXERCISES: CHAPTER 5

1. Name and describe the five kinds of ionizing radiation.

2. Define external and internal radiation hazards and indicate what materials are and are not a source of each type.

3. Describe the various effects that can result from exposure to ionizing radiation.

4. Compare the following radiation-measuring and monitoring devices, with special attention to mode of operation and limitations:
   a. pocket ionization chambers
   b. dosimeters
   c. Geiger counters
   d. scintillation counters

6. Name the three types of monitoring involved in any ionizing radiation monitoring survey and describe what is involved in each.

7. Discuss the safety training program for persons handling or otherwise involved with ionizing radiation sources, comparing the training given personnel who routinely handle such sources and that given such special groups as medical, industrial hygiene, and plant protection personnel.

8. Summarize the general handling precautions that should be followed when radioactive isotopes are handled.

9. Name the types of personal protective devices utilized in handling radioactive isotopes.

10. Contrast the types of shielding required for different types and strengths of ionizing radiation.

11. Contrast the procedures involved in disposing of solid and liquid radioactive wastes.

12. Define major and minor radioactive material spills and summarize the procedure for decontaminating each.

6

Light

Visible light is the radiation in the segment of the electro-magnetic spectrum that lies between the infrared and ultraviolet segments (see Chapter 4). Visible light has a wavelength of 10^{-6} to 10^{-7} centimeters and a frequency of 3×10^{14} to 3×10^{15} cycles per second.

Three sources of artificial light are of importance today: incandescent lamps, fluorescent lamps, and mercury vapor lamps. An incandescent lamp essentially comprises a tungsten filament inside a partially evacuated glass globe. Friction created by electrical current passing through the filament causes it to become heated and to give off light. The first lamp, invented by Thomas Edison in 1879, was of the incandescent type.

A fluorescent lamp consists of a glass tube exhausted of air, containing a small amount of mercury and some inert gas, and with an electrical element, called a cathode, at each end. The inside of the lamp is coated with phosphors. Current generates an electric arc within the mercury vapor between the two cathodes and causes the mercury to radiate ultraviolet radiation. This radiation in turn activates the phosphor coating, causing it to emit visible light.

The mercury vapor lamp has an arc tube containing mercury vapor at much higher temperature and pressure than in the fluorescent lamp, with an electrode sealed into each end. The arc tube assembly is contained in a glass bulb that is exhausted or air and usually filled with a controlled amount of nitrogen. It finds widespread usage in industrial high bay areas, as the unit produces more light for its size than the other sources.

Light may be controlled by reflection, refraction, and polarization. It is reflected in various ways from different surfaces. If a beam of light falls on a surface which breaks up the beam and scatters it in all directions so that the shape of the beam is lost, this is called diffuse reflection. If a beam of light falls on a flat, glossy surface, such as a mirror or polished metal, the beam will be reflected without being significantly changed in shape. This is called regular or specular reflection. Most surfaces have a combination of diffuse and specular reflection. The fact that light can be reflected makes it possible to control its distribution.

Refraction is the process by which the direction of the ray of light changes as it passes obliquely from one medium to another. This property of light makes it possible to control light with lenses and prisms. Lenses and prisms may be formed in glass plates that direct and diffuse the light passing through them. These are used in lighting fixtures (luminaires).

In an ordinary beam of light the waves vibrate in all directions at right angles to the direction of travel. A piece of polarizing material placed in the beam will confine the vibrations to one plane. A second piece of polarizing material can be placed in the beam to stop the remaining vibrations and eliminate all the light. Properly adjusted polarized glasses can thereby reduce bright reflections.

DEFINITIONS

The following definitions are important to any consideration of light:

Lumen: The amount of light from a standard candle that falls on a card one square foot in area and at a distance of one foot.

Luminous Flux: The quantity of light measured in terms of lumens.

Luminous Intensity: The concentration of lumens coming from a light source when observed in a given direction.

Candle: The unit of luminous intensity. Originally, the standard candle was made following rigid specifications as to length, diameter, tallow, and wick construction. Today the standard candle has been replaced with the term "candela," which is defined as the luminous intensity of one-sixtieth of one square centimeter of the projected area of a black-body radiator operating at the temperature of solidification of platinum.

Illumination: The amount of light incident upon a surface.

Foot-Candle: The unit of illumination. It is equal to the illumination on a surface one foot square in area on which there is a uniformly distributed flux of one lumen. A foot-candle is therefore equal to one lumen per square foot.

Brightness: The intensity of light emitted by, or reflected from, a surface in a given direction.

Foot-Lambert: The unit of brightness. The foot-lambert is equal to the brightness of a perfectly diffusing surface emitting or reflecting light at a rate of one lumen per square foot.

Reflectance: The percentage of incident light a surface is capable of reflecting. Plain white paper may have a reflectance of 75 to 80 percent. The reflectance of black paper is around 1 or 2 percent.

Transmission: The percentage of incident light passing through a material such as glass, plastic, or paper.

FACTORS AFFECTING QUALITY OF VISION

Effects of Bad and Good Lighting

Poor lighting does not result in direct injury to the eye. Rather, it produces eye discomfort and fatigue, and can cause headaches, nausea, and unnecessary expenditure of nervous energy in seeing. Lighting of too high an intensity is also undesirable; it can produce glare, which can also create eye strain. Good lighting eliminates these problems and provides a variety of desirable advantages.

Increased production is an important advantage of good lighting. The more light, the greater the speed and accuracy of seeing. The extra seconds an operator spends to be sure of a reading or an inspector takes to examine a part may not seem important, but if that process is repeated many times a day, the time lost can be quite significant. Greater ease in seeing means less expenditure of nervous energy, and therefore more useful energy for productive work.

Good lighting not only results in increased production, but also in better workmanship. People can see what they are doing if there is good lighting. There is no necessity for guesswork and therefore less excuse for avoidable mistakes. Since defects can be detected more readily, they are much more likely to be caught before final inspection of the work. Inspecting and checking operations can be accomplished much more quickly if both supervisor and worker can easily see the job at hand.

Along with increased production and better workmanship, reduction in worker stress can be achieved through good lighting. The eyes of workers become fatigued and strained when they are required to perform work in areas where illumination is not adequate. Eye fatigue and strain are frequent causes of headaches and general discomfort. Good lighting can alleviate these problems and also compensate for the decrease in visual acuity in older employees. Good lighting also provides a more pleasing environment, which contributes to better morale.

Accident frequency and fire hazards are certain to be higher in a cluttered and dirty plant. When illumination is poor, it is difficult to see clearly under machines and benches and into dark corners. Such places readily can be dumping grounds for debris and dirt. Good lighting shows up the accumulation of rubbish and provides a better incentive for keeping an area clean.

A uniform level of general lighting makes possible a more efficient arrangement of machinery and equipment, and a more efficient work flow.

Specific Factors

The factors that affect the quality of vision on any particular job are numerous and related in a complex way. The more important of these are discussed below.

Character of Workpiece

The character of the workpiece has an important effect on vision quality. In general, less light is required for seeing comfort with large workpieces that reflect large amounts of light. Thus, sorting large castings is a simple task even in very dim light, whereas reading micrometer calipers is quite another story. Likewise, the task of inspecting highly reflective aluminum castings is much easier than that of inspecting dark iron castings.

The extent to which the workpiece differs in brightness or color contrast from the surrounding background is another important factor. It is much easier to see a piece of white thread on black cloth than to see black thread on black cloth. Providing recommended brightness ratios (see Table 13) are followed, the contrast between an object and its immediate background should be made as high as possible.

Character of Light

The character of light is determined by quantity, or the amount of illumination that produces the brightness of the task and surroundings, and quality, which includes the color of the light, its direction, diffusion, and the amount and type of glare.

Quantity. The degree of accuracy required, the detail to be observed, the color and reflectivity of the work, as well as the immediate surroundings, materially affect the brightness requirements that will produce maximum seeing conditions. As illumination of the task is increased, the ease, speed, and accuracy with which the task can be accomplished are also increased.

Studies show that thousands of foot-candles of illumination are required to see dark, low-contrast tasks, whereas light-colored, high-contrast tasks require very low levels. However, because of other factors, a minimum of 30 foot-candles is recommended for even the simplest seeing tasks.

Quality. Direct glare, reflected glare, and shadows are the prime determiners of the quality of light. Glare may be defined as unwanted light. The more formal definition is "any brightness within the field of vision of such a character to cause discomfort, annoyance, fatigue and/or interference with vision." When the condition is caused directly by the source of lighting, whether natural or artificial, it is described as direct glare.

Direct glare may be reduced by:

1. Decreasing the area of high brightness causing the glare condition.
2. Increasing the angle between the glare source and line of vision.
3. Increasing the brightness of the area surrounding the glare source.

Reflected glare is caused by high brightness images or brightness contrast reflected from shiny ceilings, walls, desk tops, materials, and machines or surfaces within the visual field. It is frequently more annoying than direct glare because it is so close to the line of vision that the eye cannot avoid it.

Reflected glare may be reduced by:

1. Reducing the brightness of the light source. (If such brightness cannot be reduced to a desirable value, attempt to position either the lighting equipment or the task so that the reflected image will be directed from the eye of the observer.)
2. Changing the character of the offending surface (for example, by placing a matte-type pad on a highly reflective desk top).
3. Increasing the general illumination. (Reflected glare will be reduced because of the reduction in the contrast of the reflected image and its immediate surroundings.)

Ordinarily shadows are an undesirable interference in the performance of an industrial task. However, some tasks can best be seen where shadows are created. Thus glossy black thread on black matte cloth is more clearly delineated with than without shadows, and scratches on sheet metal are more easily seen.

Evenly distributed and diffused illumination is desirable for industrial interiors if unwanted shadows are to be avoided. Spot or local lighting, restricted to a small work area, can cause shadows and is unsatisfactory unless there is sufficient general illumination in the room.

Character of Environment

This is determined largely by brightness and brightness ratios, reflectance values, and light distribution.

Brightness and Brightness Ratios. The ability to see detail depends on the brightness differences between the detail and its background. The eyes function most efficiently when the brightness within other areas of view is relatively uniform.

Brightness ratios in industrial areas should not exceed those given in Table 13.

TABLE 13
Recommended Brightness Ratios

	Environmental Classification		
	A	B	C
Between tasks and adjacent darker surroundings	3 to 1	3 to 1	5 to 1
Between tasks and adjacent lighter surroundings	1 to 3	1 to 3	1 to 5
Between tasks and more remote darker surfaces	10 to 1	20 to 1	*
Between tasks and more remote lighter surfaces	1 to 10	1 to 20	*
Between luminaires (or windows, skylights) and surfaces adjacent to them	20 to 1	*	*
Anywhere within normal field of people	40 to 1	*	*

* Luminant ratio control not practical.
A Interior areas where reflectances of entire space can be controlled in line with recommendations for optimum seeing conditions.
B Areas where reflectances of entire work area can be controlled but control of remote surroundings is limited.
C Areas where it is completely impractical to control reflectances and difficult to alter environmental conditions.

Reflectance Values. The color and reflectance of room walls, ceiling, and floor, as well as equipment, determines the brightness pattern and thus influences seeing as part of the environment. Darkly painted walls, floors, and ceilings can reduce the effectiveness of lighting installations by 50 percent. Recommended reflectance values are given in Table 14.

TABLE 14

Recommended Reflectance Values

Description	Reflectance (%)
Ceiling	80 - 90
Walls	40 - 60
Desks	
Benches	25 - 45
Machines	
Floors	20

Light Distribution. The luminaire, commonly known as the lighting fixture, is the complete lighting unit including lamp, sockets, and equipment for distributing and controlling light. It is essential that lighting equipment be spaced in accordance with its lighting distribution characteristics. Recommended values for mounting height and luminaire spacing must be observed. Alternate dark and light areas that create high brightness differences are undesirable.

Luminaires are classified according to the manner in which they distribute light.

Direct Lighting: Practically all of the light emitted by the luminaires (90 to 100 percent) is directed downward. This type is most efficient from the standpoint of getting the maximum amount of light from the source to the working plane.

Semidirect Lighting: Distribution is predominantly downward (60 to 90 percent).

General Diffuse: Distribution approximately 40 to 60 percent down, 40 to 60 percent up.

Semiindirect Light: Distribution 60 to 90 percent up, 10 to 40 percent down. The ceiling should be of high reflectance to reflect the light downward.

Indirect: Distribution 90 to 100 percent up. The ceiling becomes, in effect, the light source. The total light efficiency is decreased but shadows and reflected glare are minimized.

In selecting and installing luminaires, some light should be directed to illuminate the ceiling or upper structure to eliminate the "dungeon" effect. Top openings in luminaires will minimize dirt collection on the reflectors and lamps. The outside of the fixture should be light-colored to reduce brightness ratios between the outside of the luminaire and the inner reflector surface and light sources. Luminaires should be mounted high enough to raise them out of the normal field of view, and the light source should be shielded with deep reflectors, cross-baffles, or louvers. The latter is particularly important with the high wattage filaments or mercury sources and the higher output fluorescent lamps.

EVALUATING INDUSTRIAL LIGHTING

Types of Industrial Lighting

General lighting, localized general lighting, and supplementary lighting may be employed in an industrial establishment. General lighting should produce uniform illumination on the work plane, usually considered to be 30 to 36 inches above the floor. The maximum and minimum illumination at any point should not be more than one-sixth above or below the average in the area. Although it appears to be well established that no permanent physical damage is caused to the eye when it is exposed to an improper lighting environment, a general illumination level of at least 30 foot-candles is a recommended minimum.

Industrial plants may have machinery or equipment located in such a manner that a uniform intensity of illumination is not necessary throughout the entire area. Where specific areas require higher than the general levels of illumination in the surrounding area, additional luminaires can be mounted above the task to supply the adequate level of illumination.

Supplementary lighting augments the general lighting by providing light for difficult seeing tasks or inspection processes. The choice and use of supplementary lighting must be coordinated with the general lighting. Problems of glare and extreme brightness differences can be encountered if the proper supplementary equipment is not used.

Value of Lighting Surveys

Lighting surveys are made to determine compliance with recommended practices and/or to reveal the need for maintenance, modifications, or replacement of lighting equipment. The tangible benefits derived from increased illumination are difficult to assess. Each individual case must be considered separately after a lighting survey has been made and levels are found below standard. Supervision must decide what part such factors as greater safety, decreased work spoilage, decreased fatigue, or improved morale play in relation to the economic aspect. Installation of additional lighting is not always the answer because many buildings are so designed that increased lighting will alter the brightness relationship. Many times the lighting systems in older buildings can be changed to meet present recommended levels only at considerable expense.

After a lighting survey has been made and the facts presented, it is up to supervision, together with the electrical or lighting engineer, to evaluate the cost in relation to the benefits that might be gained. In most cases some sort of compromise will be in order.

Actually, a light survey of sorts can be taken every time a visit is made to a plant or building area by keeping in mind the following questions:

1. What is the overall appearance of the area from the standpoint of pleasantness and visual comfort?
2. Does there appear to be enough general lighting? What about supplementary lighting?
3. How dirty are the lamps, fixtures, and windows? When were they last cleaned?
4. How many lamps are burned out?
5. What colors are the walls, ceiling, equipment? Are the colors acceptable from the standpoint of reflectance values? How dirty are the walls and ceilings and when were they last painted or cleaned?
6. Are there any obvious sources of direct glare or reflected glare?
7. Are any workers subjected to bothersome shadows?
8. What change would make the most obvious improvement in the visual environment?

If a thorough survey is requested or considered necessary, the form shown in Figure 6.1 outlines the basic factors to be considered.

Light-Measuring Instruments

There are three types of light-measuring instruments: light-intensity meters, brightness meters, and visibility meters.

Light Intensity Meters

Several kinds of light meters are commercially available. The ones most commonly used for lighting surveys are the General Electric pocket-sized meters and the Weston hinged-cell meters. Both are available with cosine correcting elements. With cosine correction, a light meter responds equally to illumination from all directions instead of primarily to vertically incident radiation. The Weston hinged-cell meters have wider ranges than do the pocket-sized meters,

FIGURE 6.1

General Survey Work

File No. _____ Bldg. _____ Date _____ File _____

Day _____ Night _____

    Cloudy _____

    Sunny _____

Surveyor _____ Instruments Used _____

CHARACTERISTICS OF AREA

Identification of Area _____

Length _____ Width _____ Height _____

(If area is irregular, show sketch) _____

DESCRIPTION OF WALLS, FLOOR AND CEILING

				Surface Condition		
Description	*Material*	*Color*	*Texture*	*Clean*	*Average*	*Dirty*
Sidewalls						
Ceiling						
Floor						
Work Surfaces						
Equipment						

Classification of Equipment _____

Luminaire (Desc.) _____

Lamp Specifications _____

Lamps per Luminaire _____

Number of Luminaires _____

Number of Rows _____

Luminaires per Row _____

Mounting Height _____

Luminaire Spacing _____

    (If irregular, show sketch on back) _____

Condition of Equipment   Clean _____

                        Average _____

                        Dirty _____

Description of supplementary or local lighting (Use back if necessary)

_____

_____

_____

163

but are more expensive. For highly accurate illumination measurements, special low-range sensitive and multicell light meters are commercially available. Figure 6.2 shows a light intensity meter.

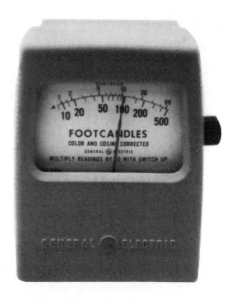

FIGURE 6.2   Light intensity meter.
(Photo by Michael D. Ells).

Brightness Meters

The most commonly used brightness meter is the Luckiesh-Taylor instrument made by General Electric. It has a battery-operated internal light source, making it a self-contained unit. With it, the brightness of any surface can be measured within the range 0.0025 to 50,000 foot-lamberts. The instrument has two eye-pieces. One is a focusing telescope, which is sighted on the surface to be measured. In the center of this field are a pair of silvered trapezoids illuminated to a constant brightness by the internal light source. The brightness of the background is adjusted with a circular gradient filter until it matches the trapezoids in brightness. The second eyepiece views a scale that rotates with the gradient filter. At the proper setting of the filter, the brightness can then be read directly from the scale.

Visibility Meters

The visibility meter is a device for determining the relative visibilities of seeing tasks under various conditions of lighting and levels of illumination. It also can be used to determine the relative foot-candles required by a variety of tasks to make them equal in visibility. The object whose visibility is being measured is viewed through variable filters, which are adjusted until the object is just barely seen. This is the point of threshold visibility.

Measurement of glare, either reflected from surfaces or direct from luminaires, has posed a problem because satisfactory field instruments are not commercially available. Research in this field has progressed, however, and several meters have been developed which should be available in the near future.

Whatever the type of measuring instrument utilized, readings should be taken at night. If day lighting supplements the artificial illumination, the evaluation of the lighting should be made during both cloudy and sunny days. The surveyor should wear dark-colored clothes to eliminate reflectance interferences in the readings, and should stand two or three feet away to prevent his or her shadow from being cast upon the cells.

Types of Measurement

Measurements Utilizing Instruments

General Illumination Measurements. In areas where the luminaires are evenly spaced, the average illumination can be determined by taking a minimum number of readings in specific sites and using the flux of light equation. The form in Figure 6.3 can be used in connection with this survey. In irregular shaped areas, or where the lighting is nonsymmetrical, a number of random readings must be taken throughout the room. The average of the sum of such readings will give the average level. Readings are taken 30 inches above the floor.

Point-of-Work Measurements. Readings should be taken at the point of work with the worker in the normal working position. The meter should be held perpendicular to the line of sight. The form in Figure 6.4 will prove useful in connection with such measurements.

Brightness Measurements. Brightness measurements may be made with a brightness meter or with a light meter. Multiply the average illumination levels (foot-candles) in the area by the reflectance value of the surface, and divide by 100.

Reflectance values can be obtained from reflectance tables or charts. These values can also be determined with a light meter. First, record the illumination (foot-candles) impinging on the area to be evaluated. Turn the cell over and slowly begin to raise it vertically above the surface. The meter will respond to the reflected light from

FIGURE 6.3

Measurement Data

Building _____ Area _____

Average Horizontal Ft.-C in Regular Areas _____

1. Flux of Light Method _____

   Average Illumination= $\dfrac{R\ (N\text{-}1)\ (M\text{-}1) + Q\ (N\text{-}1) + T\ (M\text{-}1) + P}{NM}$

   Where:
   N = Number of Luminaires/Row
   M = Number of Rows

Inner Bay R (ft.-C)	*Side Q (ft.-C)*	*End T (ft.-C)*	*Corner P (ft.-C)*
R - 1 _____	Q - 1 _____	T - 1 _____	P - 1 _____
R - 2 _____	Q - 2 _____	T - 2 _____	P - 2 _____
R - 3 _____		T - 3 _____	
R - 4 _____		T - 4 _____	
R - 5 _____			
R - 6 _____			
R - 7 _____			
R - 8 _____			

Total _____ _____ _____ _____
(Average) R = _____ Q = _____ T = _____ P = _____

2. General Avg. Method (Attach sketch and show values) _____
3. Irregular Area (Attach sketch and show values) _____
4. Nonsymmetrical Lighting (Attach sketch, show values) _____
   Remarks _____
5. Supplementary Lighting   Yes _____ No _____
   Was area measured with supplementary lighting off? _____ On? _____
   Explain _____

6. Measurement of Ft.-C in Plane of Work   Yes _____ No _____
   (If Yes, report on Plane of Work Form or sketch)

the surface. When there is no further observable increase in meter reading, record that reading as the null point. Divide the first reading by the second and multiply by 100 to obtain the surface reflectance.

The form in Figure 6.5 is designed for use with brightness surveys.

Measurements Without Instruments

General foot-candle levels can also be determined without a light meter if certain room characteristics and lamp and luminaire specifications are known. Specifically, the following factors must be determined or known before lighting calculations can be made:

1. Type of luminaire.
2. Type of lamp.
3. Mounting height of the luminaire.

FIGURE 6.4
Lighting Survey Foot-Candle Plane of Work Form

Building _____ Area _____ Date _____ Time _____

Reading	Location Description	Ht. Above Floor	Plane[1]			Foot-Candles					
			V	H	A	General		General and Supplementary			
						Average	Range	Average	Range		

[1] V = Vertical, H = Horizontal, A = Angle

167

FIGURE 6.5

Average Brightness Calculation Sheet

Building _____ Date _____ File Number _____

Time _____ Instrument Used _____ Observer _____

Weather Conditions _____

Average maintained illumination _____ Foot-Candles

Type of Lighting:

Direct _____ Uniformly Diffusing _____

Indirect or Luminous Ceiling _____

Average Reflectance:

Task _____ Floor _____ Ceiling _____

Working Plane _____ Wall _____

Windows: R.F. _____ T.F. _____

Room Dimensions: Length _____ ft. Width _____ ft. Height _____ ft.

BRIGHTNESS

Task: _____ ft.-C × Task Reflectance _____ ft.-Lambert (F.L.)

Working Place: _____ ft.-C × Work-Place Reflectance _____ (F.L.)

Walls (Midway to Ceiling) _____ ft.-C × Wall Reflectance (A*) _____ (F.L.)

Ceiling: _____ ft.-L (Luminaire Upward) × Ceiling Reflectance (B*) _____ (F.L.)

Floor: (Immediate Surround) _____ ft.C × Floor Reflectance (C*) _____ (F.L.)

Luminaire: _____ Element Efficiency (downwards) × Initial Lumens/Lamp × Number of Lamps × M.F. _____

_____ Luminous Area sq. ft.

Windows: _____ ft.-C × Window Reflectance or Transmittance _____

*If estimated, use Figure 9-22, p. 9-64 and 9-65, *IES Lighting Handbook*, 5th ed., 1972.

FIGURE 6.5 (continued)

Room coefficient = $\dfrac{\text{Ht. } (1 \times w)}{2 \times 1 \times w}$

	Standard	*Ratios*
Task to Desk	1 - 1/3 Minimum	_____
Task to Wall	1 - 1/5 Minimum	_____
Task to Ceiling	1 - 10 Maximum	_____
Task to Floor	1 - 1/10 Minimum	_____
Task to Luminaire	1 - 50 Maximum	_____
Luminaire to Ceiling	20 - 1 Maximum	_____
Window to Wall	20 - 1 Maximum	_____

*(A = Figure 9-22A; B = Figure 9-22B; C = Figure 9-22C)

169

4. Reflectance factors of ceiling, walls, and floors.

5. The room length, width, and ceiling height.

This data can be utilized to determine the number of foot-candles resulting from a predetermined or existing fixture layout, or the number of fixtures to give the desired foot-candles. The following formulas are used for these purposes:

$$FC = \frac{\text{total initial lamp lumens} \times CU \times MF}{\text{Area (square feet)}}$$

$$\text{No. of Fixtures Required} = \frac{FC \text{ desired} \times \text{area (square feet)}}{CU \times MF \times \text{total lumens/fixture}}$$

Where:

FC = Foot-candles

CU = Coefficient of utilization for fixture desired. The CU is determined from the Coefficient of Utilization Tables specified by the manufacturer with each fixture. The correct factor is chosen by reading into the table the room index figure obtained from the individual room measurements and the anticipated wall, ceiling, and floor reflectances. Room index tables may be obtained from lighting handbooks or the lamp manufacturer.

MF = Maintenance Factor (Manufacturer's specification).

Total Initial Lamp Lumens = Product of the number of fixtures per room $\times$ the number of lamps per fixture $\times$ the initial lumens per lamp (total lumens obtained from the lamp manufacturer or lighting handbooks).

Total Lumens/Fixture = Number of lamps per fixture $\times$ initial lumens per lamp.

FC Desired = Amount of light desired at the working plane.

Number of Fixtures Required = Total number of fixtures per room required to give the foot-candles required.

Area (Square Feet) = The room width $\times$ room length.

The form in Figure 6.6 can be used in conjunction with surveys conducted without a light meter.

FIGURE 6.6

Lighting Analysis Sheet

		Color	*Reflectance Factor*
Building	Ceiling		
Kind of Work	Wall		
Ceiling Height	Floor		
Area			

Foot-Candles	Outlet Spacing
Classification of Equipment	Side Wall Spacing
Luminaire	Mounting Height
Luminaire Efficiency	Lamp Specs.
	Rated Lamp Lumens
Room Index	No. of Lamps
Maintenance Factor	Lamps/Luminaires
Coefficient of Utilization	No. of Luminaires
Total Number of Lumens	No. of Rows
	Luminaires/Row
Average Illumination	Area/Luminaire
Initial Illumination	Lumens/sq. ft.
	Lumens/Outlet

$$FC = \frac{\text{Total Lumens} \times CU \times MF}{\text{Area (sq. ft.)}}$$

$$\text{No. of Fixtures} = \frac{\text{(FC Desired)} \times \text{Area (sq. ft.)}}{CU \times MF \times \text{Total Lumens/Fixture}}$$

RECOMMENDED ILLUMINATION LEVELS

Table 15* shows recommended illumination levels for various installations. To use these levels of illumination properly, additional information is necessary, as found in the *IES Lighting Handbook* within specific application sections.

*Adapted from *IES Lighting Handbook*, 5th ed. Illuminating Enginering Society, New York, 1972.

TABLE 15

Recommended Levels of Illumination

Shops and Services

Facility	Recommended Illumination Foot-Candles
Boiler Shop	
Structural Steel Fabrication	50
Drilling, riveting, and screw fastening.	70
Sheet aluminum layout and template work; shaping and smoothing small parts etc.	100
Welding	
General	50
Supplementary	1,000
Assembly	
Rough, easy seeing	30
Rough, difficult seeing	50
Medium	100
Fine	500
Extra Fine	1,000[1]
Miscellaneous machines, ordinary bench work	50
Presses, shears, stamps, spinning, medium bench work	50
Puncher	50
Tin Plate inspection, galvanized	200[1]
Scribing	200[1]
Cafeteria	
Dining area	
Cashier	50
Normal Surroundings	30 - 50
Food cleaning	20
Food Displays—twice general levels but not under	50
Kitchen	
Cooking—inspection	70
Other areas	30
Electrical Shop	
Impregnating	50
Insulating: coilwinding	100
Testing	100
Fire Department	
Wagon Room	30
Garage	
Body Repairs	100
Washing and servicing	100
Active traffic area	20
Parking garages—entrance	50
Traffic lanes	10
Storage	5
Glass Fabrication Department	
Grinding, cutting glass to size, silvering	50
Fire grinding, blowing, polishing and bending	100
Inspection, etching, and decorating	200
Instrument Department	
Assembly	
Rough, easy seeing	30
Rough, difficult seeing	50
Medium	100
Fine	500
Extra Fine	1,000
Laundry	
General	30
Washing	30
Sorting	70
Pressing	70

172

Lead Shop	
Lead Burning	
General	50
Supplementary	1,000
Moulding	
Medium	100
Large	50
Pouring	50
Library	
General	10
Reading	30 - 70
Stacks	30
Card Filing	100
Mail Department	
Mail sorting	100
Storage	20
General	30
Material Handling Department	
Wrapping, Packing, Labeling	50
Packing stock, classifying	30
Loading, trucking	20
Inside truck bodies and freight cars	10
Paint Shop	
Dipping, simple spraying, firing	50
Rubbing, ordinary hand painting and finishing;	
art, stencil and special spraying	50
Fine hand painting and finishing	100
Comparing mix and standard	200
Extra-fine hand painting and finishing	300
Photographic Department	
Engraving, all types	50 - 100
Power Plant	
Boiler Room	10
Machine Room	20
Turbine Room	30
Switchboard and control rooms	30
Transformer room	10
Printing Department	
Color inspection and appraisal	200[1]
Machine	100
Composing room	100
Presses	70
Imposing stones	150
Proof Reading	150
Railroad	
Car weigh area (scales)	10
Yards	1
Round House	30
Scheduling and Inspection	
Ordinary	50
Difficult	100[1]
Highly difficult	200[1]
Very difficult	500[1]
Most difficult	1,000[1]
Quality Control Department	
General testing	50
Extra-fine instruments, scales, etc.	200[1]
Color grading, comparison	200[1]
Sewing	
General	20
Work area	100
Knitting and weaving	100
Sewing and inspection	200
Stockroom	
Rough, bulky	10
Medium	20
Fine	50
Tool Department	
Machine tool repairs	100
General shop	30

173

Woodworking shop

Rough sawing and bench work	30
Sizing, planing, rough sanding; medium machine and bench work; blueing, veneering, cooperage	50
Fine bench and machine work, fine sanding and finishing	100

Offices

Task	Recommended Illumination Foot-Candle
Cartography, designing, detailed drafting	200
Accounting, auditing, tabulating, bookkeeping, business machine operations, reading poor reproductions, rough layout drafting	150
Regular office work, reading good reproductions, reading or transcribing handwriting in hard pencil or on poor paper, active filling, index references, mail sorting	100
Reading or transcribing handwriting in ink or medium pencil on good quality paper, intermittent filing	70
Reading high-contrast or well-printed material; tasks and areas not involving critical or prolonged seeing such as conferring, interviewing, inactive files and washrooms	30
Corridors, elevators, escalators, stairways	20

Laboratories

Areas and Devices	Recommended Illumination Foot-Candle
Corridors	20
Exits	5
Lobby	10 - 30
Receptionist Desk	50 - 30
Research Laboratory	70 - 100
Scales and balances (on face)	50
Service Areas	
Basements, air conditioning, heating equipment, etc.	10
Switchboard room	30
Tunnels, galleries, pipes	10
Shop Maintenance	
General	30
Work benches	100
Stockrooms	
Medium	20
Fine	50
Storage	
Chemical	10
General	15
Office	70
Thermometers, manometers, etc. (on face)	50

Production and Pilot Plants

Areas, Devices, Activities	Recommended Illumination Foot-Candles
Control rooms	50
Drum-filling areas	30 - 50
Gages, manometers, etc. (on face)	50

General Maintenance Shop	30
Grinding filter presses	30
Work benches	100
Hand furnaces, boiling tanks, stationary driers, *stationary and gravity crystallizers*	30
Labeling and cartoning	30
Man lifts, aisleways, walkways	30
Mechanical furnaces, generators and stills, mechanical driers, *evaporators, filtration, mechanical crystallizers, bleaching*	30
Meter panels (on face)	50
Molding and extruding operators	50
Packing, sifting, purifying operators	100
Product control	50
Production and Pilot Plant laboratories	50
Rear of switchboard panels (vertical)	10
Service areas Heating equipment, air conditioning, etc.	10
Tunnels or galleries, piping	10
Tanks for cooking, extractors, percolators, *nitrators, electrolytic cells*	30
Thermometers (on face)	50
Warehouses Inactive	5
Active	
Rough bulky	10
Medium	20
Fine	50
Loading and unloading platforms	20
Weighing Rooms Scales	30
Large	10
Small	50

Medical Department

Areas	Recommended Illumination Foot-Candles
Corridors	10
Emergency Room General	100
Local	2,000
EKG, BMR Room	30
Examination and treatment rooms General	50
Examining table	100
Dark room	10
Eye examination room	50
Ear, nose, and throat	50
Exits	5
Fracture Room General	50
Table	200
Kitchen	70
Laboratories Assay	30
Work tables	50
Close work	100
Linen closet	10
Medical records room	100

Nurses' Station
General 70
Desk and charts 70
Medicine counter 100

Rooms
General 10
Reading 30
Recovery 30

Therapy
Physical 20
Occupational 30

X-Ray Room and Facilities
Radiography and fluoroscopy 50
Viewing room 30
Filing room, developed films 30
Storage, undeveloped films 10

Exteriors

Areas	Recommended Illumination Foot-Candles
Active shipping areas, surround	5.0
Bulding surrounds	1.0
Catwalks (general)	2.0
Cinder dumps	0.1
Coal Unloading	
Docks (loading or unloading)	5.0
Barge storage areas	0.5
Car dumper	0.5
Tipple	5
Conveyors	2.0
Delivery headers	5.0
Entrances - Buildings	
Main	10
Secondary	2.0
Clockrooms	10.0
Fence	0.2
Gates	
Active	5.0
Inactive (normally locked)	1.0
Open yards	0.2
Platforms, loading and unloading	20
Roadway	
Between or along buildings	1.0
Not bordered by buildings	0.5
Storage areas	
Active	20
large (coal yards)	0.1
tank	1
Vital locations or structures	5.0

Miscellaneous

Areas	Recommended Illumination Foot-Candles
Auditoriums	
Assembly only	15
Exhibitions	30
Building entrance foyers	30
Conference and interview rooms	30

176

Corridors, elevators, stairs	20
Dishwashing	30
Front office Receptionist	50
General storage rooms	15 - 20
Lobby	10 - 30
Reading	30
Locker Rooms	20
Wash Rooms	30

[1] Supplementary lighting necessary.

MAINTENANCE

Good lighting maintenance is good economy and insures the user that he or she actually gets the light paid for from the cost of electrical energy. The four principal factors that cause light losses in industrial lighting installations are:

Dirt on Lamps and Luminaires. Lamps and lighting fixtures collect dust, which reduces the lumen output from the equipment.

Dirt on Room Surfaces. Dirt on interior surfaces decreases the reflectance characteristics of the surfaces and consequently reduces the effectiveness of the lighting system.

Lamp Lumen Depreciation. The lumen output of electric light sources diminishes with use and, in most cases, it is economical to replace lamps before they finally fail. Individual replacement is called "spot replacement." Mass replacement is called "group relamping." Labor costs saved by group relamping usually more than compensates for the value of the depreciated lamps that are thrown away before they burn out. Lamp manufacturers supply booklets or brochures describing group relamping programs.

Dirt on Windows. Dirt on the inside and outside of windows can absorb as much as 50 percent of the light that would otherwise enter the area. At the same time, the diffusing effect of such dirt can result in very high brightness, causing serious direct glare.

There should be a regularly scheduled system of maintaining the skylight, windows, lighting equipment, lamps, and room surfaces to keep them clean. Luminaires should be kept in proper adjustment and in good repair. Even if this is done, illumination levels will drop 25 to 35 percent below the original under normal operating conditions. When illumination has decreased to two-thirds of its initial value, the lighting equipment should be washed. A thorough washing at least twice a year is recommended. In addition, the useful output of some lamps decreases with age long before they burn out. Therefore it is important that a regular maintenance program be set up to clean lamps and fixtures and replace dim and burned-out lamps at suitable intervals.

EXERCISES: CHAPTER 6

1. Differentiate between the three types of lamps in common use today.

2. Discuss the three ways in which light can be controlled.

3. Distinguish between the following terms:
   a. lumen and luminous flux
   b. foot-candle and foot-lambert
   c. reflectance and transmission

4. Describe the effects of poor lighting on workers and the advantages that are achieved through good lighting.

5. Define two types of glare and indicate how each can be reduced.

6. Name the various classes of luminaires and indicate the light-distribution pattern of each.

7. Describe the mode of operation of brightness meters and visibility meters.

8. Indicate the general procedure followed in making general illumination measurements and point-of-work measurements with light-measuring instruments.

9. Name the factors that must be known or determined before lighting calculations without instruments can be made.

10. Name the four factors that cause light losses in industrial installations and indicate the countermeasures necessary to prevent such losses.

7

Heat

Whenever persons must work in extremely hot or extremely cold surroundings, the possibility of stress arises. The problems of heat are in general more serious than those of cold and have received more attention. This chapter will, therefore, be primarily concerned with heat stress, and will touch only lightly on cold problems.

Heat stress has long been recognized as a serious problem in many industrial operations, and much serious attention has been given to techniques of evaluating the exposure and attaching physiological significance to the data.

Two sources of heat are important to persons working or living in a warm or hot environment: internally generated metabolic (body) heat and externally imposed environmental heat.

Metabolic heat is a by-product of the chemical processes occurring within the cells, tissues, and organs. Under resting conditions the metabolic heat production of an adult is about 75 kilocalories per hour (300 British thermal units—Btu). Muscular activity is the major source of increased heat production. During very hard physical work, heat production may reach 600 to 750 kilocalories per hour (2,400-3,000 Btu). Thus under conditions of physical work, large quantities of heat must be removed from the body if an increase in body temperature is to be prevented.

Environmental heat is important because it influences the rate at which body heat can be exchanged with the environment, and consequently the ease with which the body can regulate and maintain a normal temperature. Other environmental factors influencing the exchange of heat between the body and the environment include humidity, air movement, and radiation.

DEFINITIONS

The following definitions are important to any consideration of heat stresses:

Dry-Bulb Temperature: The temperature of a gas as indicated by an accurate thermometer after correction for radiation.

Wet-Bulb Temperature: The temperature at which water, by evaporation into the air, can bring the air to saturation adiabatically at the same temperature. It is the temperature indicated by a wet-bulb psychrometer when properly used and shielded from radiation.

Humidity: The water vapor within a space.

Absolute Humidity: The weight of water vapor per unit volume of space.

Relative Humidity: The ratio of the quantity of water vapor in the air to the quantity that would saturate it at any specific temperature.

British Thermal Unit (Btu): The amount of heat required to raise one pound of water one degree Fahrenheit in temperature.

METHODS OF HEAT EXCHANGE

Metabolic heat—the heat produced by the body—is exchanged with the surrounding environment by the processes of conduction-convection, evaporation-convection, and radiation.

If the contact substance, whether air, water, clothing, or an external object, is at a lower temperature than the skin, heat will be lost; but if the contacting substance is at a higher temperature, heat will be gained. The rate at which transfer takes place is determined basically by the difference between the two temperatures, but if the contacting substance is fluid, such as air or water, movement in the fluid accelerates the transfer.

Exchange by Conduction-Convection

In this type of exchange, heat is transferred from the body surface to the surrounding air or a fluid by forced or natural motion of the air or fluid. Some transfer also occurs by intermolecular action without gross displacement of the particles. Nearly all the transfer of "sensible" heat from the skin to the air is by this process.

To this exchange between the skin and the air must be added heat exchanged between the respiratory tract and the inspired air, since the former behaves in this respect simply as an inward extension of the body surface, with a special mechanism—respiration—moving the air away when it is heated.

Exchange by Evaporation-Convection

Heat may also be lost from the body surface to the air by evaporation of water diffusing through the skin from deeper tissues (perspiration), produced by sweat glands (sweating), or applied from without. This is the major cooling mechanism of the body. The rate of evaporative heat loss is determined basically by the difference between the effective vapor pressure of the water on the skin and that of the air, but once again movement of the air greatly accelerates the rate of loss, so that the combined process is properly termed *evaporation-convection.*

The vapor pressure of such water as is present on the skin is determined by the temperature of the skin, but the extent of the water film varies between something less than 10 percent and 95 percent of the maximum. The effective vapor pressure of water on the skin is thus a function of these two factors. The extent of the water film is termed "skin wetness" or "skin relative humidity." It represents a balance between evaporation on the one hand and addition of water on the other. It is high only when the sweat glands are active, evaporation is inhibited, or water is applied from without.

The vapor pressure of the air is determined by the amount of water vapor present per unit volume of the air and corresponds closely to the *absolute humidity* of the air. Unfortunately, atmospheric humidity is usually expressed in terms of *relative humidity,* which is something quite different.

To determine the vapor pressure from the relative humidity one also needs to know the air temperature and to have tables or a graph by which to make the transformation.

Exchange by Radiation

Radiation is the transmission of energy by means of electromagnetic waves. Heat is exchanged by radiation between the surface of a body and all of the surfaces in its surroundings, which are at temperatures different from its own. For such things as the sky, the "surface" means that hypothetical surface that would exhibit the same radiation behavior as the sky is observed to exhibit.

The intensity of the energy emitted from a surface by radiation increases as the fourth power of its absolute temperature. The intensity is usually diminished below the theoretical maximum however, by the physical nature of the surface, the relative effect being known as its emissivity.

Radiation incident upon a surface is absorbed by it in proportion to its emissivity for the wavelength involved. The absorptivity for a particular wavelength is the same as the emissivity for that wavelength. From an opaque surface, the incident radiation which is not absorbed must be reflected, so that its reflectivity is the converse of its absorptivity and thus of its emissivity for the particular wavelength involved.

Net Heat Exchange

The net heat exchange between a person and his or her environment is the sum of the metabolic heat production, heat exchange by conduction-convection, heat exchange by radiation, solar heat gain, and heat loss by evaporation-convection. Heat exchange by conduction-convection and by radiation may be negative or positive, depending upon whether heat is lost or gained. Metabolic heat production can be calculated, since about 5 kilocalories are liberated for each liter of oxygen used by the body cells. Heat exchange by conduction-convection and radiation, and heat loss by evaporation-convection, can be determined by special formulas.

EFFECTS OF HEAT STRESS

The human body has two physiological responses that come into play whenever it becomes necessary to step up heat loss. If these fail to alleviate the situation a number of physiological disturbances can occur. These are discussed below:

Increased Flow of Blood

When the heat loss from the body becomes less than the metabolic heat production, the first corrective action taken by the body is the dilation of the blood vessels near the surface of the skin, causing an increased flow of blood to the area and an increase in skin temperature. This in turn increases both the convective and radiative heat loss from the body when the ambient air temperature and the average radiant temperature of the surroundings are less than the skin temperature. If these are higher than the skin temperature, the heat gain through the channels is decreased. The flow of blood carries heat from the interior of the body to the surface, where heat transfer occurs.

Sweating

Sweating, the second defense mechanism, is brought into action when there is an insufficient flow of blood to the skin to meet the requirements for heat loss. This occurs usually when there is anything more than a minor thermal imbalance. The number of sweat glands activated and the rate of secretion of sweat are graded to meet the magnitude of the imbalance. Sweat production of more than two liters per hour has been observed, but continuous sweat rates of about one liter-per-hour over several hours each day are considered to be maximum production rates. This means that, except for short periods of time, about 600 kilocalories per hour is the maximum amount of heat that will be lost from the body surface by sweat evaporation.

Sweat production results in a drain on the water and salt in the body. The water lost in sweat is usually replaced by an increase in water intake. In most situations the thirst mechanism is sufficient to keep the water intake and water loss in balance. However, under conditions of heat stress with large sweat production (6 to 12 liters a day), enough fluids are not voluntarily consumed to replace the water lost in the sweat. This "voluntary" dehydration may amount to two to three liters or more during an eight-hour working day. The "voluntary" water deficit is usually replaced during meals and nonworking hours if an adequate supply of drinking water is available. Dehydration in excess of three liters may have serious physiological and clinical consequences.

Heat-Induced Illnesses

If the normal responses of increased skin blood flow and sweat production are not adequate to meet the needs for body heat loss or if the mechanisms fail to function properly, physiological breakdown may occur. There are four major categories of heat-induced illnesses: heat exhaustion, dehydration, heat cramps, and heat stroke.

Heat Exhaustion

This is a state of collapse brought about by an insufficient blood supply to the cerebral cortex as a result of dilation of blood

184   *Chapter 7*

vessels in response to hot conditions. The failure is not one of heat
regulation itself, but an inability to meet the price of heat regulation.

The symptoms of heat exhaustion include weakness or extreme
fatigue, giddiness, nausea, and headache. The skin is clammy and
moist, indicating that sweating remains active. The complexion may
be pale, muddy, or flushed. Oral temperatures may be normal or low,
but rectal temperature is usually slightly elevated. If sitting, the
patient may faint when he or she stands.

Dehydration

In its early stages, dehydration acts mainly by reducing the
blood volume and promoting heat exhaustion. But in extreme cases
it brings about disturbances of cell function, which increase and
reinforce each other with worsening deterioration of the organism.
Muscular inefficiency, reduced secretion (especially of the salivary
glands), loss of appetite, difficulty in swallowing, acid accumulation in
the tissues, and nervous irritability followed by depression will occur
in increasing strength. Uremia, fever, and death will terminate the
picture.

Heat Cramps

This is a condition of cramp-like spasms in the voluntary
muscles following a reduction of the concentration of sodium
chloride in the blood below a certain critical level. The abdominal as
well as the limb muscles may be affected. A high sodium chloride
loss is brought about by high sweating rates and by high intake of
water, which dilutes the salt remaining in the body. Heat cramps can
be prevented by taking extra salt whenever heavy work is to be
carried out in hot, dry environments, especially by unacclimatized
persons.

Heat Stroke (Hyperpyrexia)

This is reached when the mean temperature of the body is such
that the continued functioning of some vital tissue is thereby
endangered. It represents, of course, a marked failure of the
heat-regulating mechanism to maintain a proper balance between the
two sides of the heat balance.

There are three chief signs of heat stroke. First, the skin is hot
and dry, with a red, mottled, or bluish appearance. Second, the body
temperature is usually $106°$ F. or higher, and rising. And third, the
victim exhibits such evidences of brain disorder as mental confusion,
delirium, loss of consciousness, convulsions, and coma.

It is unlikely that this crisis will be reached in a healthy, acclimatized person carrying out normal activities in a natural climate. But under severe emotional and physical stress and very hot conditions, heat production may reach a level high enough to produce heat stroke without prior onset of the usual escape—that is, heat exhaustion.

Relief is obtained only by early and effective reduction of body temperature. This is usually done by wrapping the patient in wet sheets and playing a fan on him or her, but it is sometimes necessary to resort to packing the patient in ice. When drastic cooling is used, care has to be taken that the temperature is not lowered too fast or too far.

Acclimatization to Heat

Persons who encounter a heat wave or who are transported to a hot climate commonly find that their performance capacities are noticeably impaired at the onset. Tasks that were easily performed in a cool environment become difficult, and heat discomfort may interfere with sleeping and eating.

If, however, the exposure to the heat is continued for several days, performance gradually returns to normal and heat discomfort subsides at least to some extent. This acclimitization is the result of certain physiological adaptations, notably increased protection against hyperthermia, improved cardiovascular function, and increased sweating.

EVALUATING HAZARDS

Heat-Measuring Instruments

Human comfort is influenced by the following environmental factors:

1. Ambient air temperature (dry-bulb)
2. Humidity of the ambient air
3. Mean radiant temperature (MRT)
4. Air velocity

All four of these factors must be measured in any assessment of the thermal environment. The measurement of these variables at some convenient point within a space usually presents little difficulty provided conditions in the space are reasonably uniform and thermal radiation is not excessive. In hot industries the dry-bulb and MRT may vary widely from place to place, and near hot surfaces the radiation intensity is often high enough to complicate the accurate determination of air temperatures.

Instruments for Measuring Dry-Bulb Temperatures

Dry-bulb temperatures usually are measured with an ordinary mercury-in-glass thermometer. This thermometer should have a range of 30°F to 120°F in 0.5°F graduations. Thermocouples, thermistors, or resistance thermometers also may be used. To obtain accurate results it is necessary to shield the sensing element from radiation from nearby surfaces that are either appreciably hotter or cooler than the air.

Instruments for Measuring Humidity

The simplest and most common type of instrument used to determine the humidity of the air is the sling psychrometer. This instrument consists of a frame on which two thermometers are mounted. At one end of the frame is a handle by which the device may be whirled through the air to obtain an air velocity of 1,000 feet per minute or more across the bulbs of the thermometers. A cloth wick covers the bulb of one of the thermometers and is wetted with distilled water prior to whirling the psychrometer. Due to evaporation from the wick, the wet-bulb thermometer will indicate a lower temperature than the dry-bulb thermometer, the difference being known as the wet-bulb depression. The whirling of the psychrometer is interrupted at intervals to read the thermometers, and is continued until both thermometer readings become steady. By use of a psychrometric chart or table the relative humidity and other thermodynamic properties of the air can be established from the dry-bulb and wet-bulb temperatures thus obtained.

Often a small fan or blower is used to produce air movements across stationary thermometers. This arrangement is termed an aspirated psychrometer, and several commercial models are available. Thermistors or thermocouples can be used as the sensing elements for dry- and wet-bulb temperatures, and by feeding the outputs to suitable recorders, a record can be made of temperature fluctuations.

Relative humidity can be determined with good accuracy with a dry- and wet-bulb type psychrometer if proper techniques are followed. Otherwise results can be grossly in error. A study of a psychrometric chart or table will show that a rather small error in reading either the dry- or wet-bulb temperature can result in a rather large error in the relative humidity. A matched pair of accurate thermometers, preferably with 1/2 degree graduations, should be used.

Where radiation is high, the wet-bulb temperatures obtained with a sling psychrometer may be seriously affected, even when there is a high-velocity air flow over the bulbs. For this reason, where radiant heat is likely to be encountered, it is recommended that a properly shielded, aspirated psychrometer, rather than a sling psychrometer, be used.

Instruments for Measuring Mean Radiant Temperature

The globe thermometer is often used to measure the mean radiant temperature. A conventional globe thermometer consists of a 6-inch hollow copper sphere painted on the outside and inside with a matte black finish. For ordinary plant measurements a common copper toilet float painted a flat black gives a sufficiently accurate reading, and is cheaper and easier to obtain. A thermocouple, thermistor, or thermometer bulb is fixed at the center of the sphere with the wires or stem protruding to the outside through a sealed opening. The temperature of the air inside the globe at equilibrium is the result of a balance between the heat gained or lost by radiation and the heat gained or lost by convection. In terms of heat transfer relationships:

$$\text{MRT}^4 = T_g^4 + 0.103 \times 10^9 \sqrt{V} \ (t_g - t_a)$$

where MRT = mean radiant temperature, deg. F,
 T_g = globe temperature, deg. F,
 V = air velocity at the globe, ft/min,
 t_a = ambient air temperature, deg. F.

The MRT may be approximated by the simplified equation:

$$\text{MRT} = t_g + C \sqrt{V} \ (t_g - t_a)$$

where MRT = mean radiant temperature, deg. F,

 C = a convection coefficient = 0.17 for a 6-in. globe,

 V, t_g, and t_a are as defined above.

It will be noted from the above equations that the air temperature and air velocity around the globe must be determined. It is the determination of this velocity that presents the greatest problem in the use of a globe thermometer. A period of 15 to 25 minutes usually is required for the globe to reach equilibrium.

The two-sphere radiometer is a more recent development that also may be used to measure MRT. Two small spheres, approximately two inches in diameter, one gold plated and the other black, are heated electrically to the same temperature. Since there is no difference in convection losses, the difference in energy input required by the two spheres is determined, and from this difference the MRT may be calculated or read from calibration curves.

The MRT at any point in a space can be calculated if the temperature, emissivity, and area of all surfaces, and their orientation with respect to the point of calculation, are known. Surface temperatures for such a calculation can be obtained by means of thermocouples, or suitable radiometers or pyrometers.

Instruments for Measuring Air Velocity

Measurement of low air velocity is always difficult. In addition to being low, the air velocity in an open room is usually quite random in direction, and a nondirectional instrument should be used for its measurement. Thermal anemometers of the heated-thermocouple, hot-wire, or heated-thermometer types are best suited for such measurements. They will produce useful data if they are maintained in calibration, and the user understands their operation and limitations.

Methods of Assessing Stress

Many attempts have been made to devise a system to express as a single number, or at least as a semiquantitative expression, the significance of a given set of environmental conditions. Two of the most widely known and employed are discussed below, as is the standard recommended by the National Institute for Occupational Safety and Health.

Effective Temperature (ET)

The earliest and most widely known scheme for indicating the thermal significance of environments based upon physiological reactions is the "Effective Temperature" scheme, developed by a research team in the American Society of Heating and Ventilating Engineers. Test subjects were exposed serially to atmospheres with different temperatures, humidities, and air movements, and asked to rate their comparative sensations of warmth or coolness. From a large number of records two nomograms were drawn up indicating the combinations that produced comparable sensations on persons stripped to the waist and on normally clothed persons.

The first of these is seldom used since workers are not usually stripped to the waist; the second is shown in Figure 7.1. For simplicity of understanding, the significance of any combination of environmental conditions is expressed as the "effective temperature"; that is, the temperature of a still, saturated atmosphere that would produce the same effect. From Figure 7.1, the effective temperature for any combination of dry-bulb temperature, wet-bulb temperature, and air movement can be determined.

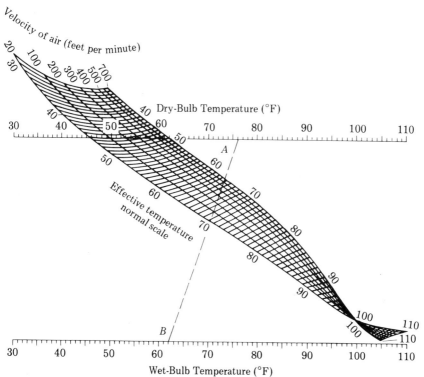

HOW TO USE THE CHART: Draw line A-B through measured dry-bulb and wet-bulb temperature. Read effective temperature or velocity at desired intersections with line A-B. EXAMPLE: Given 76°F (dry bulb) and 62°F (wet bulb) read: 69 ET at 100 fpm velocity, or 340 fpm required for 66 ET.

FIGURE 7.1. Chart showing normal scale of corrected effective (or effective) temperature. (Source: Reprinted by permission from *ASHRAE Handbook of Fundamentals*, 1967.)

This, like any other scheme, has certain limitations:

1. The basic observations were made on a specific group of subjects—healthy, young, white men and women, living under the current American conditions of climate, housing, clothing. The results are not necessarily representative of other groups and should not be applied with abandon.
2. The observations relate only to sedentary conditions, although some later modifications were drawn up to make the scheme applicable to greater activities.
3. The scheme applies only to persons in "normal" clothing. Heavy clothing could be expected to reduce the effect of air movement still further.
4. The original scheme applied only to situations in which the mean temperature of the surroundings was substantially the same as the air temperature. It has been shown, however, that the scheme can continue to be used where the temperatures of the surrounding surfaces differ from the air temperature, provided that they are fairly uniform and that the glove thermometer temperature is substituted for the air temperature.
5. It does not apply at air movements below 25 feet per minute.
6. It is not reliable at the upper end of the scale—effective temperature above $90°$F.—since sensations of heat cease to be very good guides under conditions as hot as this.
7. The scheme in its original form gives too much weight to changes in humidity at the lower end of the scale.
8. The scheme provides no scale for evaluating the physiological significance of stresses above or below the comfort zone.

Heat-Stress Index (HSI)

The HSI involves the concepts of stress and strain. Heat stress is considered to exist whenever, despite vasomotor adjustment, metabolic heat production exceeds the combined losses by radiation and convection. Strain is considered to be the probable consequence of exposure to stress, and is expressed as the ratio between the rate at which evaporative cooling is needed to balance the stress and the maximum rate at which evaporative cooling can be maintained in that particular environment. Mathematically,

$$\text{HSI} = \frac{E_{\text{req}}}{E_{\text{max}}} \times 100$$

For values below 100, the HSI is approximately the percentage of a person's total sweating capacity that is being used.

The flow charts shown in Figure 7.2 can be used to determine HSI. To illustrate, let us assume a situation in which the globe temperature is $110°F$, the dry-bulb temperature is $90°F$, the wet-bulb temperature is $75°F$, the air speed is 100 feet per minute, and the metabolism is 600 Btus per hour.

The first step is to find the globe temperature on the vertical scale of Chart A, draw a horizontal line to the air-speed line, then drop a vertical line to the horizontal scale. The value found on the horizontal scale represents the combined radiation and connection heat load.

The second step is to extend the vertical line of Chart A into Chart B to a point corresponding to the value for metabolism. From this point, a horizontal line should be drawn to the vertical scale of Chart B to obtain the net heat load in terms of the evaporation required for heat balance (E req).

For the third step, go to Chart X and draw a horizontal line from the intercept of the wet-bulb and dry-bulb temperatures to the right-hand vertical scale to obtain the vapor-pressure difference between saturated skin at $95°F$ and ambient air.

To carry out step four, extend the horizontal line of Chart X into Chart Y until it intercepts the line corresponding to the air speed. From this point, draw a vertical line to the upper scale of Chart Y, which represents E_{max}, and into Chart C. Continue extending this line and also extend the horizontal line of Chart B until the two intersect. The point of intersection represents the HSI, in this case 90. If the E_{max} reading on the upper scale of Chart Y exceeds 2,400, then the horizontal line within Chart C is extended to the vertical scale to determine the HSI.

Table 16* lists the physiological and hygienic implications of exposures to various heat-stress levels.

Under conditions of high heat stress (HSI $\geqslant$ 100) people will usually not work at a rate that will lead to collapse from heat exhaustion. Instead, they compensate by lowering their rates of activity, thereby reducing their metabolic rates.

Wet Bulb Globe Temperature (WBGT)†

Most of the methods suggested for evaluating heat stress,

---

*From H. S. Belding and T. F. Hatch, *Heating, Piping, and Air Conditioning*, 27 (1955), 129.

†For further details concerning this standard see "Criteria for a Recommended Standard . . . Occupational Exposure to Hot Environments," U.S. Department of Health, Education, and Welfare, National Institute for Occupational Safety and Health, 1972.

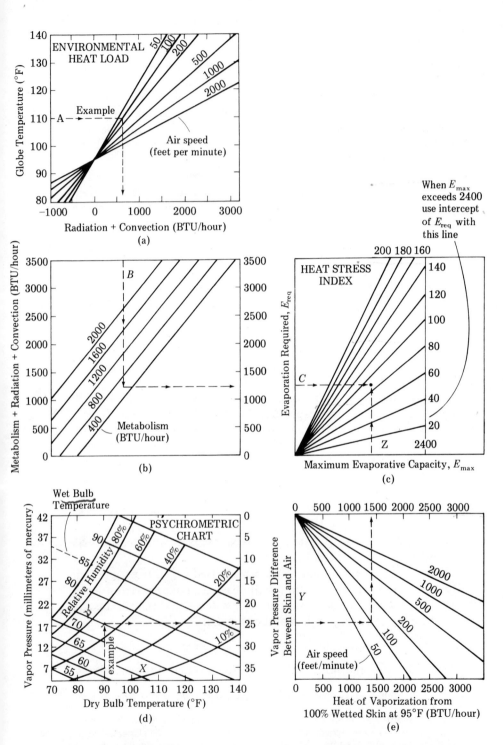

FIGURE 7.2. Flow charts for determining heat-stress index. (From H.S. Belding and T.F. Hatch, *Heating, Piping, and Air Conditioning*, **27** (1955), p. 129.

TABLE 16

Effects of Various Heat-Stress Levels

Index of Heat Stress	Effect of 8-hour Exposure
0	No thermal strain
+10 20 30	Mild to moderate heat strain. Where a job involves higher intellectual functions, dexterity, or alertness subtle to substantial decrements in performance may be expected. In performance of heavy physical work, little decrement should be expected unless the ability of individuals to perform such work under no thermal stress is marginal.
40 50 60	Severe heat strain, involving a threat to health unless people are physically fit. Break-in period will be required for those not previously acclimatized. Some decrement in performance of physical work is to be expected. Medical selection of personnel is desirable because these conditions are unsuitable for those with cardiovascular or respiratory impairment or with chronic dermatitis. These working conditions are also unsuitable for activities requiring sustained mental effort.
70 80 90	Very severe heat strain. Only a small percentage of the population may be expected to qualify for this work. Personnel should be selected (1) by medical examination and (2) by trial on the job (after acclimatization). Special measures are needed to assure adequate water and salt intake. Amelioration of working conditions by any feasible means is highly desirable, and may be expected to decrease the health hazard while increasing efficiency on the job. Slight "indisposition," which in most jobs would be insufficient to affect performance, may render workers unfit for this exposure.
100	The maximum strain tolerated daily by fit, acclimatized young men. (Tests have not yet been conducted on women.)

including the two discussed above, involve consideration of wind velocity. In view of the problem associated with the accurate measurement of wind velocity, the National Institute for Occupational Safety and Health (NIOSH) has recommended a standard not involving this factor as the standard for exposure to hot environments.

The numerical value of the index is calculated by the following equations:

1. Indoors or outdoors with no solar load.
   WBGT = 0.7WB + 0.3GT

2. Outdoor with solar load.
   WBGT = 0.7WB + 0.2GT + 0.1DB

where WB = Natural wet-bulb temperature obtained with a wetted sensor exposed to the natural air movement (unaspirated)

GT = globe thermometer temperature

DB = dry-bulb temperature

To determine the time-weighted average WBGT, the following equation is used:

193

$$\text{Avg. WBGT} = \frac{(\text{WBGT}_1) \times t_1 + (\text{WBGT}_2) \times t_2 + \cdots + (\text{WBGT}_n) \times t_n}{t_1 + t_2 + \cdots + t_n}$$

where WBGT_1, WBGT_2, WBGT_n are calculated WBGT values for the various rest or work areas occupied during the total time period. t_1, t_2, $\cdots$, t_n are the elapsed times in minutes spent in the corresponding areas.

If the exposure to heat is continuous, time-weighted average WBGT values should be calculated hourly. If exposure is intermittent, the WBGT values should be calculated on a two-hour basis.

The wet-bulb thermometer used for calculating the WBGT index should be a mercury-in-glass thermometer with a range of 30 to 120°F in 0.5°F graduations. The wick covering the bulb should be highly absorbent cotton and should extend to 1¼-inch above the bulb. The lower end of the wick should be immersed in a reservoir of distilled water. Figure 7.3 illustrates a WBGT-measuring instrument. The globe thermometer and the dry-bulb thermometer should be the types described earlier in this chapter.

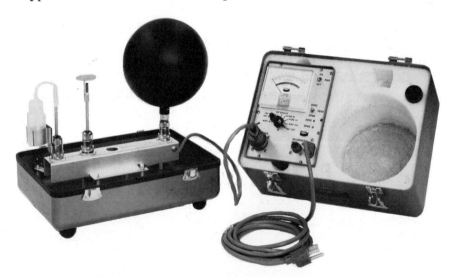

FIGURE 7.3   **Wet bulb globe temperature measuring instrument.** *(Photo courtesy Bendix Corporation Environmental Science Division.)*

The American Conference of Governmental Industrial Hygienists has recommended the following work-rest regimens for workers exposed to various WBGTs.*

*From "Threshold Limit Values for Chemical Substances and Physical Agents in the Workroom Environment with Intended Changes for 1975," American Conference of Governmental Industrial Hygienists, 1975, p. 60.

TABLE 17

Permissible Heat Exposure Threshold Limit Values
(Values are given in °C. WBGT)

Work-Rest Regimen	Light	Work Load Moderate	Heavy
Continuous Work	30	26.7	25.0
75% Work, 25% rest each hour	30.6	28.0	25.9
50% Work, 50% rest each hour	31.4	29.4	27.9
25% Work, 75% rest each hour	32.2	31.1	30.0

Typical examples of light, moderate, and heavy work operations of various types are as follows: †

Light hand work: writing, hand knitting

Heavy hand work: typewriting

Heavy work with one arm: hammering in nails

Light work with two arms: filing metal, planing wood, raking a garden

Moderate work with the body: cleaning a floor

Heavy work with the body: laying railroad track, barking trees, digging

COMBATING HAZARDS

Hazards may be combated by use of personal protective devices and/or engineering controls, by following recommended medical and work practices, and by posting warning signs, monitoring the work place, and keeping proper records.

Personal Protective Equipment and Clothing

In particularly hot areas, vortex tubes can provide very effective cooling. A vortex tube is a long tube into which compressed air is fed tangentially at a high pressure and temperature, and at a point near one end of the tube. The bulk of the air spirals inward, expanding and cooling greatly. This cool air exhausts from the vortex tube at a point close to the point of entrance. The small portion of the air that is not cooled churns around in the tube, becomes heated, and exhausts at the end opposite the cool-air exhaust. The vortex tube device is carried on the belt. Such devices can be used with head protection alone, with jackets, and with full-suit protection. Temperature decreases of 40 to 50°F. may be obtained. A vortex tube is shown in Figure 7.4.

Self-contained sources of air that can be backpacked have also been developed. One involves a liquid refrigerant, which is sealed into

†*Ibid.*, p. 63

FIGURE 7.4   Vortex tube.
(Photo courtesy 3M Company.)

a finned container. After being cooled in a deep-freezer, the container is placed in the pack. A small battery-driven fan circulates air across the fins and into the suit. More sophisticated devices employ a closed system with liquid as the coolant and a fairly elaborate network of small tubes for distribution.

Infrared reflecting face shields may be worn when radiant heat is high. When hot materials are frequently handled, workers often use oversized insulated gloves with wide gauntlets so they can be easily put on.

For bodily exposures to furnaces and other sources of very high heat, thick insulated clothing is appropriate. Such clothing acts as a heat "sponge." The most effective insulation is provided by high-density materials such as asbestos, but insulation with medium weight is best imparted by a thickness of still, trapped air.

When shielding against radiant heat cannot be accomplished by fixed barriers, aluminized clothing components may often be used to advantage. The aluminum is vacuum-deposited on the surface of the fabric. In intermittent work at high humidities, a full aluminized suit can be a handicap by interfering with the evaporation of sweat. The use of such suits can sometimes be avoided. For example, if fixed shielding to the waist is employed, workers may sometimes need only an aluminized jacket. Often, too, a worker who faces the heat source may be protected adequately by a long metallized apron. Figures 7.5, 7.6, and 7.7 show various types of heat-protective clothing.

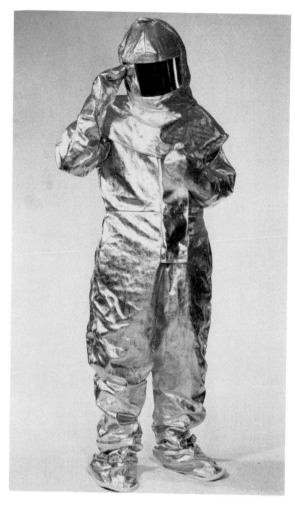

FIGURE 7.5   Aluminized suit. *(Photo courtesy
Mine Safety Appliances Company.)*

Engineering Controls

The engineering controls available for controlling heat include
insulation of hot equipment, shielding of hot surfaces, and local
exhaust ventilation. The most suitable type of control will depend
upon the means by which the heat is being transferred.

197

FIGURE 7.6   Aluminized apron.
*(Photo courtesy Wheeler Protective
Apparel, Inc.)*

Insulation of Hot Equipment

Insulating hot surfaces will decrease the amount of sensible and radiant heat lost to the environment. Likewise, enclosing hot water tanks, covering hot water drains, and maintaining joints and valves on steam lines point out the best practice of controlling heat at its

FIGURE 7.7 Aluminized apron,
gloves, and safety helmet.
*(Photo courtesy Wheeler
Protective Apparel, Inc.)*

source. In this light, it is advantageous to isolate hot processes, even putting them outdoors where practical.

Shielding of Hot Surfaces

Shielding is limited to control of radiant energy and is very effective in this application. One great advantage is that it can be designed to protect the worker specifically, rather than minimize general loss of heat as with insulating materials. An inexpensive, reflective shield consists of aluminum foil backed by light-weight plastic insulating board. Other types of shields include (1) absorbing shields made of sheet-steel panels separated by air spaces, (2) water-cooled shields where the absorbed heat is carried away with a moving film of water, and (3) "transparent" shields made of wire mesh, chains, or heat-reflecting glass.

Local Exhaust Ventilation

Although limited in use, local exhaust ventilation can be used in some areas to reduce heat exposure. Canopy hoods, using either natural or mechanically induced drafts, are in common use over furnaces and hot, open mix tanks. It should be remembered, however, that ventilation is only effective in removing convective heat; normally the use of insulation and/or shielding will further reduce the heat exposure.

A detailed discussion of exhaust ventilation may be found in Chapter 10.

Medical Monitoring

Pre-employment Physical Examination

According to the National Institute for Occupational Safety and Health Standard, any prospective employee for a hot job should be given a thorough pre-employment physical examination to determine his or her mental, physical, and emotional qualifications to perform the assignment efficiently and safely.

Both during the history taking and the physical examination itself, the examiner should be especially alert for evidence of chronic functional or organic impairments of the cardiovascular system as well as of the kidneys, liver, endocrines, lungs, and skin. Significant disease of any of these systems should disqualify the applicant for jobs involving severe heat exposure.

Careful inquiry should be made concerning use of drugs, particularly hypotensive agents, diuretics, antispasmodics, sedatives, tranquilizers, and antidepressants, as well as the abuse of drugs, particularly amphetamines, hard narcotics, and alcohol. Many of these drugs impair normal response to heat stress and others alter behavior, thus exposing the employee or other workers to health and safety hazards. Evidence of therapeutic use of one or more of these drugs or personal abuse of alcohol or other drugs should be disqualifying.

A glucose tolerance test, renal clearance studies, chest X rays, X rays of the renal pelvis and biliary system with contrast media, pulmonary function tests, and the usual blood and urine analyses are also recommended.

A history of successful adaptation to heat exposure on previous jobs is perhaps the best basis for predicting an effective future performance under heat stress, provided that work demands and heat exposures on the old and new jobs are equivalent, and the worker's health has not changed significantly.

Periodic Physical Examination

The National Institute for Occupational Safety and Health Standard stipulates that all employees exposed to hot environmental conditions should be given a periodic physical examination. For employees under 45 years old, the examination should occur every two years, and for employees over 45 years old it should be given every year. The examination should include all components of the pre-employment examination.

Examinations of employees over 45 years old should be designed to detect the onset of chronic impairment of the cardio-circulatory and cardiorespiratory systems and also to detect metabolic, skin, and renal diseases. For all employees, any history of acute illness, either occupational or nonoccupational, during the period between examinations should be noted. Repeated accidental injuries on the job or repeated sick absences should alert the physician to possible heat intolerance of the employee. Nutritional status should be noted and advice given to correct overweight.

In industrial establishments where the heat stress approaches or equals permissible limits only in the summer, periodic examination should be administered during the summer. Where heat stress at the permissible level occurs throughout the year, the examination can be administered in any season.

Safe Work Practices

The standard recommended by the National Institute for Occupational Safety and Health for exposure to hot environments sets exposure limits for new employees without previous occupational exposure to heat. These limits stipulate that such new employees not be assigned to jobs where the time-weighted average WBGT exceeds 79°F for men and 76°F for women until they are acclimatized.

The acclimatization schedule calls for acclimatization over a 6-day work period. The schedule should begin with 50 percent of the anticipated total work load and time exposure on the first day, followed by daily 10 percent increments building up to 100 percent total exposure on the sixth day.

Regular acclimatized employees who return from 9 or more consecutive days of illness should have medical permission to return to the job and undergo the 4-day acclimatization period.

Employees exposed to environmental conditions exceeding the prescribed limits should be trained in health and safety procedures through a program that includes the following:

1. Information on water intake for replacement purposes.
2. Information on salt intake.
3. Importance of weighing each day before and after the day's work.
4. Instructions on recognizing the symptoms of heat disorders and illnesses, including dehydration, heat exhaustion, heat cramps, and heat stroke.
5. Information on special cautions to be exercised in situations where employees are exposed to toxic agents and/or other stressful physical agents simultaneously with heat.
6. Information concerning heat acclimatization. The information should be kept on file and readily accessible to the worker.

A minimum of one break per hour should be provided to allow employees to get water and replacement salt. The employer should provide a minimum of 8 quarts of cool 0.1 percent salted drinking water or a minimum of 8 quarts of cool drinking water and salt tablets per man or woman per shift. The water supply should be located as near as possible to the employee's work station, but never more than 200 feet away.

Appropriate protective clothing and equipment should be provided and used, and appropriate controls to reduce environmental heat should be utilized.

Emergency Procedures

Supervisors and selected personnel should be trained in recognizing the symptoms of heat disorder and in administering first aid. First aid treatment requires immediate removal of the victim to a cooler area, soaking the clothing in cold water, and fanning vigorously. Medical facilities to treat heat disorder should be as close as possible to work areas.

In treating heat stroke, an air-conditioned room should be available and provided with a tub and ice for immersion and massage treatment or a suitable table on which the patient can be placed, wrapped in wet sheets, and fanned vigorously. To reduce shivering and increase the rate of heat dissipation, chlorpromazine can be administered intravenously in dosages of 25 to 50 milligrams. Body temperature should be measured every 3 to 5 minutes and cooling interrupted when the rectal temperature reaches 100 to 101°F. Monitoring should then be continued to prevent the temperature from rising to dangerously high levels or dropping too low.

If the patient is in shock when admitted to the medical facility and the shock persists after cooling, it can be treated by oxygen inhalation, administration or intravenous fluids, or the use of pressor agents.

Heat exhaustion is treated by oral administration of salted liquids or by intravenous infusion of normal saline if the patient is unconscious or vomiting. Heat cramps are likewise treated by intravenous infusion of normal saline, with rapid administration of 250 milliliters within 5 to 10 minutes.

Warning Signs, Monitoring, Record Keeping

The National Institute for Occupational Safety and Health standard calls for warning signs bearing the legend "Warning—Heat Stress Area" to be prominently located in one or more spots at the edges or entrances to areas where environmental conditions are 86°F WBGT or above.

A WBGT profile should be established for each work place for winter and summer seasons to serve as a guide for deciding when work practices should be initiated to conform with the requirements of the standard. After profiles are established, monitoring should be conducted once during July and August of each year.

Medical and acclimatization records should be maintained for the period of the person's employment and for one year thereafter.

NOTE ON COLD STRESS

Although the problems of cold stress have not received nearly so much attention as those of heat stress, some effort has been made to determine maximum time limits that workers can be safely exposed to different low temperatures. The findings are shown in Table 18.

Workers with outside jobs should be provided with the proper

TABLE 18

Low-Temperature Time Limits

Temperature Range	Maximum Daily Exposure
30°F to 0°F	No exposure time limit, if the person is properly clothed.
0°F to −30°F	Total cold-room work time: 4 hours. Alternate 1 hour in and 1 hour out of the chamber.
−30°F to −70°F	Two periods of 30 minutes each, at least 4 hours apart. Total cold-room work time allowed: 1 hour. (Note: Some difference exists among individuals: one report recommends 15-minute periods—not over 4 periods per work shift; another limits periods to 1 hour out of every 4, with a low chill factor, i.e., no wind; a third says that continuous operation for 3 hours at −65°F has been experienced without ill effect.)
−70°F to −100°F	Maximum permissible cold-room work time: 5 times over an 8-hour working day. For these extreme temperatures, the wearing of a completely enclosed headgear, equipped with a breathing tube running under the clothing and down the leg to preheat the air, is recommended.

Source: *N.S.C. Data Sheet 465*, Cold Room Testing of Gasoline and Diesel Engines. Published by National Safety Council, Chicago, Illinois.

clothing for the environment. Clothing may include freeze jackets, thermal underwear, special cold- and wind-resistant trousers and jackets worn over regular clothing, hoods, parkas, and insulated gloves.

Workers should avoid touching metal tools and equipment with bare hands, since the metal will conduct heat rapidly away and hands may become frozen to the metal.

In case of frostbite, the victim should be removed to a warm area immediately and proper medical care obtained.

Several articles containing detailed instructions for coping with cold-weather problems can be found in the March 1960 issue of the *Journal of Occupational Medicine*. The author and page numbers of these articles are as follows: E. A. Sellers, 115-17; L. H. Turl, 123-28; L. D. Carlson, 129-31; K. A. Provins and R. S. J. Clark, 169-76; and K. Rodahl, 177-82.

EXERCISES: CHAPTER 7

1. Name the factors that influence the exchange of heat between the body and the environment.

2. Discuss the mechanisms involved in the following methods of heat exchange:
   a. conduction-convection
   b. evaporation-convection
   c. radiation

3. Describe the two physiological mechanisms used by the body to step up heat loss.

4. Contrast the effects of the following types of heat-induced illnesses:
   a. heat exhaustion
   b. dehydration
   c. heat cramps
   d. heat stroke

5. Describe the sling psychrometer and globe thermometer, and indicate what they are used to measure.

6. Define the terms "effective temperature" and "heat stress index," and list the limitations of the effective temperature system for evaluating heat stress.

7. Describe the wet-bulb globe temperature (WBGT) system for evaluating heat stress, indicating the methods for calculating the numerical value of the index and the time-weighted average WBGT.

8. Discuss the four types of engineering controls used to combat heat.

9. Describe the medical monitoring program for persons working in hot jobs, with special attention to:
   a. the scope of the pre-employment physical examination
   b. medical conditions that should exclude persons from hot jobs
   c. the scope of follow-up physical examinations

10. Summarize the acclimatization schedules for persons working in hot jobs.

11. Discuss the procedures for treating heat exhaustion, heat cramps, and heat stroke.

8

Noise

Noise is unwanted sound. There are three basic types of noise: wide-band, narrow-band, and impulse.

Wide-band noise is, as the name implies, noise that is distributed across a broad range of frequencies. Narrow-band noises have their energies confined to a narrow range of frequencies or concentrated about a single frequency. The noise created by circular saws, planers, and similar power tools is narrow-band noise. Impulse-type noise is comprised of transient pulses which can occur repetitively or nonrepetitively. The noise associated with a jack hammer is an example of repetitive noise; the firing of a gun is a nonrepetitive noise.

DEFINITIONS

To understand the subject of noise, it is necessary to know the meaning of a number of terms.

Decibel (dB): A unit for measuring the relative sound pressure levels, and thus the relative intensity of sounds. It is approximately equal to the smallest degree of difference of loudness ordinarily detectable by the human ear. On the decibel scale, 1 represents the faintest audible sound; 120 is generally considered to be the threshold of pain.

The decibel scale is logarithmic. Therefore, values cannot be added or subtracted arithmetically. Table 19 can be used to determine the approximate decibel level that will result when the values for two sounds measured separately are combined.

TABLE 19

Difference Between Decibel Levels	Number of Decibels to be Added to Higher Level
0	3
1	2.6
2	2.1
3	1.8
4	1.5
5	1.2
6	1.0
7	0.8
8	0.6
10	0.4
12	0.3
14	0.2
16	0.1

In combining more than two decibel levels, the two highest levels should be combined first. The total obtained should then be combined with the highest remaining level and the procedure continued to completion.

It is important to realize that sounds of equal sound pressure may not be equally loud. At pressures near 100 decibels, frequencies between 20 and 1,000 cycles per second sound equally loud. At lower sound pressure levels, the lower-frequency sounds do not seem as loud as the 1,000 cycles per second tone.

Frequency: This is the number of variations in sound pressure per second. It is usually expressed in Hertz (Hz) units or in cycles per second (cps). The healthy young ear can detect sounds in the 20 to 20,000 cycles per second range. As aging occurs, some loss of hearing takes place. The frequencies comprising speech are found principally between 250 and 3,000 cycles per second.

Hearing sensitivity: An individual's threshold of hearing; that is, the lowest intensity of a given tone that he or she can detect through air.

EFFECTS OF NOISE

Noise may cause loss of hearing, disrupt speech communication and hearing, cause annoyance, and impair performance to some extent.

Loss in hearing, also known as threshold shift, may be temporary or permanent, depending upon the length and severity of the exposure. Temporary threshold shift, also called auditory fatigue, is a loss which is recoverable after a period of time away from the noise. Such losses may occur after only a few minutes of exposure to intense noise. With prolonged exposures (months or years) to the same noise, hearing recovery may be only partial. In this case, the remaining loss represents permanent hearing impairment.

Noise that is not intense enough to cause hearing damage may still disrupt speech communication and the hearing of other desired sounds. While communication is essential in many places of employment, the levels of acceptability vary with the nature of the work involved. Being able to communicate by shouting, for example, may be satisfactory when doing maintenance work on certain machinery. On the other hand, raising the voice slightly to overcome typical office noises may be quite undesirable for a conference room.

Perhaps the most widespread reaction to noise is annoyance. Some characteristics of noise seem to be more annoying than others. Thus, an intense and therefore loud noise is more annoying than one that is less intense Similarly, a high-pitched noise—one containing mostly frequencies above 1,500 cycles per second—is more annoying than a low-pitch noise of equal loudness. Sounds that occur randomly, vary in intensity or pitch, or repeatedly seem to change their location are more annoying than those which are continuous, unchanging, or stationary.

The effects of excessive noise on efficiency and work output seem to be relatively slight. Performance on tasks involving simple repetitive operations does not appear to be affected by noise, whereas losses in efficiency on most complex tasks tend to become dissipated with increased exposure time. The relation between excessive noise and such factors as accident rate, absenteeism, and employee turnover has not as yet been clearly determined.

EVALUATING HAZARDS

The industrial hygienist is being called upon more and more frequently to make noise measurements and to evaluate noise problems. In conducting an evaluation, the industrial hygienist seeks to determine the total sound-pressure level and the distribution of the sound pressure with frequency and with time. The specific situation will dictate the equipment to be utilized and the procedure to be employed.

Noise-Measuring Instruments

A wide variety of equipment is commercially available for evaluating noise. The choice of equipment will depend on the type of noise being dealt with and the purpose of the survey.

Sound Survey Meter

This is a small instrument that usually consists of a nondirectional microphone, an adjustable calibrated attenuator, an amplifier with three weighting networks, and an indicating meter.

The instrument has a function switch with an off position, A, B, and C weighting network positions, and a battery check position. It has either a continuously variable or a stepped attenuator. The sound pressure level is the sum of the attenuator setting and the reading from the indicating meter.

Even though the sound survey meter does not meet American Standards Association specifications for sound level meters, it is a good instrument for rapid screening and surveying of a large area or number of locations.

Sound Level Meter

This is the basic instrument used for noise measurements. It comprises a microphone and an electronic circuit including an attenuator, an amplifier, three frequency response networks (weighting networks), and an indicating meter. The attenuator in the circuit controls the current within limits that can be handled by the indicating meter, which is calibrated in decibels. Sound level meters are available in the range of 20 to 180 decibels. The newer meters have an amplifier with a flat frequency response from 20 to 20,000 cycles per second. The overall response depends on the microphone used.

Three weighting networks, A, B, and C, are included. Their purpose is to give a number that is an appropriate evaluation of the total loudness level. Human response to sound varies with its frequency and intensity. The three weighting networks also provide a means of compensating for these variations in human response.

The A network is less sensitive to low frequencies and is intended for use at sound pressure level below 55 decibels. Human responses can be predicted quite accurately from this level. The B network is an intermediate step for the range of from 55 to 85 decibels. The C network has a flat response and is used for everything above 85 decibels. The C network is the only one that indicates the actual sound pressure level, since the sound pressure level is based on a flat response.

Figure 8.1 shows several commercially available units, and Figure 8.2 shows the use of a unit in checking sound levels.

FIGURE 8.1 Commercially available sound level meters.
*(Photos courtesy B & K Instruments, Inc.; Bendix
Corporation Environmental Science Division;
Bausch & Lomb; and General Radio Co.)*

FIGURE 8.2   Use of sound level meter in checking
sound levels. *(Photo courtesy General Radio Co.)*

Microphones

The microphone picks up the sound energy and converts it to electrical energy, which is fed to the electronic circuit of the measuring device. The frequency response, sensitivity, directionality, and range of the meter are primarily determined by the microphone; therefore, its importance cannot be overemphasized. The four major types of microphones in general use are described below.

Rochelle Salt Crystal Microphone. The Rochelle salt crystal microphone was supplied as standard equipment on sound level meters for many years. The reasons were its low cost, ruggedness, availability, and high sensitivity. It has a frequency range of 20 to 8,000 cycles per second and an intensity range of 24 to 150 decibels. Its frequency response varies widely with the incidence of the sound, and the response is far from flat above 2,000 cycles per second.

One of the disadvantages of the Rochelle salt crystal microphone is its sensitivity to heat and humidity. Temperatures above 115°F may damage the crystal, and temperatures above 125°F will destroy it. Also, there is a reversible change in sensitivity produced by temperature change. When an extension cable is used and the input impedance of the meter is high, a temperature correction is necessary.

Ceramic Crystal Microphone. Since 1963 a PZT (lead titanate-lead zirconate) piezo-electric, ceramic diaphragm-type microphone

211

has been regularly supplied with many of the sound level meters. This is a stable and rugged microphone with a smooth frequency response, and it is relatively unaffected by normal temperature and humidity changes. It has a frequency range of 20 to 9,000 cycles per second and an intensity range of 24 to 150 decibels. It will withstand temperatures of 22 to 205°F without damage. A correction is necessary when the PZT microphone is used with an extension cable. A correction of about 7 decibels is made when a 25-foot cable is used between the microphone and the instrument.

Dynamic Microphone. This is probably the second most commonly used type. It operates on the principle of a coil moving in a magnetic field resulting in an induced voltage. This type of microphone has a dynamic range of 20 to 140 decibels and a frequency response in the range 40 to 10,000 cycles per second ±5 decibels. Its frequency response is less erratic than that of the crystal microphone, but the frequency response drops off sharply below 40 cycles per second. The dynamic microphone has the advantage of low self-noise, permitting noise measurements down to less than 20 decibels in an octave band. Sensitivity to temperature changes is small up to 180°F, and it can be used with long extension cables without correction. This microphone cannot be used in strong magnetic fields such as those encountered around large transformers, generators, or electric arc welders. Care must be taken to avoid getting the microphone wet because of the possibility of internal shorts.

Condenser Microphone. This is the least commonly used of the four microphones. It consists of two plates that act as a condenser; one is a thin diaphragm, the other a solid backing. Movement of the diaphragm changes the capacitance of the condenser in proportion to the displacement of the diaphragm. This change in capacitance produces an electrical signal proportional to the sound pressure level.

This microphone has the best frequency response of all types in current use. It is available in models that are essentially flat in response to frequencies from 20 to 8,000 cycles per second and ±3 decibels at 15,000 cycles per second to models that are essentially flat from 20 to 40,000 cycles per second. The dynamic range is from 40 to 145 decibels in the standard models, with special models available with an upper limit of 200 decibels. Its temperature sensitivity corresponds to that of a dynamic microphone. It can be used at higher temperatures and with longer cables without correction.

Disadvantages are its high cost, high self-noise, limited humidity range, and additional electronic complexity due to the preamplifier that is required. High humidities may increase internal noise, but this can be corrected by keeping the microphone in a dessicator when not in use. Figure 8.3 shows three condenser-type microphones.

FIGURE 8.3   Condenser-type microphones.
(Photo courtesy General Radio Company.)

Noise Analyzers

When the sound to be measured is complex— that is, consisting of a number of tones or having a continuous spectrum—the single value obtained from a sound level meter reading often is not sufficient for analytical purposes. It may be necessary to determine the sound pressure distribution according to frequency. Such information can be obtained by means of a noise analyzer connected to the sound level meter.

In general, there are three types of noise analyzers:

1. Octave, half-octave, and third-octave bandwidth.
2. Constant bandwidth.
3. Constant-percentage narrow band.

Octave Band Analyzer. The most practical and widely used analyzer for industrial noise studies is the octave band analyzer. As indicated by the name, the upper cutoff frequency is twice the lower cutoff frequency. For a half-octave analyzer the upper cutoff frequency is $\sqrt{2}$ times the lower, and for a third-octave analyzer the ratio is $\sqrt[3]{2}$. A commonly used set of octave pass bands in commercial equipment is 20 to 75, 75 to 150, 150 to 300, 300 to 600, 600 to 1,200, 1,200 to 2,400, 2,400 to 4,800, and in excess of 4,800 cycles. This analyzer provides sufficient data on industrial noise to evaluate its physiologic significance and to provide a basis for noise control. The small number of bands makes it possible to perform an analysis in a reasonable length of time, and the bands are wide enough to reduce the problem of transient components affecting stability of the meter.

This analyzer may be equipped with one of two types of filters. In one type there is a choice of eight pass bands and the selection is made by a single switch. The second type consists of a low cutoff filter, which filters out all frequencies below the setting of the dial; and a high cutoff filter, which filters out all frequencies above the setting of the dial. Measurements can be made in bands of one octave or multiples of an octave to cover the complete spectrum, using the second type of analyzer. The attenuator, which is 50 decibels in 10 decibel steps, and the amplifier serve the same purpose as in the sound level meter. The meter has a 16-decibel range.

All octave band analyzers are equipped with an output jack permitting the use of headphones, a recorder, or an oscilloscope.

Some octave band analyzers using the American Standards Association's preferred frequencies are on the market. In these analyzers the octave bands are centered around the following mid frequencies: 63, 125, 250, 500, 1,000, 2,000, 4,000, and 8,000 cycles per second. Although there is a shift in octave bands utilized on the standard analyzers, there is no practical difference in the interpretation of data of industrial hygiene significance. Figure 8.4 shows two commercial octave band analyzers.

Constant Bandwidth. This instrument has a fixed bandwidth that is a specific number of cycles wide. Common bandwidths range between 5 and 200 cycles per second. These are essentially laboratory instruments requiring line voltage for operation. Because of the transient nature of industrial noise and the narrow bandwidth of the analyzer, this instrument is seldom used in industrial noise studies.

Constant-Percentage Bandwidth. The constant-percentage narrow-band analyzer has a bandwidth which is a fixed percentage of

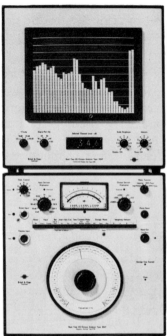

FIGURE 8.4   Octave band analyzers. *(Photos courtesy General Radio Co. and B & K Instruments, Inc.)*

the mid-band frequency. Thus, at low frequencies the bandwidth will be only a few cycles, while at high frequencies the bandwidth will cover a wide range of frequencies. This instrument has essentially the same limitations as the constant bandwidth analyzer but it can be used for some noise control problems.

Accessories

A number of accessories are available for use with sound level meters and frequency analyzers to give additional information about the noise situation under study.

Impact Noise Analyzer. This instrument operates directly from the output of the sound level meter or frequency analyzer. It can be used to measure the peak level and duration of impact noise.

Through the use of electrical storage systems, three characteristics are measured by the analyzer for every impact noise. These are: the maximum instantaneous level, the average level, and a continuously indicating measure of peak sound levels. The values can be read individually on the same meter by means of a selector switch. A reset position of the selector switch restores the meter to its prereading condition.

Cathode-Ray Oscillograph. An oscillograph affords a useful means for observing the wave form of a noise and also for observing many general noise patterns, particularly short duration or impact noises. The oscillograph can be operated from the output signal of a sound level meter, octave band analyzer, or magnetic tape recorder. It can be used to measure the peak amplitude, the rate of decay, and the shape of a wave.

A cathode-ray oscillograph with a sweep rate extending down to at least two seconds and a long-persistence screen is preferable for acoustic measurements. If the wave form is to be photographed, a short-persistence screen should be employed.

Magnetic Tape Recorder. The magnetic tape recorder can be used for recording noise in the field for subsequent analysis in the laboratory. Detailed and repeated studies can be made immediately or the tape can be kept for later analysis or for comparison purposes.

If meaningful measurements are to be made from a recording, a high quality recorder must be used. It should have a flat frequency characteristic over a wide range, low hum and low noise levels, low nonlinear distortion, a constant speed drive, and it should be of good mechanical construction. A tape speed of 15 inches per second is recommended, since these characteristics are more readily obtained and maintained at high tape speeds. The gain control should be set according to instructions supplied with the recorder. After it is set a reference signal should be recorded, such as the signal from an acoustical calibrator. Whenever the gain control is reset, a reference signal should again be recorded.

When an analysis is desired, a representative sample of the tape should be selected and that portion of the tape spliced to form a loop so it can be repeated continuously. If the absolute level needs to be known also, a sample of the recorded reference signal should be measured with the same control settings as were used for the original noise recordings.

Graphic Level Recorder. The graphic level recorder provides a permanent record of the noise characteristics by tracing a graph on a moving paper. The recorder can be operated from the output of the sound level meter or the octave band analyzer to record the level of a noise as a function of time. It is subject to the same limitations as the sound level meter in the analysis of noise of short duration.

When using the graphic level recorder it must be calibrated with an acoustical calibrator. The gain control should be adjusted according to instructions supplied by the manufacturer. Figure 8.5 shows such a unit.

FIGURE 8.5   Graphic level recorder.
(Photo courtesy General Radio Co.)

Acoustic Calibrator. An overall acoustical check can be made of the sensitivity of a sound level meter, including its microphone, by use of an acoustic calibrator. Calibration may be accomplished either by electronic or mechanical devices. The electronic calibrator consists of a small speaker mounted in an enclosure which fits over the microphone of the sound level meter. The enclosure is designed in such a way that the measurements can be repeated with a high degree of accuracy. Signals may be supplied to the calibrator from either an audio oscillator or random noise generator depending on which calibrator is being used. The designated operating voltage of the supplied signal and the sound level output can be found in the operating instructions furnished by the manufacturer.

Two types of mechanical acoustical calibrators are available commercially. One is a piston-activated diaphragm mounted in an enclosure placed over the microphone. The other type consists of a diaphragm activated by steel balls. It is located a specified distance from the microphone to be calibrated. Each type of calibrator, when operated according to the manufacturer's instructions, will produce sound of a specified level.

Dosimeters. Three different types of noise dosimeters are available at present. These are designed to indicate in some manner the total noise dose during a specified time interval. Each of these uses a different approach. One measures the total amount of sound energy to which a worker is exposed for a work day; another measures the amount of time that a specified decibel level is exceeded; and the third measures the rate at which sound energy impinges on the exposed person for selected short periods of time. Correlation of dosimeter readings with hearing loss is lacking, so they cannot be used to evaluate hearing risks. Figure 8.6 shows a personal noise dosimeter.

FIGURE 8.6   Personal noise dosimeter. *(Photo courtesy B & K Instruments, Inc.)*

Factors in Conducting Evaluation

The factors of concern in conducting any evaluation include the purpose of the evaluation and sources of error.

Purpose of Evaluations

There are three basic reasons for conducting noise evaluations: to determine if there is a problem, to check exposures of one or more individuals, and to survey for noise control.

Surveying to Determine Existence of Problem. This relatively simple survey merely involves taking a large number of measurements throughout the area with the sound survey meter or sound level meter, paying particular attention to the noise sources.

Surveying to Determine Individual Exposure. This survey requires the use of a noise analyzer to determine the distribution of sound pressure with frequency. In making the evaluation, the microphone should be placed as near to the individual as possible without interfering with his or her work. The microphone also should be positioned so that it is not between the noise source and the observer or operator.

It is important to relate the actual time the worker spends in each work area to the level of noise measured. If the worker spends most of his or her day operating a certain piece of equipment, measurements must be made that will adequately represent the exposure. If the worker spends some time each day in another and less noisy area, measurements must be made there also.

The period of time over which measurements should be made is also of importance. If the noise is steady and the operation is continuous, one or more measurements may be sufficient. Conversely, if the machine or operation produces an impulse-type noise, measurements should be made periodically over a period of time. If the plant operates more than one shift, it may be necessary to make noise measurements during each one. In addition, if a given piece of equipment is used on different products at different production rates, data for all rates and products are required. There is no "cookbook" procedure available that covers the selection of sampling sites or suggests the number of measurements to be made. Decisions on these questions have to be based on the experience and judgment of the individual conducting the study. The object, however, is to obtain data that are truly representative so that the worker's exposure can be evaluated.

Once the survey is completed, the job may be classified by some job profile system which indicates its noise level and which can be used for making job assignments. Once such job profile system and its use are discussed in the medical monitoring section of this chapter.

Surveying for Noise Control. In this type of survey measurements are made to determine the total acoustical output and directional characteristics of the source. Measurements at one point yield information about that point only and in that environment. In order to determine how a source differs in another environment, the acoustic power characteristics must be determined. A number of measurements must be made in a geometrical pattern about the source and at some distance from it. Care must be taken to insure that the measurements are representative. Thus, the possibility that obstacles may upset the distribution of sound, particularly at higher frequencies, must be kept in mind.

This type of survey is basically an engineering problem and the measurements are made with particular attention to the physical characteristics of the source and environment and without regard to personnel in the area.

Sources of Error

Wind. When a microphone that is very sensitive to low frequencies is used, wind will produce an appreciable low-frequency noise. If it is necessary to sample in the wind, a wind screen forming a sphere about 18 inches in diameter can be made out of wire mesh. Polyurethane foam screens are also commercially available.

Circuit Noise. All vacuum tubes are affected by mechanical vibration. The tubes used in sound-measuring equipment are less sensitive to shock than usual tubes. But at high noise levels some of these tubes may be vibrated to such an extent as to produce extraneous noises in the equipment. To test for this, disconnect the microphone from the sound-level meter and observe whether there is a reading on the meter. If there is a reading, the meter can be placed on a rubber pad or rubber-tired cart, which should eliminate the trouble. In very intense noise fields it may be necessary to remove the meter from the area and use an extension cable to the microphone.

When making measurements around electrical equipment, a check should be made to see if there is an appreciable pickup of the electromagnetic field by the sound-measuring equipment. This check is particularly important when using a dynamic microphone. A pair of good quality earphones can be used to monitor the output of the sound level meter or octave band analyzer. The hum produced by the electromagnetic field will be of the same frequency as the power source, usually 60 cycles per second. The orientation of the instruments and microphone should be changed to determine if there is a difference in the meter reading; also, the microphone should be disconnected to check if the pickup is in the microphone or the instruments. If the pickup is in the instruments they can be removed from the area under study and an extension cable used; if it is caused by the dynamic microphone, a crystal or condenser microphone may have to be used.

When making noise measurements at low levels, the inherent circuit noise may be the limiting factor. A check for this is made by removing the microphone and substituting a capacitor of the proper size. The lowest reading that can be obtained with the sound level meter or analyzer is determined. This will be the lower limit of measurement.

Background. In some cases, when making measurements to determine the amount of noise generated by a particular source, it may be necessary to make corrections for background noise. If there is any evidence that background noise is contributing to the noise being measured, a background measurement should be made with the noise source turned off.

Recording Data

Certain basic data are essential and must be recorded in almost all noise surveys. These data include:

1. Equipment used for measurements: the type and serial numbers of all microphones, sound level meters, and analyzers.
2. Corrections for measured values, such as cable, temperature, and acoustical calibration.
3. The time and date that measurements are made and name of person conducting the study.
4. Description of space in which the measurements were made, such as dimensions and nature of ceiling, walls, and floor, and locations of windows and doors.
5. Description of the primary noise source under test. This should include a clear description of the machine as to size, nameplate data, speed, and power rating; also types of operations and operating conditions, number of machines in operation, locations of the machines, and types of mounting.
6. Description of secondary noise sources, including location and types of operations.
7. Noise control measures instituted, including the types and effectiveness of ear protectors.

8. Overall and band levels at each microphone position and the extent of meter fluctuation.

9. The meter speed and weighting network.

10. Position of the microphone and the direction of the sound with respect to the microphone, tests for standing wave patterns and the decay of sound level with distance.

11. Time pattern of the noise; that is, whether continuous, intermittent, or impact.

12. Personnel exposed, directly and indirectly.

In recording data it is helpful to have a blueprint or sketch of the building on which locations of measurements can be noted. Frequently, location of columns in a building can be used to form a grid system such as shown in Figure 8.7.

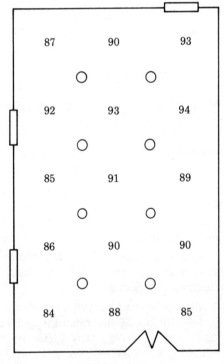

FIGURE 8.7.

The data obtained should be recorded on a form such as that shown in Figure 8.8.

FIGURE 8.8

Sound Survey

Date: _____   Page _____ of _____

Time: _____

Wind Velocity: _____   Wind Direction: _____

Sound Level Meter: Type _____   Model _____   Serial No. _____

Microphone: Type _____   Cable Length _____

Analyzer: Type _____   Model _____   Serial No. _____

Other Equipment: _____

Location: _____   *Sketch*

_____

_____

_____

Calibrated: 60 cps: Acoustic: Corr. Factor:

Location	Weighting Network	Overall Level	Octave-Band Pressure Levels							
			200-75	75-150	150-300	300-600	600-1,200	1,200-2,400	2,400-4,800	Above 4,800
1										
2										
3										
4										
5										
6										
7										
8										
9										
10										
11										
12										

Remarks: _____

Recorded By: _____

The type of report made will depend upon the purpose of the survey and who will receive the report. In any event, the criteria that apply to the situation should be presented and appropriate recommendations made.

COMBATING HAZARDS

OSHA standards require employers to provide protection against the effects of noise exposure when the sound levels exceed

those shown in Table 20, when measured on the A scale of a
standard sound level meter at slow response.

TABLE 20

Permissible Noise Exposures[1]

Duration Per Day, Hours	Sound Level, Decibels
16	80
8	85
4	90
2	95
1	100
1/2	105
1/4	110
1/8	115*

[1] When the daily noise exposure is comprised of two or
more periods of noise exposure of different levels. their
combined effect should be considered, rather than the
individual effect of each.
If the sum of the following fractions: $C1/T1 + C2/T2 \ldots$
Cn/Tn exceeds unity, then the mixed exposure should be
considered to exceed the limit value. Cn indicates the total
time of exposure at specified noise level, and Tn indicates
the total time of exposure permitted at that level.
Exposure to impulsive or impact noise should not exceed
140 decibels peak sound pressure level.
*No exposure to continuous or intermittent in excess of
115 decibels.
Source: "Threshold Limit Values for Physical Agents,"
American Conference of Governmental Industrial Hygienists, Cincinnati, Ohio, 1975.

When noise levels are determined by octave band analyzers, the
levels may be converted to the equivalent A-weighted sound level by
plotting them on Figure 8.9 and noting the A-weighted sound levels
corresponding to the point of highest penetration into the sound
level contours. This level is used to determine exposure limits from
Table 20.

When the noise levels exceed those listed in the table, feasible
administrative or engineering controls must be utilized. If these
controls fail to reduce the noise to the levels stipulated in the table,
personal protective equipment must be provided to do so.

It is the responsibility of the employer to maintain, and have
available for OSHA compliance officers, records that show when
noise surveys were taken and the results. The form in Figure 8:10
can be used for this purpose.

Hazards may be reduced by a safety education program, proper
plant design, use of personal protective equipment, reduction of
sound at the source, and medical monitoring. These are discussed
below.

Safety Education Program

To have an effective hearing conservation program, employees
must be informed of the hazards of noise exposure. Specific topics

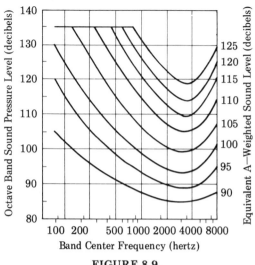

FIGURE 8.9.

that should be discussed in any educational program include the effects of noise on humans, the various types of exposures, the proper action to protect oneself from such exposures, and the engineering methods by which noise can be controlled.

Any effective educational program must be continuous; providing a single session at the beginning of the employee's service is not enough. Only through repeated discussions and continued emphasis will employees come to appreciate the seriousness of a hearing loss from noise exposure.

Safety engineers, because of their frequent visits to all operations, can provide a service to supervisors and the educational program. They can advise in the training of supervisors and employees so that employees will know how to avoid exposure, why it is necessary to wear protective equipment, how to recognize high noise levels so they can report them to supervision, to report to medical if they detect any change in their hearing, and the importance of audiometric testing of their hearing.

Design of Facility

Selection of Building Site

Several factors should be taken into consideration in the selection of a plant site. One of the most important is the existing noise level of the surroundings. A noisy facility should be located in an area of the plant with a high background noise level rather than in a quiet area.

FIGURE 8.10

Occupational Safety and Health Act Noise Summary

Superintendent _____ Buildings _____ Plant _____

Job Classification _____ Index Number _____ No. of Employees _____

Date of Survey	Area	Db A Scale	Exposure Minutes	Exposure Rating	Dosimeter Reading	Engineering and Administrative Program

Advantage often can be taken of the topography and prevailing winds in controlling noise. Sound waves tend to be deflected upward, so it is better to locate a facility on top of a hill rather than in a low spot if there is a possibility of creating a noise problem in adjoining installations. Further, while sound waves traveling into the wind tend to be deflected upward, those traveling with the wind tend to be deflected downward and carried for longer distances than if no wind were present. In some situations it may be possible to take advantage of this by considering the prevailing winds.

It is also possible to reduce noise by proper landscaping, using embankments, terraces, or shrubbery as sound barriers. They can isolate a building from existing noise or prevent noise occurring in a building from creating a neighborhood problem.

Building Layout

Noisy operations should be isolated from the rest of the work room. When a quiet area is desired or necessary, it is better to locate offices, conference rooms, libraries, and other like areas away from plant noises.

Choice of Equipment

When purchasing new equipment, serious attention should be given to procuring the type with the lowest possible noise level. Engineering specifications should incorporate the tolerable noise levels and the requirements for noise performance data.

The form in Figure 8.11 is used by The Dow Chemical Company in buying new equipment that might create a noise hazard. The engineer in charge of ordering fills in the "Dow Required" columns with the sound pressure levels that he or she obtains from the noise control engineer. The prospective supplier then fills in the applicable remaining columns, and describes any special design or acoustical treatment required to achieve the specified sound levels, and their costs. The Dow form and accompanying instructions specify the material or product that is to be utilized in the machine when tests are carried out, along with the proper test procedures and microphone locations. If necessary, Dow supplies the product or material to the equipment manufacturer.

If there are a number of suppliers of the equipment, the form is sent to each, and noise level data is considered along with other factors when Dow makes a final choice of supplier.

Personal Protective Equipment

Such equipment should be used only when administrative and engineering controls are unfeasible, never as a substitute for such

FIGURE 8.11

		FILE NO.
PLANT	B M NO.	JOB NO.
MANUFACTURER	P.O. NO.	CHARGE NO.
NAME OF EQUIPMENT		

Sound pressure level (SPL) limits are specified below for each piece of equipment or integrated unit (such as compressor, speed changer, and driver).

INSTRUCTIONS TO MANUFACTURERS' SUPPLIERS OR VENDORS:

Refer to the attached "INSTRUCTIONS TO MANUFACTURER" for required measurement procedures and microphone locations.

1. Complete stock item (i.e. your Standard Design) column in the table below, and complete all applicable items below the table. If the SPL of the stock item column do not exceed the SPL required in a frequency, Omit Step 2.

2. Complete special design and acoustic treatment columns in the table below and describe design and treatment, and the cost thereof below the table.

SOUND PRESSURE LEVELS IN DECIBELS RE. 0.0002 MICROBAR

CENTER FREQUENCY (HERTZ)	DOW REQ'D.		MANUFACTURERS' MAXIMUM EQUIPMENT NOISE LEVEL							
	Pres. or Expected Background	Max. Sound Level	Stock Item		Special Design		Acoustic Treatment			
			Test Back-ground	Equip. Level	Test Back-ground	Equip. Level	Test Back-ground	Equip. Level	Test Back-ground	Equip. Level
All Pass										
"A"										
31.5										
63										
125										
250										
500										
1,000										
2,000										
4,000										
8,000										
16,000										

SPL ARE CONTINUOUS BROAD BAND IN NATURE AND NON-DIRECTIONAL

EXCEPTIONS: _____

DESCRIBE SPECIAL DESIGN _____ ADDED COST: _____

DESCRIBE ACOUSTIC TREATMENT _____ ADDED COST: _____

THE MACHINERY WILL NOT EXCEED THE NOISE LEVELS QUOTED ABOVE:
NOTE: The supplier *shall* furnish data of the above before shipment. All measurements
are to be taken while the equipment is running under as normal conditions as
possible at full production rates.

This completed form will become a part of the Purchase Order.
(NA) = Not Applicable.

Spec. By	THE DOW CHEMICAL COMPANY			Equip. No.
Checked:				Spec. No.
App'd:				
Date:	REVISION DATE	A	B	C
Vendor to complete all applicable information not given by purchaser			Sheet	of

NEW EQUIPMENT
NOISE LEVEL
SPECIFICATIONS

Form M-45820

controls. The two usual types of ear protectors are earplugs and earmuffs.

Earplugs

These are inserted into the ear canal and are intended to reduce the airborne sound just before it impinges on the eardrum. They may be made of rubber, plastic, neoprene, or cotton impregnated with wax. Material and shape have very little to do with the effectiveness of commercially available plugs. Contrary to popular opinion, dry cotton gives very little protection. Figure 8.12 illustrates two types of commercially available plugs.

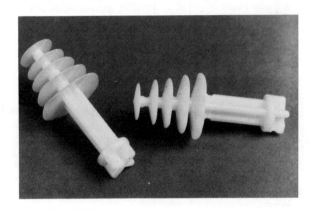

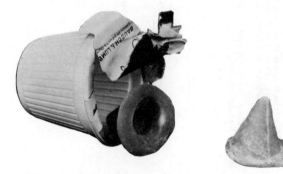

FIGURE 8.12   Ear plugs. *(Photos courtesy Bausch & Lomb and 3M Company.)*

Earmuffs

The muff-type protector consists of a set of cups designed to cover both ears and held snugly to the head by a headband. The cups may be made of plastic or rubber and encompass the ear without compressing it, yet at the same time fit so closely to the head that very little sound leaks through. Figure 8.13 shows a set of ear muffs in use.

FIGURE 8.13   Ear muffs. *(Photo courtesy Bausch & Lomb.)*

The amount of protection that can be expected from plugs or muffs in current use is shown in Table 21 as the average attenuation in decibels at different frequencies. The table clearly indicates that the typical muff affords better protection than the typical plug. Muffs also have the advantage of eliminating the complex fitting problems of earplugs and the many complaints associated with them.

The ear protection program should be under the supervision of the medical department. An employee's ears should be examined and his or her hearing tested at the time he or she is fitted with ear protectors. A number of types and sizes of plugs should be available, and plugs should be fitted individually for each ear. Most of the available ear protectors, when properly fitted, provide about the

TABLE 21

Noise Protection Afforded by Earplugs and Earmuffs

Ear Protector	Frequency in Cycles per Second					
	250	*500*	*1,000*	*2,000*	*4,000*	*6,000*
Typical plug	15	17	20	25	30	30
Typical muff	20	22	30	37	45	40
Typical muff plus plug	33	35	42	47	50	45

same amount of protection; therefore, the one that fits correctly and is the most comfortable for a particular employee is the best for him or her. Properly fitted protectors can be worn continuously by most persons and will provide adequate protection against most industrial noises.

Employees often do not understand why it is necessary to wear ear protection. Therefore, an educational program is needed to convince them of the importance of ear protection and the personal benefits to be gained from consistent use. The employees should be informed that:

1. Good protection depends on a seal between the surface of the skin and the surface of the ear protector. A very small leak can destroy the effectiveness of the protection. Protectors have a tendency to work loose as a result of talking, chewing, or the like, and they must be resealed from time to time during the work day.
2. A good seal cannot be obtained without some initial discomfort due to pressure on the skin, perspiration, and so on.
3. There will be no health problems or effects on the ears as a result of using ear protectors if they are kept reasonably clean.
4. The use of ear protection will not make it more difficult to understand speech or to hear warning signals when worn in a noisy environment.

Where earplugs are the protective device, employees should be instructed on how to insert them and keep them clean.

In larger plants, the safety engineer or industrial hygienist should supervise the use of such devices.

Reduction of Sound at Source

Reduction may be accomplished by equipment substitution, modification of the noise source, and modification of the sound waves generated.

Equipment Substitution

Replacement of noisy equipment and work processes with quieter equipment and processes should always be a consideration in any noise-reduction program. Several possibilities are:

1. Compression riveting or welding instead of pneumatic riveting.
2. Hot instead of cold working of metals.
3. Grinding instead of chipping.
4. Pressing instead of rolling or forging.
5. Using squirrel cage blowers instead of propellor or axial-type fans.
6. Using turbine drive units instead of gear increasers.
7. Using spiral or helical gears in place of spur gears.
8. Using belt drives instead of gear drives.

Modifying the Noise Source

If at all practicable, it is usually more economical to modify the noise source than to substitute equipment. The major types of possible modifications are discussed below.

Eliminating Rattle. Noises of this nature often are attributed to poor maintenance. Replacement of worn parts, securing, stiffening, or damping loose panels usually will eliminate a large amount of this noise.

Eliminating Friction. Noise is produced by frictional forces generated by grinding and cutting operations, or from parts of machines such as clutches and brakes. Some of this noise can be eliminated by proper lubrication of moving parts on the machine and the use of cutting oils or coolants on the products being handled. The use of sharp and properly shaped cutting tools will also usually reduce noise at such operations. Substitution of materials used in the manufacture of brakes or clutches may give a reduction in noise.

Reducing Impact Noise. Impact noise is generated when materials are forged, hammered, sheared, tumbled, or peened. Noise from such sources is generally difficult to control. One approach is to spread the impact out over a longer time interval. This can be done in some cases by the use of stepped dies, which produce a number of small impacts rather than one large impact. Other approaches may be clamping a weight on the piece being hammered, coating tumbling barrels with a mastic material, or using suitable innerliners.

Reducing Gas-Flow Noises. If the flow of gas varies irregularly in a piece of equipment, noise is usually generated. This may occur, for example, in an internal combustion engine, from the use of a high-speed fan or blower, or from jets of compressed air, steam, and air-driven tools. Noise from such sources can be controlled by installing proper intake and exhaust mufflers on internal combustion engines, equipping compressed air jets with mufflers or redesigning the jet nozzle, redesigning the exhaust parts of tools that produce a sirenlike noise, or slowing down the speed of operation.

Reducing Solid-Borne Transmission. Vibrations can be transmitted for considerable distances through solid materials with very little attenuation. If some break can be made in this path, the flow of energy will be reduced. This can be done by using soft, flexible material, such as vibration mounts, flexible couplings on shafts or pipes, flexible connectors on ventilation ducts, belt drives, and mastic materials at joints when fastening materials together.

Energy may be transmitted to structural members or panels, which in turn are set into vibration and radiate the energy. Structural members and panels can be stiffened to reduce vibration. Vibration-damping materials resist motion in any direction during every cycle of vibration. Thus, they provide definite energy losses during each cycle of vibration. Materials such as mastic, and vegetable and mineral fibers impregnated with mastic, have been used successfully to reduce vibration and, in turn, noise levels.

Other possible ways of modifying the noise source include installing silencers and increasing housing clearances.

Modifying Sound Waves Generated

Sound waves are usually modified by the use of an enclosure or by absorption.

Enclosure. It is usually more economical to construct a partial or total enclosure around the machine rather than to install sound-absorbing materials throughout the entire work area involved.

Partial enclosures can be of a number of shapes or designs. They may consist of single panels beside the source, panels on both sides of the source, or tunnels with the work flowing through them. Sound-absorbing linings must be used inside these enclosures. The barrier can be constructed by using plywood, wallboard, sheet metal, or almost any material with the requisite mass and strength. This type of enclosure usually does not protect the operator but does produce a shadow effect to give protection outside of the enclosure up to about 15 decibels.

Machines or noisy operations can be totally enclosed except for small access openings, which are treated by the use of short acoustical ducts. These should be made of an impervious material to stop the flow of energy and should be provided with an acoustical lining to absorb the energy within the enclosure. Properly designed enclosures will give a 20- to 30-decibel noise reduction. They do have the disadvantage of interfering with ventilation, handling of materials, and maintenance. When it is impractical to enclose the noise-producing device, it is sometimes possible to enclose the employee. Thus, a desk area in a very noisy factory may be enclosed with a suitable wall.

Absorption. The most common use of sound-absorbing material is in general or overall noise control in enclosed spaces. Sound-absorbing material has very little effect on the noise level close to a source, but it will give a considerable reduction some distance from the source in the reverberant field. The limit for most practical installations is a reduction of 10 decibels. Sound-absorbing materials are usually porous and may be made of vegetable or asbestos fibers, glass, wool, or mineral wool. They are available in a number of forms, including blocks, panels, or blankets, which can be fastened to the walls or ceiling, and space absorbers, which can be suspended from the ceiling.

Besides enclosures and absorption materials, devices such as baffles and sound traps may be placed around noisy parts of equipment to modify the generated sound wave.

Medical Monitoring

Pre-Employment Audiometric Examination

An evaluation of hearing ability should be part of any prospective employee's pre-employment physical examination. The evaluation is carried out with an audiometer, a device that produces pure tones at specified frequencies (see Figure 8.14). It consists of measuring the individual's sensitivity for pure tones in each ear at each of the following test frequencies: 500, 1,000, 1,500, 2,000, 3,000, 4,000, 6,000, and 8,000 cycles per second. The "hearing sensitivity" of the individual is his or her threshold of hearing; that is, the lowest intensity of a given tone that he or she can detect through air.

The tests are carried out in special audiometric rooms. These rooms are located and constructed so as to reduce background noises

FIGURE 8.14   Audiometer. *(Photo courtesy Bausch & Lomb.)*

to levels that will not mask or interfere with the pure tone thresholds of the test frequencies.

Table 22 lists the allowable background noise levels for audiometry rooms. This table, compiled by the American Academy of Opthalmology and Otolaryngology, is based on data developed by the American National Standards Institute.

TABLE 22

Allowed Background Noise Levels

It is assumed that (1) no frequencies below 500 c/s will be measured, and (2) well-fitted binaural earphones will be worn.

Octave Band *Cycles Per Second*	300 to 600	600 to 1200	1200 to 2400	2400 to 4800	4800 to 9600
Level in B (C scale)	40	40	47	57	62

The tests may be administered by an industrial nurse or an audiometrician, a person who has received special training in conducting the test. Interpretation of the test results, however, should be done only by a physician.

The results of the audiometric examination, along with the medical history and the results of an ear examination of the prospective employee, should be recorded on a form like that shown in Figure 8.15. This information is then used to make the final job assignment. In one such assignment procedure, the person tested is placed in one of five hearing profile categories (H-1, H-2, H-3, H-4, and H-5), which are utilized in conjunction with four job profile categories (A, B, C, and D).

FIGURE 8.15

AUDIOMETRIC RECORD   DATE _____ HOUR _____

MAN NO.	LAST NAME	FIRST NAME	MIDDLE NAME	SEX	AGE

DOW WORK HISTORY PRIOR TO FIRST AUDIOGRAM

INDEX	DEPARTMENT	JOB CLASSIFICATION	CODE NO.	HEARING PROFILE CLASS	DEPARTMENT YEARS	10 THS

PRE-DOW EXPOSURE TO NOISE	YEARS	10 THS
INDUSTRIAL		
MILITARY		
CIVILIAN GUNFIRE		

WEARS EAR PROTECTION DURING EXPOSURE TO NOISE			
ALWAYS	1	SELDOM	2
FREQUENTLY	3	NEVER	4

MEDICAL HISTORY		R	L
EARACHES (INCLUDING CHILDHOOD)	1		
DRAINAGE	2		
EAR INJURY	3		
SURGERY (EAR OR MASTOID)	4		
RINGING IN THE EARS	5		

MOST FREQUENTLY USED EAR PROTECTION			
COTTON	1	HELMET	2
WAXED COTTON	3	WAX COTTON & MUFF	4
PLUGS	5	PLUGS & MUFF	6
MUFF	7		8
	9	OTHER	0

		YES	NO
HEAD INJURY: WITH UNCONSCIOUSNESS	6		
HEARING LOSS IN THE IMMEDIATE FAMILY (UNDER 60)	7		
RECEIVED DRUGS FOR AN INFECTION	8		
HAD SCARLET FEVER	9		
MUMPS	0		

LAST NOISE EXPOSURE					
TIME SINCE	10 THS	LENGTH OF	10 THS	USED EAR PROTECTION	
				YES	NO

PHYSICAL FINDINGS		R	L
OBSTRUCTION OF, OR DRAINAGE FROM CANAL	1		
PERFORATIONS OF DRUMHEAD	2		
SCARRING OF DRUMHEAD	3		
DO YOU FEEL YOUR HEARING IS NORMAL?	4	YES	NO

AIR CONDUCTION AUDIOGRAM - ISO

BY	AUDIO NO.	RIGHT								LEFT							
		500	1000	1500	2000	3000	4000	6000	8000	500	1000	1500	2000	3000	4000	6000	8000

BONE CONDUCTION AUDIOGRAM

BY	AUDIO NO.	RIGHT								LEFT							
		500	1000	1500	2000	3000	4000	6000	8000	500	1000	1500	2000	3000	4000	6000	8000

GENERAL REMARKS - TECHNICAL:

DISPOSITION OR TREATMENT:

FORM 5820 PRINTED IN U.S.A. R4-67

A job profiled "A" is a quiet job in a quiet department. A job profiled "B" is a quiet job in a noisy department, with 10 percent of the working time spent in noise approaching basic damage-risk noise levels, as shown in Figure 8.16. A job profiled "C" is a noisy job with 10 percent of the working time spent in noise that exceeds the basic damage-risk noise levels and 10 percent or more spent in noise within a 10-percent decibel range below these levels. A job profiled "D" is a noisy job, with 10 percent or more of the working time spent in noise that exceeds the basic damage-risk noise levels.

A person profiled "H-1" has normal hearing and can work in any job profiled "A," "B," "C," or "D." A person profiled "H-2" has only a minor deviation in hearing and can work in jobs profiled "A," "B," and "C." A person profiled "H-3" has a moderate or functional loss of hearing and can work in any job profiled "A" or "B." A person profiled "H-4" has a serious hearing defect and can work only in a job profiled "A." A person profiled "H-5" is essentially deaf and unemployable in these jobs because of hearing defects.

Periodic Audiometric Examination

Every employee on a noisy job should have an audiometric examination at least once a year. When an employee is being considered for a transfer from a quiet to a noisy job, a hearing test should be run to determine if the employee is qualified for the job.

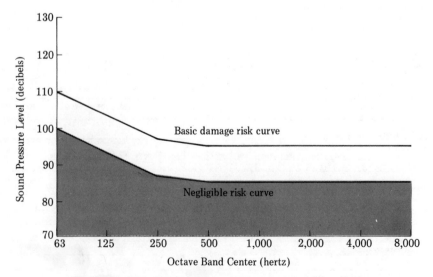

FIGURE 8.16. Hearing conservation damage risk criteria.

EXERCISES: CHAPTER 8

1. Differentiate between wide-band, narrow-band, and impulse-type noises.

2. Define the term "threshold shift" and name the two types.

3. Describe the effects of noise cn workers.

4. Contrast the purposes for which the following instruments are used:
   a. sound level meters
   b. octave band analyzers
   c. cathode-ray oscillographs
   d. dosimeters

5. Name the sources of error that may be encountered in making noise evaluations and tell how they may be compensated for.

6. List the basic data that must be recorded in making noise evaluations.

7. Discuss the elements involved in any good hearing conservation program.

8. Name the two major types of ear protectors and summarize what workers must be told concerning their use.

9. Discuss the three ways by which sound may be reduced at its source.

10. Describe the use of hearing profile categories in audiometric examinations.

9

Microbiological Hazards

Although laboratory workers long have been acquiring accidental infections from pathogenic organisms, the subject of safety in
biological laboratories has never received the attention it deserves.
The magnitude of the problem is apparent from the results of a
number of studies and surveys. A 1961 study published by the
American Public Health Association reported 2,262 cases of accidental infection resulting in 96 deaths.* The Bureau of Labor Statistics
has reported that about 23 percent of the lost-time accidents in
hospital clinical laboratories are due to infections.

The agents that cause laboratory infections include bacteria,
viruses, fungi, rickettsiae, and protozoa. Three diseases—brucellosis,
tuberculosis, and hepatitis—account for some 33 percent of all
infections. The number of infections caused by a particular organism
is not related to the number of persons handling them. Thus, the
organisms of tularemia, typhus, and Q fever are responsible for a
relatively large number of infections, although very few laboratory
personnel work with them. The fatality rates of infected personnel
differ rather widely between the different classes of organisms. Of
those infected with viruses, some 7.3 percent die. For bacterial,
rickettsial, and fungal infections, fatality rates of 4.0, 2.6, and 2.3
percents, respectively, have been recorded.†

*S. E. Sulkin, and others, "Diagnostic Procedures and Practices," American
Public Health Association, 1961.
†S. E. Sulkin, Bact. Rev. 25, September 1961, pp. 203-9.

240

Office workers, janitors, dishwashers, and other personnel not closely associated with infectious organisms account for an appreciable percentage of all infection cases.

HOW ACCIDENTAL INFECTION OCCURS

Only 16 to 20 percent of all cases of accidental infection can be related to known unsafe acts and conditions. The remainder is usually listed as cause unknown—probably the result of faulty manipulation (malmanipulation) of test tubes, flasks, beakers, syringes, pipettes, and inoculating loops. In such cases, the act causing microorganisms to escape may have occurred in a fleeting second and not been noted or remembered by the laboratory worker. Examples of malmanipulations include:

1. A film of culture breaking on an inoculating loop.
2. A drop of culture escaping from the tip of a pipette or syringe.
3. A drop of culture running down the outside of a test tube or flask.
4. A surface bubble breaking when a culture is stirred.

Whenever such a malmanipulation occurs, an aerosol—that is, a dispersion in the air of liquid droplets ranging in diameter from 0.01 to 100 microns—is created. The larger droplets soon settle out and become incorporated with dust particles. These dust particles are resuspended by any activity that stirs up air currents, or they may be picked up by direct contact. The smaller droplets do not settle out. Instead, they quickly dehydrate to leave nuclei which contain high concentrations of the infectious agent and which remain in the air almost indefinitely, being wafted about by the air currents until removed by the room ventilation. These particles are small enough so that they can be inhaled all the way to the aleveolar spaces of the lung. This distinction gives rise to two basic varieties of airborne infectious agents: droplet nucleus-borne and dust-borne.

Laboratory animals contaminated either naturally or experimentally with pathogenic agents may be a second significant source of laboratory infection. Pathogens may be transferred from experimentally infected animals to handlers and other personnel via aerosols, biting, scratching, and contaminated cages and litter. The infection of cagemates or other animals in the same room by experimentally infected animals creates yet another avenue by which laboratory personnel can become infected. Experiments involving monkeys, guinea pigs, and a number of bacteria species show that when contaminated animals are placed with unexposed cagemates, the latter rapidly become contaminated by test organisms coming from the fur, saliva, and excretions of the infected animals. Animal autopsies sometimes reveal cross-infection between two groups of animals, each inoculated with a different organism and caged in the same room or airstream. These findings provide powerful evidence of the potential danger for laboratory personnel from supposedly uncontaminated animals.

EVALUATING HAZARDS*

The basic methods for sampling airborne bacteria include impingement in liquids, impaction on solid surfaces, filtration, sedimentation, centrifugation, electrostatic precipitation, and thermal precipitation. Occasionally, cotton swabs may be utilized for sampling surfaces, and the use of control animals can provide an estimate of the risk of infection. All of these methods of evaluating hazards are discussed below. Discussions of particle sizing, selection of samples, and operating samplers are also included.

Air Sampling Methods

Impingement in Liquids

Samplers that use the principle of impingement and washing of air, in which organisms are entrapped in a liquid medium, will be referred to as liquid samplers. In high-velocity liquid impingers, air is drawn through a small jet and directed against a liquid surface, and the suspended particles are collected in the liquid. High-velocity impingers are constructed so that the air inlet tube projects into the liquid or is in close proximity to the liquid surface. The airstream approaches sonic velocity in the high-velocity liquid samplers, resulting in almost complete collection of suspended particles, but this condition tends to cause the destruction of vegetative cells. For the recovery of some vegetative cells, the highest efficiency appears to occur when the jet velocity is approximately 70 percent of sonic velocity. Due to the agitation of the collected particles in the collecting liquid, bacterial clumps are likely to be broken into smaller fragments or even into unicellular particles. For this reason, the bacterial counts obtained with this type of sampler tend to reflect the number of bacterial cells suspended in the air.

*See H. W. Wolf, and others, "Sampling Microbiological Aerosols," *Public Health Monograph No. 60*, U.S. Dept. of Health, Education and Welfare, 1964.

In liquid samplers that utilize the air-washing principle, air enters the sampler at a low velocity through either a large jet, a fritted glass disk, or a perforated tube. The airborne particles are wetted and entrapped in the collecting medium. Since the air is bubbled through the liquid, fragmentation of bacterial clumps will occur, but this may not be as extensive as in the high-velocity impingers. There is also less likelihood of destruction of the collected bacteria as compared with the high-velocity liquid impingers. The bubbler or air-washing type of sampler is not efficient for the collection of small (less than 5 microns) airborne particles, since these small particles tend to remain in the air bubbles and are carried out of the sampler with the discharged air.

The selection of a collecting medium for liquid samplers depends on the particular organism being isolated. The primary consideration in determining a suitable medium is the preservation of the recovered bacteria. In quantitative studies, a medium must be used that will minimize the multiplication or death of the organism. The more common collecting media include buffered gelatin, tryptose saline, peptone water, and nutrient broth. Under certain conditions it may be desirable to use an antifoam agent in the collecting medium to prevent excessive foaming and carry-over of the liquid, since this may result in the loss of entrapped bacteria.

Atmospheric conditions must be considered when utilizing liquid impingers for recovery of bacteria in air. If the temperature is high and the relative humidity low, rapid evaporation of the collecting liquid will result, thus limiting the length of the sampling period. If the ambient temperature is below 40°F, freezing of the collecting liquid is likely to occur. A nontoxic, low freezing-point diluent such as glycerol may be added, or some means of temperature control may be used to prevent freezing during the sampling period.

In quantitative studies, careful measurement of the total airflow is required to determine the number of organisms per specific volume of air. The volume of collecting fluid at the completion of a sampling interval also must be measured to determine the number of cells collected.

Impaction on Solid Surfaces

Impactors deposit particles suspended in the air directly on solid surfaces. The bacterial count obtained with this type of sampler reflects the number of airborne particles carrying bacteria. Certain samplers operated by this principle also have been designed to determine the particle size distribution of airborne microorganisms.

If the volume of the sampled air is accurately metered, a quantitative estimate of the airborne bacteria can be determined. The solid surface on which the particles are collected may be a dry, coated, or agar surface. Samplers that collect particles on dry or coated surfaces, such as the cascade impactor, are used for determining the number of particles and the size distribution by means of microscopic examination of the deposits. The number and size distribution of viable organisms then can be approximated by washing the particles from the collection surfaces and enumerating by bacteriological methods. However, only resistant organisms, such as spores, will remain viable on dry or coated surfaces for an appreciable length of time. In the surface-washing process, large particles that may contain many viable organisms may be broken into individual organisms. The result would then indicate the number of viable organisms present but not the number of particles collected.

Samplers that collect particles on agar surfaces are used to determine the number of viable particles present, since bacterial colonies grow directly on the collecting surface. The sieve, slit, Andersen, and single-stage impactor samplers, when used with an agar collecting surface, are examples of solid-medium impactors. The Andersen, the cascade slit, and the single-stage impactor also can be used to determine the size distribtuion of the viable particles in an aerosol.

The sieve and Andersen samplers are operated by drawing air at a rate of 28.3 liters per minute through a large number of small, evenly spaced holes drilled in a metal plate. The suspended particles are impacted on an agar surface located a few millimeters below the perforated plate. The Andersen sampler consists of a series of six sieve-type samplers with holes of smaller diameter in each succeeding plate. The velocity of the incoming air increases for each succeeding plate, resulting in a separation of the particles into six size ranges. From the number of colonies growing on each agar plate, the particle count as well as the particle size distribution can be determined. In the operation of slit samplers, air is drawn through a narrow inlet slit, and particles are impacted on an agar surface contained in a moving dish or tray. The slit samplers permit determination of a time and viable-particle-concentration relationship. Samples from the sieve, slit, and Andersen-type samplers are assayed by incubating the exposed plates and counting the developed colonies.

These samplers require a vacuum source sufficient to draw a constant flow of air through the sampler—usually 28.3 liters per minute, although the airflow through the slit sampler may be changed by altering the dimensions of the slit.

In the operation of a single-stage impactor, the air is drawn through a small orifice at a constant airflow of 12.5 liters per minute, which impacts the larger particles on the surface of a small dish of agar or on a glass disk. The smaller particles are carried on through the sampler and are collected in a suitable final collector, such as a filter or liquid impinger. By changing the diameter of the inlet orifice, the size range of particles collected on the agar or glass disk may be changed. Results of samples collected indicate the number of viable organisms in particles of a size range that are above and below the cutoff point of the sampler.

The cascade impactor consists of a series of four or five stages, each with a small air inlet slit that causes the particles to be impacted against a glass slide. In each successive stage, the slit area decreases, resulting in a higher velocity of the passing airstream, so that a separation of particles into size ranges results. The addition of a thin layer of adhesive to the slides aids the collection of particles. Long sampling periods may cause desiccation of certain organisms and overcrowding of particles so that microscopic sizing is difficult. The total number of viable organisms can be determined by washing the particles from each of the slide samples into sterile water and assaying by standard bacteriological techniques.

Filtration

In the filtration of bacterial aerosols, the particles either strike the fibers of the filter directly or are attracted to them and held in place by electrostatic and other forces. The particles are not normally removed from the airstream by a simple screening or sieving action, except in the case of membrane filters. An efficient filter is usually composed of fibers, generally perpendicular to the direction of the airflow and having a diameter equal to or less than the diameter of the particles collected. The mesh or opening between the fibers may be quite large compared to the size of the particle. Filtration efficiency will increase if the filter thickness or number of fibers is increased, but the resistance to airflow also will increase. Materials that have been used as filter media include glass wool, alginate wool, cotton, cellulose-asbestos paper, gelatin foam, and membrane.

Membrane filters differ from the other types in that they mainly sieve the particles from an aerosol. The pores of the membrane are smaller than the diameter of the collected particles, and therefore the particles do not penetrate into the filter but are retained on or within a few microns of the filter surface.

When assaying most filter samples, the complete filter or a section of it is agitated in a suitable liquid until the particles are uniformly dispersed. Aliquots of the suspension are then assayed by appropriate bacteriological techniques. The bacterial counts obtained are usually higher than the actual number of viable particles in the aerosol, since clumps of bacteria may be fragmented during agitation of the sample.

The viability of some organisms is detrimentally affected if the bacteria remain on the filter during long sampling periods, because the continuous passage of air over organisms will result in desiccation. This effect is reduced at high relative humidities.

Sedimentation

In the sedimentation process, particles suspended in the air are allowed to settle on plain surfaces or on surfaces coated with a nutrient medium. This method can be used to determine the number of viable particles, the total number of viable organisms, and the total number and size of all particles that settle in a given time. This sampling method is obviously more suitable for the enumeration of larger-sized particles, since they settle more rapidly than the smaller-sized particles. Table 23 gives the settling velocities of various sizes of particles of unit density in still air.

TABLE 23

Settling Rates of Airborne Particles

Diameter of Particles (microns)	Velocity of settling	
	Feet per minute	Inches per hour
0.1	0.00016	0.115
0.2	0.00036	0.259
0.4	0.0013	0.936
0.6	0.002	1.44
0.8	0.005	3.60
1.0	0.007	5.04
2.0	0.024	17.3
4.0	0.095	68.4
6.0	0.21	
8.0	0.38	
10	0.59	
20	2.4	
40	9.5	
60	21.3	
80	37.9	
100	59.2	
200	352	
400	498	

Note: This table has been compiled from "Size and Characteristics of Air-Borne Solids," by W. G. Frank, published in the Smithsonian Meteorological Tables. Rates are for particles in the shape of spheres, having a specific gravity of 1.0 and settling in air at a temperature of 70° F.

Samples collected by sedimentation provide only a qualitative index of the particles suspended in the air. Air movement will influence the deposition of the particles so that the particle size distribution may indicate a higher percentage of large particles than is actually present in the air. When an aerosol is collected in a closed container of known volume and the particles are allowed to settle quiescently onto a collecting surface, a quantitative measure of the number of particles suspended in the air can be obtained bacteriologically and a true particle size distribution can be obtained by microscopic examination.

The number of viable particles that settle from the air in a given time is determined by allowing the particles to settle on a nutrient surface such as an open agar Petri dish or agar-coated slide, and the undisturbed particles are allowed to develop into visible colonies during subsequent incubation. The number of settled particles is then determined from the colony count.

The total number of viable organisms settling from the air in a given time is determined by allowing the particles to settle on surfaces from which they can be removed by washing in a suitable liquid. The collecting surface is agitated thoroughly in the wash liquid to break up clusters of organisms and to dislodge the bacteria from inert particulate matter, so that each individual cell may develop into a separate colony when the liquid is plated on a nutrient medium.

The number and the size of all particles settling from the air in a given time are determined by allowing the particles to settle on glass slides, membrane filters, or other surfaces. The particles are then counted and sized microscopically without removal from the surface on which they have been collected.

Advantages of sedimentation samplers are inexpensiveness, simplicity of use, and ability to collect the particles in their original state. The disadvantages are inability to give quantitative counts unless the aerosol sample is allowed to settle in a closed container, the long waiting period required for all particles to settle, and some possible loss in viability during the settling process.

Centrifugation

Since deposition of particles by sedimentation in the size range of bacteria is not always satisfactory or efficient, a few bacterial samplers are available which employ centrifugal force for propulsion of the particles to the collecting surface. When the aerosol moves in a circular path at high velocity, the suspended particles are impacted on the collecting surface by a force proportional to the particle velocity and mass. The particles are usually collected on glass or agar surfaces, which are removed from the sampler and assayed by appropriate methods. Two samplers of this type in use are the cyclone (aerotec tube) and the air centrifuge (Wells). In the cyclone type, the sampler remains stationary while the aerosol travels in a circular path and the large particles are collected at the bottom of the sampler. In the air centrifuge sampler, the collecting vessel and aerosol rotate at high velocity, resulting in the impaction of particles on the walls.

Electrostatic Precipitation

Samplers using this method collect particles by drawing a given amount of air at a measured rate (up to 85 liters per minute) over an electrically charged surface. In some samplers the air is ionized prior to passing over the collection surface. The electrostatically charged particles are subsequently collected on either positively or negatively charged surfaces. These samplers may employ a variety of solid collecting surfaces such as agar or glass. Although these samplers have a relatively high sampling rate, high collection efficiency, and low resistance to airflow, they are complex and must be handled carefully. Furthermore, little is known about the effect on viability and clumping of electrostatically charged particles.

Thermal Precipitation

Samplers that collect particles on surfaces by thermal gradients are classified as thermal precipitation samplers. Their design is based on the principle that airborne particles are repelled by hot surfaces and are deposited on colder surfaces by forces proportional to the temperature gradient. These samplers are used for determination of particle size distribution and are more effective when collecting small particles (1 micron and less).

The particles are usually collected on a glass coverslip or electron microscope grid, and the particles are subsequently sized and counted microscopically.

Other Sampling Methods

Surface Sampling

Surfaces may be monitored for settled particles by wiping or swabbing them with a cotton swab, then inoculating an appropriate growth media and counting the number of colonies that develop. Rodac plates may also be used for surface sampling. These are small plates of solid nutrient medium which are pressed directly on the surface being sampled. When the plate is lifted, organisms will stick to the agar medium and a colony of bacteria will develop wherever an organism was transferred.

Control Animals

A convincing method of detecting microbiological contaminants consists of placing a control animal or animals in the same cage (or an adjoining one) as the experimentally infected animal. This procedure provides some evidence of the risk to the experimenter as well as a check on the validity of the experiment. If the control animal becomes infected by association, some thought then should be given to the effect of the supplementary accidental dosage upon the experimental animals.

Particle Sizing

Consideration should be given to the determination of particle size when studying airborne microorganisms and their relation to respiratory infections, since the amount of particulate material and bacteria retained in the respiratory system is largely dependent upon the size of the inhaled particles. Particles larger than 5 microns are efficiently trapped in the upper respiratory tract and are removed by ciliary action or some other means. Particles in the 1 to 5 micron range, which reach the alveoli, are usually retained; however, retention increases as particle size decreases from 1 micron.

To facilitate reporting the size of airborne bacterial particles, it can usually be assumed that all particles in a given sample have the same density and shape. To simplify further the determination of bacterial particle size, it can be assumed that all the bacterial particles are spherical and that the mean diameter need only be approximated. Although this is not precise, it will provide the necessary information for all practical purposes.

The method most often used for size analysis of bacterial particles is direct measurement with a light microscope and a calibrated scale or graticule. The particles are collected without disintegration on a glass slide or on a membrane filter so that there are between 25 and 150 particles per field. A total of 200 to 500 particles should be counted and sized so that a representative sample of the deposited particles is obtained. The most useful graph to show size variation in particle sizing is the cumulative type, in which the cumulative percentage count, or mass less than a stated size, is plotted against that size on logarithmic probability paper. The count median diameter, mass median diameter, and geometric standard deviation are determined from this graph. An example of these determinations from hypothetical data follows.

After 200 particles from the sample slide are counted, sized, and tabulated (Table 24), the percentage of the total number is calculated for each size and these percentages are cumulated for determination of count median diameter. If the mass median diameter is desired, the total mass for each size group is calculated by cubing the diameter and multiplying by the number of particles in that size group. Since the density and the factor $4/3\ \pi$ are constant, they are eliminated from the calculation. The percentage of the total mass is calculated for each size group, and these percentages are cumulated for determination of mass median diameter. The cumulated figures for number percentage and mass percentage are plotted against the corresponding size as shown in Figure 9.1. By reading the size corresponding to the 50 percent intercept, as shown on the graph, the count median diameter (CMD) is found to be 3.02 microns and the mass median diameter (MMD) is 7.70 microns.

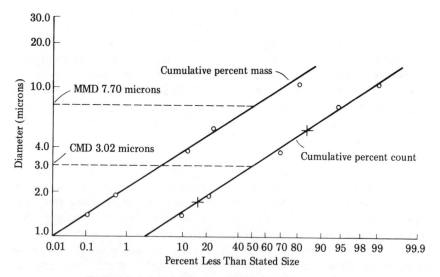

FIGURE 9.1. Particle size distribution.

The geometric standard deviation (GSD) is found by substituting values from the graph in the following relationship:

$$\text{GSD} = \frac{84.13\ \text{percent size}}{50\ \text{percent size}} = \frac{50\ \text{percent size}}{15.87\ \text{percent size}} = 1.7$$

The amount of work involved in sizing an aerosol can be reduced considerably by using such samplers as the Andersen, TDL slit, and cascade impactor, which separate the particles into size ranges as the sample is collected. However, these samplers should first be calibrated by sampling similar aerosols under the same conditions as those for which they are to be used. Particles collected on each stage should be measured microscopically to determine CMD or MMD as explained previously. When the instrument has been calibrated, it can be assumed that the CMD or MMD for each stage will remain relatively constant, providing the particulate material and sampling conditions do not change.

TABLE 24

Particle Sizing Data

Diameter (microns)	Number counted	Percent of total	Cumulative percent	Volume (diameter³ x number)	Percent of volume	Cumulative percent by volume
1	5	2.5	2.5	5	0.01	0.01
1.4	14	7	9.5	38	.10	.11
1.9	23	11.5	21	157	.43	.54
2.7	41	20.5	41.5	807	2.19	2.73
3.8	57	28.5	70	3,127	8.50	11.23
5.4	29	14.5	84.5	4,566	12.41	23.64
7.7	20	10	94.5	8,130	24.82	48.46
10.9	9	4.5	99	11,655	31.68	80.14
15.4	2	1.0	100	7,304	19.85	99.99
Totals	200	100		36,789		

Selection of Samplers

When selecting a sampler for a specific air sampling program, many factors must be considered, but first a clear understanding is needed of the type of information desired and the particular determinations that must be made. For instance, is information desired on (1) one particular organism or all organisms that may be present in the air, (2) the concentration of viable particles or of viable organisms, (3) the change of concentration with time, (4) the size distribution of the collected particles, or (5) a combination of these points? Are the results on concentrations to be qualitative or quantitative?

Factors to be considered before selecting a particular sampler are:

1. Viability and type of organism to be sampled.
2. Sensitivity of particles to sampling.

3. Assumed concentrations and particle size.
4. Volume of air to be sampled and length of time sampler is to be continuously operated.
5. Background contamination.
6. Ambient conditions.
7. Sampler collection efficiency.
8. Effort and skill required to operate sampler.
9. Availability and cost of sampler.
10. Availability of auxiliary equipment and facilities, such as vacuum pumps, electricity, and water.

When selecting a sampler for the collection of airborne bacteria, it must be kept in mind that although the bacteria may be present as individual organisms, they generally occur as clumps or mixed with or adhering to particles of foreign matter, such as dust and organic substances. Complete or partial disintegration of these clumps will usually occur when certain types of samplers (particularly those using liquid media) are used. The results would then indicate a greater number of viable clumps or particles than was actually present in the air at the time of the sampling. When the clumps or particles are completely disintegrated, the result will indicate the total number of individual organisms present in the air. When reporting the bacterial concentrations determined by various air samplers, the results should indicate whether the concentrations represent individual organisms or particles bearing bacteria.

The air sampling devices described in this chapter can be placed in the following categories according to the purpose for which they are used:

1. Determination of the concentration of viable particles.
2. Determination of the concentration of viable organisms.
3. Determination of the change of concentration with time.
4. Determination of the size of the collected particles.

For Concentration of Viable Particles

Samplers used for this purpose should deposit the particles (without disintegration) on the surface of a solid nutrient medium on which the particles can grow or multiply into discrete colonies with suitable incubation. If a qualitative estimation of the number of viable particles present in the air is desired, an open agar plate may be used. If quantitation is desired, the slit or sieve sampler may be employed, since these samplers permit the viable particles in a measured volume of air to be deposited uniformly on an agar surface. All samplers employing agar as a collecting surface are limited to use in temperatures above 32°F unless some method of heating is provided.

Membrane filters can be used to collect viable particles at temperatures below freezing. However, this method of sampling is limited to the collection of microorganisms that are resistant to drying.

For Concentration of Viable Organisms

A general requirement for the determination of the number of viable organisms present in the air is that the collected particles must be disintegrated during the sampling process or in the subsequent assessment of the collecting medium so that individual organisms, when placed in the proper growth medium, can develop into discrete colonies. Samplers generally used for this purpose are the liquid impingers, since they are very efficient in the collection of all microorganisms. The sampling time is limited to relatively short periods because of the evaporation of the collecting fluid and the killing effect upon the organism. Their use is generally limited to temperatures above 40°F unless some method of heating is applied or a suitable antifreeze agent is added to the collecting fluid.

Samplers employing filter media and nonnutrient solid impaction surfaces can be used to determine the concentration of viable organisms in the air. However, these samplers are limited to the collection of spores and other microbial forms that are resistant to the drying usually associated with this method of sampling.

For Change of Concentration with Time

The variation of concentration of viable particles or organisms with time can be determined by collecting intermittent samples with any suitable air sampling device. A simpler method of obtaining this information is by the use of automatic samplers, such as slit samplers that collect continuous samples on a rotating agar surface, the modified filter-tape sampler that collects bacterial spore samples automatically on filter media, and the liquid samplers that collect samples in a series of bubblers.

For Size of the Collected Particles

The Johnson single-stage impactor, the preimpinger, and the Casella cascade impactor are devices which can be used to separate all collected particles into either two or four general size ranges. By direct microscopic examination of the collecting surfaces, it is possible to measure the size of the collected particles. When the collecting surfaces are biologically assayed by washing in a suitable liquid, the results represent the total number of viable organisms associated with the particles recovered on the various stages of these samplers.

The TDL cascaded slit and the Andersen samplers are two devices which separate all collected viable particles into four and six general size ranges, respectively, by direct impaction on agar media. The plate counts represent the number of viable particles containing vegetative and spore-forming microorganisms.

The final selection of an air sampling device for use in specific air sampling problems must be left to the judgment of the investigator. Consideration should be given to the factors discussed in this section and to the individual characteristics of the candidate samplers.

No one type of sampler and assay procedure exists that can collect and enumerate 100 percent of the airborne organisms. Experience indicates that the most satisfactory samplers are the sieve (limited to low concentrations), slit, all-glass impinger, and the tangential-jet impinger. In any case, the sampler chosen should have an adequate sampling rate in order to collect a sufficient number of particles within a reasonable time period so that a representative sample of air is obtained for biological analysis.

Operation of Samplers

Control of Air Sampling Rate

If sampling is to be quantitative, the amount of air passing through the samplers must be known. The rate of flow for each sampler may be determined prior to the utilization of the device for collecting air samples. Many devices are available for metering the flow. Three are described here. The simplest but most expensive device is a calibrated flowmeter connected in the vacuum line between the sampler and vacuum source. A somewhat less expensive device consists of a manometer or a vacuum gauge and a regulating valve. To use this device, a calibrated flowmeter is first connected to the inlet of the sampler for calibrating the manometer. One side of a U-tube manometer or vacuum gauge is connected to the vacuum line between the sampler and the source of vacuum. The other side of the manometer is open to the atmosphere. With this system in operation, the amount of air passing through the sampler for a particular manometer reading is determined from the flowmeter and regulated by a bypass valve, preferably of the needle type, located on the vacuum inlet to the pump. In subsequent utilization of the sampler for collecting airborne microorganisms, the flowmeter is not included in the system. The rate of flow is maintained by adjusting the vacuum to the same manometer reading recorded for the desired flow rate. This calibration will be accurate as long as the resistance of the sampler does not change. Figure 9.2 illustrates the above arrangement.

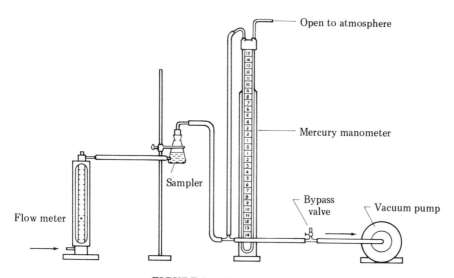

FIGURE 9.2 Calibration of airflow.

The third device for metering airflow, which eliminates the need for individual flowmeters for samplers when a number are in use, is a capillary tube in the vacuum line. This tube may be used as a critical orifice to keep the sampler flow at the desired rate. This is based on the fact that the velocity of a gas flowing through an orifice will reach a maximum or acoustical velocity when the critical pressure is reached. The theory applies to orifices in a thin, flat surface, but for all practical purposes it is also applicable to short lengths of capillary tubing. The critical pressure ratio is represented by the ratio of the downstream absolute pressure to the upstream absolute pressure. This critical ratio is approximately 0.5. When a vacuum is applied to the downstream side, the flow through the orifice will increase until this pressure ratio is reached. Increasing the vacuum further will not increase the flow no matter how high it may go, provided the upstream conditions do not change (see Figure 9.3).

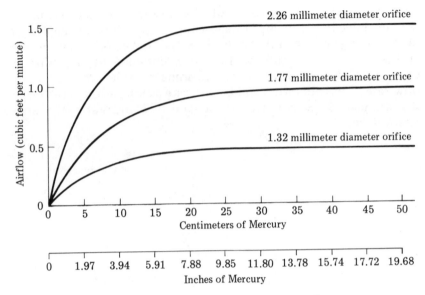

FIGURE 9.3. Critical airflow through orifices.

This means that constant flow can be maintained through a critical orifice without the necessity of having a constant vacuum as long as one uses a vacuum source of sufficient capacity so that the downstream pressure is always less than half the upstream pressure. How much less than 0.5 this ratio is does not matter.

An orifice may be cut from a glass or metal capillary tube in a convenient length (2 inches). The outside diameter should be large enough to fit snugly inside the vacuum tubing, and the inside diameter should be of a size to give approximately the desired flow. The exact flow through this orifice must be accurately measured, taking care to do this in the same arrangement in which it will be used so that there will be no change in conditions when samples are taken. To measure or calibrate the orifice, proceed as follows: Place the orifice in the vacuum line between the sampler and the vacuum source, and connect a standard flowmeter to the inlet of the sampler to measure the sampling rate accurately. Connect a mercury manometer across the orifice and adjust the vacuum so that the pressure differential is 38.0 centimeters of mercury or above. The airflow is indicated on the standard flowmeter, and if this value is too high or too low, repeat this procedure using another orifice with a different diameter until one is found giving the desired flow. During each measurement, make certain that the pressure drop is at least one-half atmosphere. After the proper orifice diameter has been determined, more orifices of the same length may be cut from the same tube, but each of these should be calibrated separately since the capillary diameter often is not constant through the length of the same tube. If a glass capillary is used and the diameter is larger than desired, it may be reduced by holding one end in a burner flame until the opening at the end is constricted. The flowmeter is not needed after the orifices have been calibrated (Figure 9.4).

Sterilization of Samplers

Most bacteriological samplers should be sterilized prior to use. Sieve samplers and filtration types that have been wrapped in kraft paper, and liquid impingers with cotton plugs inserted in the intake and exhaust ports, can be conveniently autoclaved. The sampling chamber of the Casella slit sampler can be removed and autoclaved, but in actual use it is much more convenient and adequate simply to swab the interior with a suitable disinfectant prior to each sampling period. Gaseous sterilization techniques, if available, can be used to sterilize all of the samplers.

Media

Many collecting and culturing media are available for biological air sampling. The selection of a particular nutrient medium will depend primarily on the nutritional requirements of the organism or organisms under study and, to a lesser extent, on the type of information desired from the study, as well as on the sampling method and the sampling conditions. When initial collection is in a liquid medium, the organisms must remain viable without growth or decay until aliquots are taken for culture. Some of the more common liquid media used are: tryptose saline, buffered gelatin, peptone water, buffered gelatin enriched with brain-heart infusion, buffered saline, and buffered water. These media also are used as diluting fluids to obtain suspensions suitable for plating. Buffered saline and buffered water are used only for the collection of spores and other resistant microbial forms.

When collection is directly on a solid nutrient medium, it is essential that a sufficient concentration of agar be used to produce a stable medium capable of withstanding the impingement action from a rapidly flowing airstream. Solid media containing an agar concentration of 1.5 to 2.0 percent possess adequate stability for use with all types of air sampling devices. Some of the more common solid nutrient media employed for general bacterial air sampling are: blood agar, tryptose agar, trypticase soy agar, proteose extract agar, and nutrient agar. These media are also employed for culturing the liquid

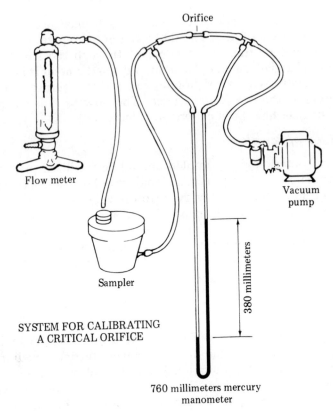

Orifice

Flow meter

Vacuum
pump

Sampler

SYSTEM FOR CALIBRATING
A CRITICAL ORIFICE

380 millimeters

760 millimeters mercury
manometer

FIGURE 9.4. System for calibrating a critical orifice.

collecting media by using surface plating methods or the conventional pour-plate method, and for culturing membrane filters by placing the filter directly on the agar surface or by placing the filter on blotter pads saturated with the broth form of these media.

Processing Samples

Samples should be processed as soon as possible after collection. Unless liquid samples can be processed shortly after being exposed, they should be kept at low temperatures, since variations may occur in the results due to growth or death of the entrapped organisms prior to plating. Long intervals between collection and incubation may also be detrimental to the organisms on agar plates. The most desirable situation would be to provide the necessary facilities at the location where samples are being collected. This can be done by using a mobile laboratory unit at the sampling station.

Computation of Sampling Data

Monitoring the levels of airborne microorganisms requires recording a substantial volume of pertinent information from a variety of sources. A comprehensive analysis of the data is difficult unless a systematic method of tabulating is available. One method of recording data, which has been found satisfactory in previous studies, is shown in Figure 9.5. Forms I and II are used at the sampling site, and Form III in the laboratory.

Assuming that each colony on the agar plate is the growth from a single bacteria-carrying particle, the contamination of the air in question is determined from the number of colonies counted. The airborne microorganisms may be reported in terms of the number per cubic foot of air sampled.

The following formulas can be applied to convert the colony count to organisms per cubic foot or air sampled.

For solid agar samplers

$$\frac{C}{R \times P} = N$$

where C = total plate count,
R = airflow rate, c.f.m.,
P = duration of sampling period (minutes),
N = number of organisms collected per cubic foot of air sampled.

For liquid impingers

$$\frac{C \times V}{Q \times P \times R} = N$$

where C = total number of colonies from all aliquots plated,
V = final volume in ml of collecting media,
Q = total number of ml plated,
P, R, and N as above.

FIGURE 9.5

Sampling Data Forms I, II, and III

Sampling Data

Data _____

Sample number	Sampling site	Type and number of sampler	Time sampling started	Duration of sampling period (min.)	Air flow rate through sampler (cu. ft./min.)	Total air volume sampled (cu.ft.)	Collecting media	Initial volume of media	Remarks

Meteorologic Conditions

Date _____

Time of recording	Temp. at sampler level	Psychrometer readings		Barometric pressure (in. Hg.)	Relative humidity (percent)	Wind		Precipitation	Sky Condition	Remarks
		Dry bulb	Wet bulb			Direction	Velocity (mph)			

Laboratory Data

Date _____

Sample number	Final broth volume	Volume plated	Culture media	Time sample placed in incubator	Total bacterial colony count			Bacterial colonies per cu. ft. of air		Remarks
					After incubation					
					24 hours	48 hours		24 hours	48 hours	

Planning a Sampling Program

When plans are being made for conducting an air sampling program, many obvious factors that may cause delays are easily overlooked. The obvious points listed below should serve only as a checklist.

1. Are enough samplers available to complete the proposed program?
2. Is the vacuum source sufficient for operation of the samplers?
3. If pumps are used, is the power line of sufficient capacity to carry the total current load?
4. Are all pumps in good operating condition, and is sufficient and proper lubricant available?
5. Are tools available for emergency repairs?
6. Are personnel instructed properly on the operation of samplers?
7. Will enough sterile media or samplers be available when needed?
8. Are facilities available for storage (refrigeration) of samples until they can be assayed?
9. Are complete laboratory facilities available for assessment of samples?
10. Are proper culture media prepared for assessment of samples?
11. Is all assay equipment sterile and ready for use?
12. Are duplicate supplies of materials on hand for items easily broken or consumed, such as fuses, glassware, thermometers?
13. Have laboratory personnel been instructed in the proper methods of assay?
14. Are personnel prepared to record all data relative to sampling and assaying?

COMBATING HAZARDS

Any effective program to combat microbiological hazards must encompass an effective safety training program, safe laboratory practices, properly designed facilities, the use of the proper protective equipment and devices, and medical monitoring. These factors are discussed below.

Safety Training Program

A rational approach to designing a safety program is first to define the specific problems and attempt to identify areas of possible hazard. This should be done by administrators and the various project leaders. The laboratory people can make assessments of risk on a technical basis, and the administrators can decide what degree of protection the program will be designed for.

In the case of some agents and procedures, the assignment of risk to the various tasks is rather easily done; with others, hazard levels cannot be readily ascertained. In the latter situation, the administrator has no alternative but to assign the highest risk and plan for maximum protection. As more knowledge of a particular risk is gained, it may be possible to scale back the safety design.

There must be an early indication from management whether the goal is to be (1) protection of the laboratory personnel, (2) protection of the population in the external environment, (3) protection of the experiment and concurrent experiments from contamination, or (4) a combination of the three. In addition, management must choose what level of protection is desired. Are all laboratory infections to be prevented, or are subclinical infections that can be detected only by serological procedures acceptable? Does management wish to prevent only incapacitating or lost-time infections—only those for which there is no treatment? To what extent is management willing to jeopardize the validity of experiments by cross-contamination?

Once these decisions have been made, management should issue a strong policy statement to serve as the basis or authority for the subsequent safety program. A safety manual should be published and made available to all laboratory personnel. Likewise, the proper procedures should be stipulated for all laboratory activities that can affect safety. A safety and environmental control officer reporting directly to laboratory management should be appointed.

The final safety training program should make provisions for both employee and supervisor training, include procedures for job safety analysis and accident prevention programs, utilize safety meetings, and establish an accident investigation procedure. (See Chapter 2 for a detailed review of safety training and investigating programs.) The following section discusses those safe laboratory practices that pertain to microbiological laboratories.

Safe Laboratory Practices

General Precautions

Only authorized employees, students, and visitors should be allowed to enter infectious disease laboratories or utility rooms and attics serving these laboratories.

Food, candy, gum, or beverages for human consumption should not be taken into infectious disease laboratories. Smoking should not be permitted in any area in which work on infectious or toxic substances is in progress. Employees who have been working with infectious materials should wash and disinfect their hands thoroughly before smoking. Workers should develop the habit of keeping hands away from mouth, nose, eyes, and face. This may prevent self-inoculation. Drinking fountains should be the sole source of water for drinking by human occupants.

Library books and journals should not be taken into rooms where work with infectious agents is in progress. An effort should be made to keep all other surplus materials and equipment out of these rooms.

According to the level of risk, the wearing of laboratory or protective clothing may be required for persons entering infectious disease laboratory rooms. Showers with germicidal soap may be required before exiting. Contaminated laboratory clothing should not be worn in clean areas or outside the building. No person should work alone on an extremely hazardous operation.

Safety Cabinets and Similar Devices

With many of the common operations carried out in micro-biological laboratories, there is real danger that infectious aerosols will be formed. A ventilated safety cabinet should be used for all procedures with infectious substances, such as opening test tubes, flasks and bottles, using pipettes, making dilutions, inoculating, necropsying animals, grinding, blending, opening lyophile tubes, operating sonic vibrators, operating standard table model centrifuges. A safety box or safety shaker tray should be used to house or safeguard all containers of infectious substances on shaking machines.

A safety centrifuge cabinet or safety centrifuge cup should be used to house or safeguard all centrifuging of infectious substances. When centrifuging is done in a ventilated cabinet, the glove panel should be in place with the glove ports covered. A centrifuge in operation creates reverse air currents that may cause the escape of the agent from an open cabinet.

A respirator or gas mask should be worn when changing a glove or gloves attached to a cabinet if an infectious aerosol may be present in the cabinet.

Pipetting

The major hazards of pipetting are (1) accidental aspiration of the fluid in the pipette, (2) contamination of the mouthpiece from the pipetter's contaminated finger, and (3) production of microbial aerosols. The use of a cotton plug in the pipette and attaching a mouthpiece to the pipette are well-known precautions against oral contamination, but they are not entirely successful in eliminating this danger. The best procedure is to eliminate all oral pipetting by using one of a variety of nonautomatic pipetting devices.

For optimum protection, pipetting pathogens should always be done in a hood. The pipette should be drained by holding the tip against the side of a tube or bottle below the level of the fluid. A disinfectant-wetted towel should be used to cover the work area to minimize the danger of splash from dropped culture contaminants. Contaminated pipettes should be placed in a tray large enough to allow their complete horizontal immersion in germicidal solution immediately after use. The pan should be equipped with a well-fitting cover that is replaced at the completion of the task, and the pipettes should then be autoclaved in the disinfectant before removal for cleaning and reuse.

Centrifuging

The greatest hazard associated with centrifuging infectious material is breakage of the culture tubes with release of extremely large numbers of aerosolized microorganisms. Two types of breakage may occur: (1) tubes may crack or break in the carrier cup, but the fluid remain in the cup by centrifugal force or (2) the tube may shatter and release the culture into the centrifuge chamber. Few aerosolized microorganisms are released into the laboratory from breaks in the carrier cup, but a large number are released by shattering tubes.

Aside from tube breakage or other malfunctions, microbial aerosols are produced during the normal operation of the centrifuge and manipulations associated with its use. If the rim of the tube is wet with culture fluid, this culture is thrown out as an aerosol from either a smooth-rimmed tube or from liquid trapped between threads of a screw-capped tube. This contamination may occur from careless filling of the container or from uncapping and recapping a container that has been shaken or inverted. Material spilled on an exposed rotor will also spin off, creating an aerosol composed predominately of particles in the size range capable of lung deposition. Further, when fluids are spun in screw-capped tubes in an unshielded angle centrifuge, some of the material can be sprayed over a wide area around the centrifuge. Manipulations of the centrifuge tube and its contents also add to aerosol production. Organisms may remain in the supernatant of centrifuged cultures.

To insure maximum safety, a biological safety cabinet should be used for filling and opening the centrifuge tubes or cups, and for continuous-flow centrifugation of infectious material.

Before centrifuging, inspect tubes for cracks or chips, and discard those that are imperfect. Use centrifuge trunnion cups with screw caps, or the equivalent. Inspect the inside of the trunnion cup for rough walls, and remove bits of glass or debris from the rubber cushion. A germicidal solution added between the tube and trunnion cup not only disinfects the surfaces of both, but also provides an additional cushion against shocks that might otherwise break the tube. Avoid decanting centrifuge tubes. Material should be withdrawn by aspiration. If the rim of the tube is accidentally wetted, it should be wiped off with a disinfectant or the rim should be flamed. If the cap is wetted, it should be removed, the tube flamed, and a sterile cap used.

Using Syringes and Needles

Two types of hazards are associated with the use of needles and syringes: (1) accidentally puncturing the skin with a contaminated needle, and (2) producing infectious aerosols.

Aerosols may be produced in a number of ways. For example, when a needle is withdrawn from the rubber stopper of a vaccine bottle, it vibrates considerably and droplets containing the test microorganism are discharged into the air. Aerosols can also be formed during actual inoculation operations. Internasal inoculation of mice with nasal suspensions has been found to be especially hazardous, producing large amounts of aerosol together with extensive contamination of the environment. An extremely dangerous situation is created when the needle is accidentally separated from a syringe containing infectious material and liquid is sprayed into the air. Accidents of this kind are usually caused by improperly affixed needles. Animal sites must be thoroughly disinfected before and after inoculation to prevent aerosol production, and even when this is done a few airborne organisms may be recovered following the injection.

Still another consideration in the use of syringes and needles is finger or hand contamination. Such contamination may occur (1) from the syringe plunger and (2) from wads of cotton held in the fingers and used to receive excess fluid and air bubbles from the syringe. The rear portion of the plunger becomes contaminated by capillary action, and organisms are transferred to the fingers. This hazard may be minimized by grinding down the rear portion of the barrel, thereby eliminating capillary action.

To reduce infection hazards from the use of syringes and

needles, use them only in a ventilated safety cabinet or hood and avoid quick and unnecessary movements of the hand holding the syringe. Examine glass syringes for chips and cracks, and needles for barbs and plugs *prior* to sterilization. Use needle-locking syringes only, and be sure that the needle is locked securely into the barrel. A disposable syringe-needle unit (where the needle is an integral part of the unit) is preferred. Wear surgical or other type of rubber gloves for all manipulations with needles and syringes.

Fill the syringe carefully to minimize air bubbles and frothing of the inoculum. Expel excess air, liquid, and bubbles from a syringe vertically into a cotton pledget soaked with the proper disinfectant, or into a small bottle of sterile cotton. Do not use the syringe to expel forcefully a stream of infectious fluid into an open vial or tube for the purpose of mixing. Mixing with a syringe is to be condoned only if the tip of the needle (or syringe tip without needle) is held below the surface of the fluid in the tube. If syringes are filled from test tubes, take care not to contaminate the hub of the needle, as this may result in transfer of infectious material to the fingers. When removing a syringe and needle from a rubber-stoppered bottle, wrap the needle and stopper in a cotton pledget soaked in 70 percent alcohol or the proper disinfectant. If there is danger of the disinfectant contaminating sensitive experiments, a sterile dry pledget may be used and discarded immediately into disinfectant solution.

Inoculate animals with the hand "behind" the needle to avoid punctures. Be sure the animal is properly restrained prior to the inoculation, and be on the alert for any unexpected movements of the animal. Before and after injection of an animal, swab the site of injection with a disinfectant.

Discard syringes into a pan of disinfectant without removing the needle. The syringe first may be filled with disinfectant by immersing the needle and slowly withdrawing the plunger, and finally removing the plunger and placing it separately into the disinfectant. The filling action clears the needle and dilutes the contents of the syringe. *Autoclave syringes and needles in the pan of disinfectant.* Note: Use separate pans of disinfectant for disposable and nondisposable syringes and needles to eliminate a sorting problem in the service area.

Using Inoculating Loops

A hot wire loop will create a bacterial aerosol when it is inserted into a liquid culture. The neck of the container and its plug may also become contaminated by this process. The amount of aerosol increases if the loop is agitated in the culture, or if it accidentally touches the side and vibrates while being withdrawn from the culture tube.

A heavily contaminated loop inserted into a flame may cause the spattering of viable organisms into the environment, depending on the heat resistance of the organisms in the material, the degree of spattering that takes place, and the flaming technique used by the individual.

When liquid cultures are being transferred, the film held by the loop will often burst accidentally and thereby produce aerial contamination. Streaking smooth agar plates rarely looses bacteria into the air, but when the agar surface is rough, or if a careless technique is used, the loop can vibrate and create an aerosol.

Splashing and concomitant aerosol formation may occur each time a loop is lifted from the drop of bacterial suspension during preparation of slide agglutination tests and films.

To prevent the formation of aerosols, allow the loop to cool before inserting it into a culture. A sterile applicator, which can be discarded into disinfectant immediately after use, avoids this error. Alternately, to insure adequate cooling, several wire loops can be held in a stand and used in rotation. Avoid rapid or jerky movements of the charged loop in the air, touching the sides of culture tubes with the loop, and agitating the loop in the culture. Use a shielded loop incinerator to sterilize contaminated loops, or rub off the excess inoculum in a disinfectant prior to flaming the contaminated loop. Use a sterile bent glass rod to spread inocula over the surface of *smooth* agar plates to reduce the aerosolization produced by loops. Work within the confines of a safety cabinet to eliminate dispersal of infectious aerosols into the room air.

Operating Blenders and Similar Devices

Aerosols are released by (1) a loose-fitting cover, (2) lack of a gasket in a tight-fitting cover or (3) a worn bearing or loosely fastened drive shaft. Removal of the cover immediately after operation of the blender can also allow the escape of an aerosol. Although the amount of aerosol liberated by removing the blender cover decreases as time passes, there can still be considerable escape of organisms as long as 1 and 1/2 hours after the blending operation.

High-speed blenders have been modified in various ways to reduce or eliminate aerosol production. One suggested modification involves a vaccine stopper-type harvesting port or a spigot and plug port, and a modified bowl containing a hermetically sealed cover for removing the fluid after emulsification without liberating the infectious, vaporized material in the blender. The suggestion also includes the use of a plastic cover placed over the blender during operation to prevent spraying of the entire room in case of a leak in the rotor bearing, careful inspection before use, and prompt autoclaving of the entire blender and its residual content after use. High-speed blender bowls have been designed that are leakproof in operation and can be unloaded without opening the bowl. These devices should be used with supplemental cooling of the bowl to prevent destruction of the bearings and to minimize the thermal effects on the products.

The effect of aerosol contamination when conventional glass blender containers are utilized can be reduced by securing a ring of nonabsorbent cotton close to the rim around the glass container with rubber bands, and then covering the container with two layers of plastic film. The film is secured over the cotton ring with rubber bands, allowing some sag in the film to obviate a pressure build up. The regular lid is placed over the entire assembly. After mixing, the film is punctured with pipettes to remove samples.

For maximum protection to the operator during the blending of infectious materials, blenders and such other devices as ultrasonic disintegrators, colloid mills, ball mills, and jet mills should be operated in a ventilated, gas-tight safety cabinet. Use safety blenders of the type described, inspect the rotor bearing at the bottom of the blender bowl for leakage prior to operation, and sterilize the blender and residual infectious contents promptly after use.

Operating Shaking Machines

With such devices, the shaking motion produces aerosols in the culture flasks. These contaminate the plugs and subsequently the room air when the flask is opened. The breakage of culture flasks on the mechanical shaker or the loss of plugs from the culture flasks being shaken presents serious contamination problems, especially since the accident may not be detected immediately and considerable aerosol may be formed. The possibility of accidentally dropping a flask during transfer to and from the shaker must not be overlooked.

Because of the obvious danger, various precautions are used for infectious agents of high virulence-low infectious dose. The shakers should be enclosed in ventilated cabinets provided with such features as temperature controls, UV irradiation, view-port lights, and a disinfectant spray device. The shaking machine and air-circulating fan are automatically stopped in some models by opening the door. The carriage tray may be kept moistened with a liquid disinfectant, and the machine may have a spray tube that permits washdown of the interior without opening the doors in the event of flask breakage. The chance of flask breakage may be reduced by the use of thick-wall Pyrex® or plastic flasks. To prevent escape of aerosols from the shake flask, parchment paper may be fastened over the plugs. Alternately, a vented container that securely holds and encloses a single culture flask, and is provided with a spun glass filter in the sealed top, may be used. The filter allows sufficient air exchange for microbial growth. In use, the containers themselves are clamped to the shaking machine. All containers and shake-culture flasks should be opened in a ventilated safety cabinet.

Another alternative to the ventilated cabinet may be considered when working with infectious agents of low virulence and concomitant high infectious dose. This alternate is comprised principally of close attention to details, which will minimize creation of aerosol and prevent spread of any viable aerosols generated. These details include no violent shaking, a leakproof pan to hold containers, a shock- and liquid-absorbent germicidal base on which the flasks or other containers rest, and holding devices that safely and effectively immobilize the container. It is also highly desirable to operate the shaking machine in an incubator, refrigerator, or small room separate from the working area. This enclosure should be equipped with a vision panel to permit inspection of the shaking machine prior to entry, a shaker electrical switch control mounted outside the enclosure, and interior ultraviolet and illuminating lights controlled from outside the enclosure.

Lyophilization

Two aspects of the process of lyophilization call for particular attention from the point of view of safety—contamination of the drying apparatus and opening or breaking of ampuls containing infectious dried material.

The lyophilizer becomes heavily contaminated during its operation. Danger to the operator lies in contamination of the hands from the manifold outlets during removal of the ampuls, unless rubber gloves are worn. Moreover, viable bacteria and viruses may be present in the vapors removed from frozen suspensions of biological material during lyophilization, and will contaminate the drying apparatus and subsequently the room air. It is essential, therefore, that the manifolds and condenser of the lyophilizer be sterilized immediately after use. Contamination of the vacuum gauge and pump by infectious vapors can be prevented by the use of a double-trap cotton filter in the vacuum line between the condenser and pump. The use of such a filter also allows safe removal of the remainder of the apparatus for sterilization.

The extent of contamination of the environment produced by opening ampuls containing lyophilized cultures depends on the consistency of the end product, that is, the nature of the suspending menstruum. Fluffier end products tend to create more concentrated aerosols, which, probably as a result of their lower density, persist for longer periods of time. Inclusion of mother liquor in the suspending menstruum can reduce the viable aerosol concentration of lyophilized cultures severalfold, while the use of a cotton pledget moistened with ethanol to surround the point of ampul breakage also greatly reduces aerosol release. Yet another aerosol reduction technique involves the use of an intense but tiny gas-oxygen flame. The researcher uses the flame to heat the tip of the hard glass ampul until the expanding internal air pressure blows a bubble. After allowing this to cool, he or she breaks the bubble while holding it in a large low-temperature flame; this immediately incinerates any infectious dust that may come from the ampul when the glass is broken.

Accidental breakage of ampuls, a relatively common laboratory accident, creates an extremely concentrated aerosol that may contaminate large areas of the laboratory. Significant contamination of laboratory air may remain one hour after breakage when a fluffy lyophilized powder is involved.

Preparing, reconstituting, and transferring lyophilized material, and opening ampuls containing lyophilized material, in a ventilated safety cabinet provides the best margin of safety for the laboratory worker. For added safety, surround the ampul with an alcohol-moistened pledget of cotton while the neck of the ampul is being broken. Wear rubber gloves to handle the drying apparatus and the ampuls during the process of lyophilization. Plug the necks of the ampuls with glass wool and install a microbiological filter in the vacuum line between the condenser and the pump to prevent contamination of the pump. Sterilize the manifolds and condenser of the lyophilizer immediately after use.

Inoculating and Harvesting Embryonated Eggs

The virus-inoculated egg represents a propagation system contained only by an extremely fragile barrier and is, therefore, an infection hazard to be handled with caution. Infectious aerosols have been produced and heavy surface contamination of the egg trays, shells, and hands of the operator has occurred during the inoculation and harvesting of infected eggs. One factor that contributes to the danger in handling infected embryonic egg tissues may be the high concentration of virus in such material.

The hazard of potentially contaminated dust generated by the tool used to cut windows in the eggs, or to open the eggs, must also be considered. The uninoculated egg may not be without danger, as fertilized eggs may harbor the avian lymphomatosis agent.

To minimize hazards from handling infected embryonated eggs, inoculate, examine, and harvest embryonated eggs in a ventilated safety cabinet. Disinfect the shell before and after inoculation. Shell contamination becomes airborne easily in the drying, and frequently turbulent, atmosphere of the incubator. Finally, wear rubber gloves to inoculate, examine, and harvest infected embryonated eggs.

Plugs and Caps

A cotton-wool plug that has become contaminated will create an aerosol when it is removed from the container, whether it is wet or dry. More aerosolized organisms are recovered when plugs are removed from containers that have been centrifuged—the process apparently creates sufficient aerosol in the tube to moisten the plugs or the threads of screw-cap tubes. Aerosols created in containers that have been shaken are fairly stable, as evidenced by the numbers of organisms that escape into the room when the caps or plugs are removed.

The rims of screw-capped bottles often become contaminated accidentally during inoculation procedures. Microorganism growth occurs in the fluid on the rims, and an aerosol is produced when the film around the cap is broken as the cap is removed. Plastic caps, after having been autoclaved, frequently become warped and cracked and are likely to leak, thus creating an additional danger. Nearly half of the infectious particles dispersed in this manner from screw caps are estimated to have a diameter of 4 μ (microns) or less. Such particles are small enough to penetrate the lungs. Intense flaming of the neck of the bottles for up to 10 seconds after inoculation does not always sterilize them, and bottles may break as a result of the heating.

To avoid producing infectious aerosols when opening culture tubes of infectious agents, open all tubes of infectious or potentially infectious cultures in a ventilated safety cabinet. Use rubber gloves to avoid contamination of the hands. Avoid wetting the plugs of culture tubes with the contents. The use of small or compartmented baskets will aid in preventing tubes from tipping and wetting the plugs. Avoid wetting the rim of the containers. To minimize aerosol production in tubes that are to be centrifuged, and also to reduce subsequent contamination of plugs or caps, fill the tubes only half full.

Using Incubators

The use of forced-air incubators with circulating and exhaust blowers that cause rapid dissemination of any spilled or overgrown material has added to the problems of spreading infectious materials. Tests which involved dropping glass and plastic Petri plates containing colonies of various bacteria to the floor have shown that infectious aerosols are produced during the rebound or breakage of the plates, during attempted decontamination of the plates with a spray apparatus, and while picking up the scattered and/or broken Petri plates. The bacteria are carried by air currents to distances remote from the accident site. Bacterial cultures in unbreakable plastic Petri dishes release less bacterial aerosol when dropped than those in glass Petri dishes. Similar aerosolization, of course, will occur when infectious cultures are spilled or plates are accidentally broken in the incubator.

To prevent accidental contamination of workers using incubators, place infectious materials to be incubated in nonbreakable containers that will minimize tipping and spilling, and use incubators that have been designed or modified so that the blower fans will automatically stop when the access door is opened.

Water Baths

The temperature of the water and the presence of organic contaminants in the water promote viability and, in some cases, multiplication of organisms that are accidentally released into a water bath. For these reasons, water baths used to inactivate, incubate, or otherwise treat infective material should contain an adequate disinfectant. Optimum levels should be maintained by frequent changes of fresh water and disinfectant.

Storage Freezers

Mechanical deep freezers, liquid-nitrogen freezers, and dry ice chests must be inspected frequently for broken containers of infectious material, to prevent airborne and direct-contact transmission of the material. If breakage does occur, the freezer must be decontaminated immediately. Since effective decontamination of all exposed surfaces is not practical without raising the temperature, it is essential that primary containers of infectious materials be stored in nonbreakable and leakproof secondary containers, which may be more conveniently and quickly decontaminated. Such protection will not only reduce the potential contamination of the freezer and its contents, but will facilitate decontamination and rapid transfer to a reserve box in the event of accidental breakage of an unprotected container. The use of secondary containment is especially important in liquid-nitrogen refrigerators where the entire nitrogen volume may become contaminated.

Under some circumstances, a liquid-nitrogen storage freezer may present an explosion hazard. This happens when the cap of a flask is left off for an extended period of time and atmospheric moisture condenses and freezes in the neck of the flask. Complete occlusion of the neck results in elimination of free venting and the explosion occurs from the pressure buildup. To prevent such explosions, cover liquid-nitrogen flasks promptly after use with the closure provided by the manufacturer.

Animal Care and Handling

All animal cages should be marked to indicate uninoculated animals, animals inoculated with noninfectious material, and animals inoculated with infectious substances.

Careful handling procedures should be employed to minimize the dissemination of dust from cage refuse and animals. Cages should be sterilized by autoclaving. Refuse, bowls, and watering devices should remain in the cage during sterilization. All watering devices should be of the nondrip type. Each cage should be examined every morning and at each feeding time so that dead animals can be removed.

Special attention should be given to the humane treatment of all laboratory animals in accordance with the Principles of Laboratory Animal Care as promulgated by the National Society for Medical Research. Monkeys should be tuberculin-tested and examined for herpetic lesions, and persons regularly handling monkeys should receive periodic chest X-ray examination and other appropriate tuberculosis detection procedures.

When animals are to be injected with pathogenic material, the animal caretaker should wear protective gloves and the laboratory workers should wear surgeons' gloves. Every effort should be made to restrain the animal to avoid accidents that may result in disseminating infectious material. Heavy gloves should be worn when feeding, watering, or removing infected animals. Under no circumstances should the bare hands be placed in the cage to move any object. Animals in cages with shavings should be transferred to clean cages once each week unless otherwise directed by the supervisor. If cages have false screen platforms, the catch pan should be replaced before it becomes full. Infected animals to be transferred between buildings should be placed in aerosol-proof containers.

Doors to animal rooms should be kept closed at all times except for necessary entrances and exits, and unauthorized persons should not be permitted entry to animal rooms.

A container of disinfectant should be kept in each animal room for disinfecting gloves, boots, and general decontamination. Floors, walls, and cage racks should be washed with disinfectant frequently. Floor drains in animal rooms should be flooded with water or disinfectant periodically to prevent backing up of sewer gases. (Drains should be avoided where possible.) Shavings or other refuse on floors should not be washed down the floor drain. An effective poison should be maintained in animal rooms to kill escaped rodents. Special care should be taken to prevent live animals, especially mice, from finding their way into disposable trash.

Necropsy of infected animals should be carried out in ventilated safety cabinets. Rubber gloves and surgeons' gowns over laboratory clothing should be worn when performing necropsies. The fur of the animal should be wet with a suitable disinfectant, and the animal should be pinned down or fastened on wood or metal in a metal tray.

Upon completion of necropsy, all potentially contaminated material should be placed in suitable disinfectant or left in the necropsy tray, and the entire tray should be autoclaved. The inside of the ventilated cabinet and other potentially contaminated surfaces should be disinfected with a suitable germicide. Grossly contaminated rubber gloves should be cleaned in disinfectant before removal from the hands, preparatory to sterilization. Dead animals should be placed in proper leakproof containers and thoroughly autoclaved before being placed outside for removal and incineration.

Disinfection and Sterilization

All infectious or toxic materials, equipment, or apparatus should be autoclaved or otherwise sterilized before being washed or disposed of. Each individual working with infectious material should be responsible for its sterilization before disposal. Infectious or toxic materials should not be placed in autoclaves overnight in anticipation of autoclaving the next day. To minimize hazard to firemen or disaster crews, at the close of each work day all infectious or toxic material should be (1) placed in the refrigerator, (2) placed in the incubator, or (3) autoclaved or otherwise sterilized before the building is closed. Autoclaves should be checked for operating efficiency by the frequent use of Diack, or the equivalent, controls.

All laboratory rooms containing infectious or toxic substances should designate separate areas or containers labeled: INFECTIOUS—TO BE AUTOCLAVED or NOT INFECTIOUS—TO BE CLEANED. All infectious disease work areas, including cabinetry, should be prominently marked with a biohazards warning symbol.

Floors, laboratory benches, and other surfaces in buildings in which infectious substances are handled should be disinfected with a suitable germicide as often as deemed necessary by the supervisors. After completion of operations involving plating, pipetting, centrifuging, and similar procedures with infectious agents, the surroundings should be disinfected. Stock solutions of suitable disinfectants should be maintained in each laboratory.

Floor drains throughout the building should be flooded with water or disinfectant at least once each week to fill traps and prevent backing up of sewer gases. (New construction plans should omit floor drains wherever possible.) No infectious substances should be allowed to enter the building drainage system without proper sterilization.

Floors should be swept with push brooms only. The use of a floor-sweeping compound is recommended because of its effectiveness in lowering the number of airborne organisms. Water used to mop floors should contain a disinfectant. (Elimination of sweeping through the use of vacuum cleaners or wet mopping is highly desirable.)

All laboratories should be sprayed with insecticides as often as necessary to control flies and other insects. Vermin proofing of all exterior building openings should be given close attention. Mechanical garbage disposal units should not be installed for use in disposing of contaminated wastes. These units release considerable amounts of aerosol.

Transporting Infectious Materials

It is essential that infectious materials in glass containers be transported in leakproof, nonbreakable containers that are large enough to contain all the infectious material in case of accidental breakage. This rule also applies to material being packed for shipment.

Posting Hazardous Areas

A number of national and international organizations have specified the use of the biological hazard warning symbol to signify the actual or potential presence of a biohazard and to identify equipment, containers, rooms, materials, experimental animals, or combinations thereof that contain or are contaminated with viable hazardous agents. These agents are identified as "agents presenting a risk or potential risk to the well-being of man, either directly through his infection or indirectly through disruption of his environment."

The symbol is shown below.

BIOHAZARD

FIGURE 9.6.

The hazard symbol is a fluorescent orange or orange-red color. Background color is optional as long as there is sufficient contrast for the symbol to be clearly defined.

Appropriate wording may be used in association with the symbol to indicate the nature or identity of the hazard, name of individual responsible for its control, and precautionary information, but this information should never be superimposed on the symbol.

Design of Facility

A properly designed and constructed laboratory is of prime importance in controlling biological contamination. Such a laboratory should also provide for flexibility and growth.

Early Planning Phase

Determination of Building Area. The building area may be

approximated on the basis of personnel, allowing 300 square feet per worker. This figure is determined by the nature of the research to be carried out and by a poll of several newer facilities designed for microbiological research. The number of personnel to occupy the new facility may be determined by extrapolation based on past growth.

Cost. The cost of the facility will include the following: (1) building shell, (2) building accessories such as heating and ventilating equipment, light, biological waste treatment system, building air filters and/or air incinerators, power wiring, plumbing, elevators, and service headers, (3) permanently installed equipment and furniture such as autoclaves, branch service piping, instrumentation, sinks, and laboratory furniture, and (4) the engineering and contractor's fees. Money for land, site work, underground services outside of the building, movable laboratory ware, and biological safety cabinets is not included in the cost of the facility.

Preliminary Planning. The scope of the preliminary planning phase normally is limited to the following: (1) an assessment of risk, (2) site location, (3) development of flow charts, floor plans, and sketch sections with approximate dimensions, (4) a brief outline of specifications, (5) notes regarding special biological barriers and controls, and (6) a review of the project with a biological safety consultant. Preliminary planning should include careful consideration of a means for future expansion. The degree of compatibility with neighboring facilities will influence site selection.

In this early stage, it is usually helpful to inspect other research laboratories in the vicinity and to study the construction reports on these buildings. In this way, many good ideas can be borrowed and previous errors avoided.

Preparation of Working Drawings and Specifications. Final drawings will be prepared by the engineering department or a selected A & E (architectural and engineering) firm. The length of time required for this phase varies considerably with the complexity of the project. For an ordinary-size laboratory, the time lapse may be from 3 to 6 months. Small projects seldom require less than 3 months. Large and complex projects may require as long as 8 to 10 months. During this period, a great many decisions must be made by the laboratory staff. Questions relating to services built into the facility, finish materials, safety provisions, biological barrier controls, laboratory and office furniture, and a multitude of other subjects must be answered promptly if the project is to move forward without interruption.

Both the laboratory director and his or her representative for the project should be readily accessible for making design decisions. The laboratory staff should have a voice in determining design requirements. It is the responsibility of the laboratory director, however, to make certain that essential features have not been omitted and also that important overall considerations are not unduly influenced by personal prejudices. When preliminary or final drawings and specifications are sent to the laboratory for review, it is extremely important that they be examined thoroughly. Approval of these documents by the laboratory is construed to mean that they have been studied in absolute detail. Requests for changes in drawings or specifications should be made in writing. Such requests should be followed up to make sure that proper action has been taken.

Meetings attended by the design engineering group and representatives of the laboratory staff should be held throughout the design period. The purpose of such meetings is the transfer of design information between the two groups.

The possibility of construction delays due to delayed ordering of purchased items can be eliminated by the preparation of a time schedule showing order and desired delivery dates. Such a schedule should be checked periodically. The contractor will do his or her own scheduling for the materials to be purchased as part of a "lump sum" contract.

Mechanical Considerations

This part of a laboratory design has become increasingly complex and expensive, averaging about 60 percent of the total cost. Early and careful attention to cooling, ventilating, plumbing, electrical, and other systems is necessary if the design is to be successful.

Cooling and Ventilating. Due to the danger of contaminated materials passing from one room to another, the airflow system should be designed with a rigidly controlled pressure zone. Air from "cold" (minimum biological contamination) areas must flow into "hot" (maximum biological contamination) areas through a mechanism of pressure differentials.

The air-conditioning system should provide a controlled environment suitable for predicted research programs and for the personnel using the facility. The facility, except for the mechanical equipment room, should be air conditioned with 100 percent outside air to satisfy year-round heating, cooling, ventilating, and humidity requirements. Specifically, the system should be capable of a design-controlled temperature range of 70 to $85° ± 2°F$ dry bulb temperature, while relative humidity is designed with a controlled range of 40 percent to 60 percent ± 5 percent year round in all research laboratories, animal rooms, and autopsy rooms. A visual indicating and recording control panel should be provided in a "cold" area to monitor the air conditioning, humidity, and air balance. Those areas not served by the year-round air-conditioning system should be provided with a 100 percent fresh air supply by means of a mechanical ventilation system comprised of both supply and exhaust ducts. This system should furnish tempered air in sufficient quantity to satisfy ventilation requirements.

An automatic control system should be installed that is capable of maintaining the previously outlined temperature and humidity conditions. While it is desired that individual thermostatic and humidity controls be provided for each laboratory and each animal room, consideration should be given to the selection of a multizone air-handling unit for each laboratory suite.

All air supplied to the building should be filtered. Minimum filtration efficiency should be 90 percent (some areas 99.97 percent) as determined by the National Bureau of Standards Dust Spot Test.

Supply and exhaust ducts should be located where they will not interfere with piping, electrical lines, other building elements, and with the work to be carried out in the laboratory. It is very important that fresh air supply and exhaust systems be completely separated. Exhaust outlets and fresh air intakes should be situated so as to minimize the possibility of exhaust effluent air being drawn into the fresh air inlets. There are a number of ways of reducing the cooling load without affecting the building design adversely. This may be accomplished by minimizing the building perimeter, by minimizing the number of windows, by using fluorescent lights, or by use of insulating and reflective materials.

The supply and exhaust plenums of the system should be designed so that maintenance of motors, bearings, control valves, and steam traps can be performed without entering any "hot" areas.

Special consideration must be given to the exhaust requirements of the "hot" areas of the building. In these areas provisions must be made for the exhausts and vents to be equipped with ultra-high-efficiency air filters. Space should also be provided for air incineration equipment, if this is needed later. The laboratory suite system should also be provided with dual exhaust fans, one as an operating unit and another as an emergency standby, suitably selected for capacity, suction, and static pressure level to maintain proper airflow conditions within the room. The methods selected for diffusing supply air into laboratories and animal rooms and exhausting spent air from these areas should provide for draft-free circulation. The supply and exhaust fans should be electrically interlocked so that a stoppage of a supply fan for any reason will also stop the exhaust fan.

Exhaust duct-work and filter plenums for all "hot" laboratory suites should be pressure tested and provided with aspirator fittings to permit the introduction of decontaminants for sterilization of the system.

Plumbing. Air, gas, steam, hot and cold water, and demineralized water outlets should be provided as required. Vacuum and distilled water should be handled by individual units within the various laboratory suites. Vacuum breakers should be provided on all water lines servicing the building. Exposed piping generally should not be permitted. Piping should be run in service chases with access from "cold" areas. Animal rooms should be equipped with floor drains of such size that waste and bedding material can be disposed of by this manner. These drains should be flush-mounted with a flush screw cap to seal the drain when it is not needed. Drains from all "hot" areas should be of a weld cast-iron type, which will insure that the waste material leaving the area is going directly to the sewage pretreatment tanks.

Electrical and Lighting. For power distribution it is becoming popular to run 440-volt lines to the laboratory rooms and reduce at those points to 110/220 volts. For some types of work, there is increasing use of 440-volt power. In addition, it is less expensive to distribute power at the higher voltage.

The minimum acceptable levels of lighting in the various areas are as follows: 100 foot-candles for laboratory suite, offices, animal rooms, glassware washing areas, "hot" and "cold" corridors, and cage washing areas; and 80 foot-candles for other areas. The use of incandescent fixtures should be kept to a minimum. Particular attention should be given to the selection of fixtures that will allow ease in maintenance and relamping. Fixtures located in "hot" areas should be of such a design as to allow for servicing and relamping from "cold" areas.

Provisions should be made to provide emergency electrical power for minimal lighting, freezers, air conditioning in animal rooms, filtered exhaust system equipment, fire alarm panels, refrigeration room compressors, and thermostatic control circuits and incubators.

Sewage Pretreatment. All liquid waste leaving "hot" areas should eventually terminate in a pressurized pretreatment vessel. After being subjected to the required temperature-time cycle, such waste can then be distributed to the normal sewage system. These vessels should be designed to permit withdrawal of samples necessary to test the effectiveness of the cycle in destroying viable microorganisms.

Location of Piping. The design of laboratory piping services involves a great number of seemingly unimportant details. However, if these details do not receive careful attention, the system may become the source of many irritations. Pipes may be hung in the upper service area and run to almost any point in the laboratory by being brought through the first-floor ceiling. This is a very convenient arrangement, as the pipes are easily accessible for maintenance and at the same time are located outside of the "hot" areas.

Special Systems and Equipment. Certain areas will require special equipment, including ultraviolet lights, moisture- and/or explosion-proof electrical devices, interlocked doors, automatic timers for lighting, and temperature-recording panels.

The building design should also include provisions for adequate raceways, conduit and telephone panels, space for the extension of the telephone system, and possibly a closed circuit television monitoring and Telefax system. Access to these panels should be from "cold" areas only. An intercommunication voice system should be provided for each room within the proposed facility, with a master control panel in a designated location somewhere in the office area.

Clock outlets should be provided in all offices, laboratories, animal rooms, locker rooms, and the lunchroom. The most recent accepted practices in safety and fire protection should be fully incorporated in the total design concept. In view of the sensitive electrical instruments required in various areas, particular attention should be given to the elimination of stray electrical currents and electrostatic and electromagnetic fields. A separate grounding system should be provided to the laboratories.

Architectural Considerations

Architectural considerations include such factors as windows, doors, and construction materials, as well as containment features.

Window, Doors, Construction Materials. The number and size of windows are dictated by several factors. If the building is air-conditioned, fewer windows are desired. Windows are generally more expensive than an equivalent area of exterior wall. Moreover, they have a much higher heat-gain factor. In buildings that will have windows, maximum wall space and light can be obtained by keeping window openings as high as possible.

The number of exterior doors is chiefly a matter of safety. In addition to the regular controlled building entrances, there should be some means of escape from areas where hazardous work may be in progress.

Construction materials should provide for maximum sanitation. They must also be insect-proof, rodent-proof, and capable of withstanding decontaminating agents. Such items may include wall-hung water closets, ceiling-hung toilet enclosures, integral wall coves, and impervious seamless wall, floor, and ceiling coverings.

Containment Features. The proposed facility should offer the best in environmental control and containment for the following areas: (1) protection of the investigator and supporting personnel from all known and potential biohazards; (2) protection of the experiment, resource materials, and animals by elimination of all possible routes of cross-contamination; and (3) absolute protection of the exterior environment against discharge of hazardous material.

Containment features necessary to attain maximum versatility in use of space and equipment in a laboratory include: (1) multiple exhaust outlets for adaption of flexible tubes to biological safety cabinets or other enclosures within a laboratory; (2) biological safety cabinets mounted on casters for relocation within the same room, suite, or other suites; (3) removable panel walls for enlargement of rooms or suites; (4) dual exhaust fans—one as an operating unit and one as an emergency standby; (5) an individual room supply and exhaust system for each suite; (6) quick-disconnect couplings for services to the individual laboratory rooms; and (7) pass-through autoclaves mounted in the walls between "hot" laboratories and "cold" corridors. Each room within a laboratory suite should be so designed that it is possible to accomplish the following steps in decontaminating the area: (1) seal the room from all adjacent areas, (2) completely shut off the air supply and exhaust, and (3) introduce a decontaminating agent directly into the room.

Contaminated material leaving the suite should be decontaminated with the proper disinfectant, autoclaved, or enclosed in sealed containers. "Hot" and "cold" change rooms, rest rooms, pass-through showers, and lockers should be provided in each suite for both men and women. This will allow laboratory personnel to change from contaminated laboratory clothes to clean garments prior to leaving their particular suite. An air-lock barrier should be provided for incoming and outgoing equipment. Visitor observation windows should be provided so that "hot" laboratories may be viewed from "cold" corridors.

Layout of Laboratory Facility*

A typical laboratory facility consists of five functional zones— clean, laboratory research, animal research, laboratory support, and engineering support. They are each discussed below.

Clean Zone. The clean zone of a laboratory includes the entrance area, the office area, conference room, library, and those functional rooms where administrative operations, conferences, reading, writing, and other tasks not involving infectious materials are carried out. Also within this zone are the transitional rooms through which personnel and materials enter and leave the potentially infectious parts of the building. In addition, the shipping, receiving, and clean storage areas are in the clean zone.

The clean-change room should have lockers for street clothing, storage shelves for laboratory clothing, and toilet and wash facilities. An air lock with an ultraviolet door barrier and ultraviolet lights mounted in the ceiling may be used to separate the clean-change room from the contaminated-change room. A shower room should be located between the two change rooms. The contaminated-change room should have a storage rack for laboratory shoes, a bag for used laboratory clothing, and toilet facilities.

Laboratory Research Zone. This zone contains laboratories for infectious microbiological operations not involving actual work on animals. The zone should be separated by at least a corridor from the zone where infected animals are used. In addition to laboratory rooms, the zone may contain potentially contaminated offices adjacent to offices in the clean zone, toilets, change rooms, constant-temperature rooms for incubation and refrigeration, and instrument rooms for ventilated safety cabinets, centrifuges, and other equipment.

Within the zone, provision should be made for the microbial filtration of nonrecirculated air exhausted from laboratory rooms and the safety cabinets and other apparatus. The air pressure in the laboratory rooms should be negative with respect to the adjacent halls; air pressure in the cabinets should be negative with respect to

*Based on G. B. Phillips and R. S. Runkle, *Applied Microbiology*, 15 (March 1967), 378-89. For a more detailed discussion of the layout, equipment, and design criteria associated with each of the functional zones, consult the original article.

the laboratories. The entire zone should be maintained at a negative pressure compared to the noncontaminated zones.

Paint and other surface coatings used within the area must be resistant to steam and to the disinfectants employed. Walls and other surfaces should be free of cracks and the coating flexible enough so that minor shifts in the structural system do not cause it to crack. Equipment should preferably be sealed to the floors and walls to lessen the danger of spreading contamination. Unsealed equipment should be easily movable to facilitate cleaning and decontamination.

Pipes, conduits, lighting fixtures, and the like should be installed in such a way as to preserve the biological separation between the clean and contaminated zones. Ultraviolet fixtures installed in the ceiling and operated from the corridor may be useful for reducing nonspecific microbial contamination in unoccupied rooms. They are also useful for decontaminating accidental spills.

Animal Research Zone. This zone ordinarily includes rooms for animal inoculation and autopsy, and infected animal holding rooms equipped with isolation equipment such as ventilated cages and ultraviolet cage racks. If one is not available elsewhere, the room should have an incinerator for disposing of dead animals and an autoclave for sterilizing cages and passing them into a cage-washing room.

Dust filters should be installed in the air-exhaust ducts of the animal rooms to prevent animal hair and dander from overloading the downstream microbial filters. The zone should also have waterproof lighting fixtures in the wash areas, floor drains, cage- and rack-washing equipment, and adequate storage. The type of cages and isolating equipment will depend upon the exact nature of the work. Thus, where a high degree of isolation is required, infected animals may be held in small, individually ventilated cages. In other instances, animals may be housed in cages under an ultraviolet barrier. Whatever isolation system and equipment is used, it should be adequate to prevent undesired cross-infection of animals and should provide safe working conditions for personnel.

Laboratory Support Zone. This zone may include rooms for washing and sterilizing glassware and animal cages, preparing culture media, storing equipment, glassware, and animal cages, and repairing laboratory devices. Sometimes it may also contain animal rooms for quarantining and acclimatizing animals before they are passed into the infectious area for use.

Since many heat- and odor-generating operations are carried out in the working area, a well-designed, adequate ventilating system is of special importance. Walls and other surfaces should be moisture-resistant. The air in the room used for preparing culture media and in the safety cabinets should be maintained at a pressure that is positive with respect to that of the other laboratory support rooms.

Engineering Support Zone. This zone contains the pipes, ducts, blowers, filters, waste-treatment system, and most of the air-handling systems. The zone may be on grounds adjacent to the laboratory, or in the attic, basement, or some other space within the laboratory building. As much as half of the total building area may be taken up by the engineering control zone.

Exhaust air ducts coming from infectious zones must be airtight to insure no leakage of infectious microorganisms before the air reaches the exhaust filter plenum. The exhaust stack must be located downwind from the air-intake grills. Exhaust air plenums should be equipped with filters to remove infectious microorganisms from air discharged from laboratory rooms. These may be high-efficiency spun glass mats or ultra high-efficiency units, depending upon the particular circumstances. In virus research laboratories inlet air is also often passed through ultra high-efficiency filters.

A central control board that provides readouts of all systems in the laboratory is an important unit in this zone. The board should be equipped with audible and visual alarms that will automatically signal the failure of any part of the system. There should also be a standby electrical generator in case of commercial power failure.

Protective Equipment and Devices

Personal Protective Devices and Clothing

Animal workers should wear skin and eye protection when working around cage racks. Ventilated personnel hoods provide excellent protection, and goggles can be worn for short work periods in the area. Proper laboratory clothing includes coveralls and a smock, plus a head and face covering of the type shown in Figure 9.7, whenever cultures are handled. In addition, rubber or heavy cloth gloves should be worn where appropriate. (See descriptions of operations in text.)

Clothing should be changed frequently, preferably each day. Street clothing should not be worn beneath laboratory clothing, since it may become contaminated if an accident occurs. Specific and specially designed receptacles should be used for soiled laundry, so that contamination of the room air is held to a minimum and there is no personal contact with the soiled clothing. Laboratory clothing

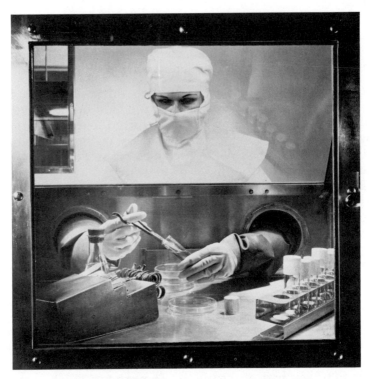

FIGURE 9.7 Biological safety cabinet in use.
(Photo courtesy The Dow Chemical Company)

worn by persons working with infectious agents must be autoclaved before it is washed. Since personnel cleanliness is an important barrier to infection and cross-contamination, change rooms should be equipped so that showers may be taken by personnel exposed to contaminated or potentially contaminated areas, and in case of accidental exposure to infectious materials. Germicidal soap should be provided.

Ultraviolet Fixtures

Ultraviolet fixtures mounted in cage areas, the ceiling between the clean and contaminated change rooms, and other strategic areas are an effective means of reducing the number of airborne vegetative organisms. The lights must be cleaned frequently and replaced when measurements with a detector (see Chapter 6) indicate a 40 percent loss of output.

Ventilated Cages and Modular Units

Infected animals may be housed in sealed cages ventilated with filtered air pulled through the cage. Such a cage shields one animal from another and protects the handler from the animal.

A cabinet system comprising one or more levels of individual modular units may also be utilized for housing animals. The levels can be connected to each other by electrically operated elevators for the transfer of animals, feed, and equipment. All procedures—including opening culture tubes, blending and grinding, inoculating, injecting and autopsying animals—are carried out within the modular system.

Biological Safety Cabinets

There are two types of biological safety cabinets in current use—the Class I and the Class III. The first is a variation of the chemical fume hood. It is a stainless steel cabinet that uses the inward flow of air to prevent the escape of microorganisms from the researcher's environment. Its major disadvantage is that the experimental materials can be exposed to large numbers of airborne contaminants brought into the cabinet by the room air. Also, the air turbulence inside the cabinet creates a potential for cross-contamination within the cabinet.

The Class III cabinet is a gas-tight enclosure fitted with arm-length gloves and is used when absolute containment of the biological materials is required. It also prevents contamination of the experiment from the room environment. Its major disadvantage is the difficulty of performing routine microbiological manipulations with any speed. Figure 9.7 illustrates use of a biological safety cabinet.

HEPA Filters

The high-efficiency particulate air (HEPA) filter, a development of the United States Army, has an efficiency of 99.7 percent for particulates 0.3 microns in diameter. It is sometimes referred to as an "absolute" filter. This device has proven to be a tremondously effective means of providing the ultraclean environment required by industry. It has made the now familiar "clean room technology" possible.

Laminar Airflow Clean Rooms

These rooms are based on the concept that if contaminant-free air is allowed to flow through a space at a uniform velocity and with an absolute minimum of turbulence it will be very efficient in capturing and removing airborne particulates. There are two basic types of such rooms: the vertical-flow type and the horizontal-flow type. The vertical or downflow type distributes air through a ceiling bank of HEPA filters and exhausts the air through a grated floor. The horizontal-or cross-flow type allows air to enter the room through a bank of HEPA filters located in one wall, with the entire opposite wall serving as the exhaust. These rooms provide very rapid removal of small particle contamination, eliminate cross-contamination between adjacent operations, and minimize the effect of personnel-generated contamination. Their cost, however, is prohibitive for many research activities.

Laminar-Flow Biological Safety Cabinets

Laminar-flow biological safety cabinets (LFBSC) provide highly localized control of the environment surrounding critical laboratory operations. They combine the benefits of high-efficiency filtration and unidirectional laminar airflow without the hazards of operator exposure.

Several types of LFBSC have been developed. One type utilizes a 10 percent free area perforated work surface and a 20 percent free area slot at the face opening. The front window is perpendicular to the cabinet back. There is a raised lip at the leading edge of the cabinet to turn the air stream entering the cabinet to more nearly parallel the flow of air recirculating within the cabinet. Figure 9.8 shows the salient features of this type.

A second type eliminates the perforated work surface and utilizes an outward-slanting window. In this case, minimum-turbulence flow is more difficult to achieve; therefore, more reliance is placed on the effects of dilution ventilation and adequate particle-capture velocities. Figure 9.9 shows this type of cabinet.

The principal design features that make the LFBSC effective are:

HEPA filtration. Air supplied to the work areas and exhausted through the cabinet is passed through HEPA filters and is essentially contaminant-free.

Dilution ventilation. The air within the work zone is changed up to 30 times per minute.

Controlled Airflow. The air recirculated within the cabinet is supplied to the work area in a laminar-flow condition. Room air entering the cabinet does not overflow onto the work surface, and air recirculating within the cabinet does not flow out at any point. This condition is achieved by effective aerodynamic

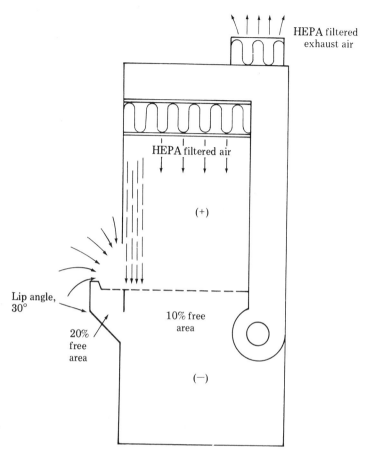

FIGURE 9.8. Laminar flow cabinet with perforated work surface. *(Reproduced by permission of the American Society for Microbiology)*

design of the cabinet work opening and proportionate balancing of the incoming airflow with the recirculated flow.

Sealed construction. The air downstream of the work surface is considered contaminated and, in many cases, is at a positive pressure. The construction joints of all air plenums are sealed to prevent the escape of potential pathogens to the room environment.

Medical Monitoring

Any prospective employee of a microbiological facility should be given a thorough pre-employment physical examination, during which a detailed medical history is taken.

Persons with active tuberculosis, bronchiolectosis, recent active rheumatic fever, or recent hepatitis should not be employed in a

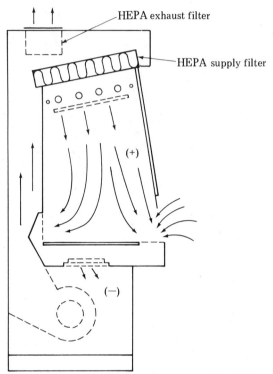

FIGURE 9.9. Laminar flow cabinet with solid
work surface.

microbiological facility, nor should chronic alcoholics or pregnant
women. Ethnic origins should be considered in those laboratories
handling coccidiodomycosis and tuberculosis, as the dark-skinned
races are quite susceptible to these organisms. As a precautionary
measure, persons whose families show a high incidence of cancer,
leukemia, or lymphoma; who have a history of allergy, including
allergy to drugs; who have been exposed to ionizing radiation or such
known or potential carcinogens as benzene, burning coke, smelter
fumes, uranium ore dusts, or mustard gas; or who have a history of
anemia, leukopenia, thrombocytopenia, collogen disease, or infectious mononucleosis should not be allowed to work in a microbiological facility.

When technically possible, periodic serological examinations are
recommended for all personnel as a means of revealing a more
accurate infection rate and gaining a truer measure of the effectiveness of control measures. Otherwise, a subclinical infection may go

unnoticed or be passed off as a severe cold or mild attack of influenza.

Immunization may be utilized as a supplement to proper personnel training, techniques, and equipment. It should not be assumed, however, that the effectiveness of a vaccine for laboratory workers will be as great as its effectiveness in preventing disease in the general population. Under laboratory conditions, infectious agents are usually handled in high concentrations, and infections may be acquired by routes uncommon in the natural mode of infection. Effective vaccines have not yet been developed for a number of human diseases occurring in the laboratory. These include blastomycosis, brucellosis, coccidiodomycosis, glanders, histoplasmosis, infectious hepatitis, leptospirosis, and toxoplasmosis.

EXERCISES: CHAPTER 9

1. Name the types of agents that can cause laboratory infections.

2. Indicate the ways in which accidental infection can occur.

3. Discuss the mode of operation of the following air-sampling devices:
   a. liquid impingers
   b. impactors
   c. centrifuges
   d. electrostatic precipitators

4. Describe the procedure for the size analysis of bacterial particles.

5. Name the types of sampling devices best suited for determining:
   a. concentration of viable particles
   b. concentration of viable organisms
   c. change of concentration with time
   d. size of collected particles

6. Discuss the elements in any effective safety training program for combating microbiological hazards.

7. Name the major hazards involved in the operations listed below and indicate how they can be combated.
   a. pipetting
   b. centrifuging
   c. using syringes and needles

   d. using inoculating loops

   e. operating blenders

   f. operating shaking machines

   g. lyophilization

   h. using incubators

   i. handling and caring for animals

8. Indicate the reason for maintaining air-pressure differentials in a microbiological laboratory facility.

9. Name the principal design features and benefits of the laminar-flow biological safety cabinet.

10. Describe the medical monitoring program for persons working in microbiological laboratories, with particular attention to:

   a. the scope of the pre-employment physical examination.

   b. medical conditions that should exclude persons from microbiological jobs.

10

Ventilation

Ventilation plays an important role in combating nonradioactive and radioactive air contaminants, heat, and microbiological hazards, as well as providing a more comfortable, and hence safer, general work environment. Thus, ventilation should be considered in some detail in any discussion of industrial hygiene. There are two major types of ventilation: dilution ventilation and local exhaust ventilation. They are discussed below.*

DILUTION VENTILATION

This type of system dilutes the concentration of the contaminant with clean and uncontaminated air before it reaches the workers' breathing zone. It does not reduce or eliminate the amount of hazardous material released into the workroom air.

When to Use Dilution Ventilation

If the contaminant escapes into the air at a comparatively low rate and more or less uniformly throughout the work area—not at one or more isolated locations—dilution ventilation may be the most practical means of controlling it at the breathing zone of the worker.

*The material in this chapter has been adapted from *The Industrial Environment—Its Evaluation and Control*, U.S. Department of Health, Education, and Welfare, 1973. For a more detailed discussion of any aspect of engineering control devices, consult Chapters 39 through 43 of this publication.

The successful application of dilution ventilation depends upon the following conditions:

The material or contaminant generated should not be in excess of the amount that can be diluted with a reasonable volume of air.

The contaminant must have a low toxicity (high T.L.V.).

The distance between the workers and the point of generation of the contaminant should be sufficient to insure that the workers will not be exposed to average concentrations in excess of the currently established T.L.V.

The material or contaminant should be generated at a reasonably uniform rate.

Dilution ventilation is used with the greatest success when the rate of consumption can be determined with some degree of accuracy and the required volume of dilution air can be calculated. This type of ventilation is not successful for controlling fumes and dusts. The amount of material generated is usually too great, and reliable data on the rate of generation are very difficult to obtain. It is, however, often used to control vapors from organic liquids of low toxicity.

Selection of the Air-Movement System

In older buildings, dilution ventilation is sometimes provided by natural means. In these buildings, heated air rises to the roof and is expelled at the top of the building, while replacement air enters at the lower perimeter. Modern industrial plants and multistory buildings, however, rely almost completely on mechanical ventilation, including mechanical exhaust of contaminated air and the mechanical supply of fresh air.

In large open industrial buildings, general ventilation can be achieved with or without gravity ventilators. Where little or no heat is available to furnish natural ventilation or where head room is low, roof fans should be used in place of gravity ventilators. The best way of assuring general ventilation in a closed building is to supply air through duct work and distribute it into the work area so as to provide both humidity and temperature control.

Great care is needed in selecting the air-movement system for a particular installation. Depending upon the installation, an air-supply system, an air-exhaust system, or a combined supply-exhaust system may be indicated.

Normally, the supply air should be introduced along one side of the room or work area and the exhaust opening located on the opposite side of the room; however, under some circumstances, the airflow pattern may be changed or short-circuited by partitions, large machines, hot processes, or other disruptions of the normal airflow. Therefore, care should be taken to avoid short-circuiting the dilution air. The supplied air must pass over the point or points where the contaminant is generated and toward the exhaust opening. In most circumstances this should assure that the contaminant is not carried to the breathing zone of the worker.

The air exhausted must be replaced by an equal amount of uncontaminated fresh air, either from a mechanical system or by entry through openings in the building. This will assure that the room is not under excessive negative pressure and thus reduce the possibility of workers being exposed to dangerous concentrations of other gases or vapors. Provision must be made to heat the makeup air during cold weather. Care must be exercised to prevent drafts around windows, doors, and other sources of makeup air. This requires special consideration, if the temperature of the makeup air is more than $10°F$ below the temperature of the workroom.

Normally, better dilution is possible with a good air-supply system because the air can be directed toward the sources of generation of the contaminant. However, if adjacent areas are occupied, the positive pressure created in the room by the excess air supplied will cause the contaminated air to flow into any adjoining occupied areas of lower pressure. This is true also of combined supply-exhaust systems when there is excessive makeup air. In the case of the combined system the adjoining area would not be contaminated if the workroom in which the ventilation was being utilized was under a slight negative pressure caused by an excess of exhaust air.

In a combined supply-exhaust system the ventilation rate is the rate of either the supply or exhaust, whichever has the greatest capacity, and is not the sum of the rates of the two systems.

Consideration must be given to any local exhaust hoods located in the work area to assure that the airflow at the face of the hoods is not disturbed by air from the supply inlets. The location of exhaust openings should be selected to keep the source of contaminant between the operator and the exhaust, and to avoid re-entrance of the contaminated exhaust air into the workroom. This usually requires that the contaminated air be exhausted well above the roof line of the building or above closely adjacent buildings.

LOCAL EXHAUST VENTILATION

The action of local exhaust ventilation is to control the contaminant at its source or point of generation, thus preventing the release of the contaminant within the general workroom. A local exhaust system consists of hoods and associated ducts, an air-moving device (fan), and an air-cleaning device. This type of ventilation is preferred over dilution ventilation in most cases for the following reasons:

1. Less air needs to be exhausted to control the contaminant. This means that a smaller fan will be used and the power consumption will probably be lower. Also, the makeup required will be less, and in cold climates or air-conditioned shops this may be an important economic factor.
2. The contaminant is captured before it enters the workers' breathing zone and not after it has been generally disseminated in the room. With some highly toxic and radioactive substances, this is the only method of exhaust possible.
3. The contaminant is more highly concentrated in the conveying airstream. This is an important factor where the air will be cleaned either to salvage some valuable product or to prevent air pollution.

Considerations in Local Exhaust Ventilation

Among the more important general considerations are capture velocity, control velocity, dispersive forces, and air distribution in hoods. Other considerations are beyond the scope of this discussion.

Capture Velocity

The capture velocity is that velocity in front of the local exhaust hood that is necessary to overcome the dispersive forces and room air currents and capture the contaminant. Table 25 gives the capture velocities required for different industrial operations and different conditions of contaminant dispersion.

Control Velocity

The control velocity produced by a hood is approximately inversely proportional to the square of the distance from the hood. Also the additions of side and back flanges to a hood reduce the

TABLE 25

Range of Capture Velocities

Condition of Dispersion of Contaminant	Examples	Capture Velocity fpm (m s⁻¹)
Released with practically no velocity into quiet air.	Evaporation from tanks; degreasing; etc.	50-100 0.25-0.5
Released at low velocity into moderately still air.	Spray booths; intermittent container filling; low-speed conveyor transfers; welding; plating; pickling.	100-200 0.5-1
Active generation into zone or rapid air motion.	Spray painting in shallow booths booths; barrel filling; conveyor loading; crushers.	200-500 1.0-2.5
Released at high initial velocity into zone of very rapid air motion.	Grinding; abrasive blasting; tumbling.	500-2000 2.5-10

amount of air required for a given velocity. These points argue strongly for good enclosure as a means of obtaining adequate control with minimum air volumes. The velocity at the entrance to the pipe connected to the hood is of comparatively little significance. For example, a 6 inch diameter open pipe with a flow of 350 cubic feet per minute will produce a velocity of 130 feet per minute on the centerline 6 inches from the open end of the pipe. If the pipe size is decreased to 4 inches, the inlet velocity increases 2½ times but the velocity 6 inches from the open end of the pipe is only 4 percent higher. However, the static pressure and power required will be 6.4 times higher.

Dispersive Forces

As previously indicated these forces are a factor influencing capture velocity. Specific dispersive forces include the rising current of hot air above a hot process, the aspiration of air downward by falling rocks from a conveyer, the rotative air current above a grinding wheel, natural air currents caused by room ventilation, and similar external forces.

In designing hoods it is preferable to take advantage of these dispersive forces whenever possible. The standard grinding wheel hood design, for example, takes advantage of the motion of the chips or fragments from the wheel. While these large particles, which are projected long distances, are of no hygienic importance, they do create an aspirated air current that carries small particles along with them into the hood, and their collection also aids in housekeeping. In woodworking operations, housekeeping and fire prevention are the major factors in control so that collection of the large, nonrespirable particles is most important.

Air Distribution in Hoods

Airflow into a hood should be uniform throughout its cross section. In the case of slots for lateral exhaust, for example, this may be done by "fishtail" design. An easier and also effective method of design, however, is to provide a velocity of 2,000 to 2,500 feet per minute into the slot, and use a low-velocity plenum or large area chamber behind it. The 2,000 feet per minute velocity will give a static pressure behind the slot of 0.4 inch of water. This pressure drop will equalize flow along the slot, since the static pressure throughout the plenum will be practically uniform. For large shallow hoods, such as paint spray booths, lab hoods, side-draft shakeout hoods and the like, the same principle may be used. In these cases, unequal flow may tend to occur, with a concentration in the neighborhood of the takeoff. Baffles provided for the hood increase velocity and pressure drop into the hood, giving the plenum effect. Where the face velocity over the whole hood is relatively high, or where the hood or booth is quite deep, no baffles will be required.

Hoods

A hood is a structure designed to enclose or partially enclose a contaminant-producing operation and to guide airflow in an efficient manner to capture a contaminant. The hood is connected to the ventilation system with a duct that removes the contaminant from the hood. There are four basic types of hoods: enclosure, booths, receiving hoods, and exterior hoods.

Enclosures

Enclosures usually surround, completely or partially, the point at which the contaminant is emitted or generated. Because all dispersive actions take place within the confines of the hood, enclosures require the lowest exhaust rate of all hood types.

Enclosure hoods are economical and efficient. They should be used whenever possible, especially when the contaminant is a hazardous material. Figure 10.1 illustrates a typical enclosed hood.

Booth-type Hoods

These are actually a special type of enclosure. Booths are typified by the common laboratory hood or spray-painting booth in which one face of an otherwise complete enclosure is open for air. Air contamination takes place inside the enclosure, and air is exhausted at such a rate as to induce an average velocity that will be sufficient to prevent air from escaping through the opening. Table 26 shows applications for enclosure-type hoods.

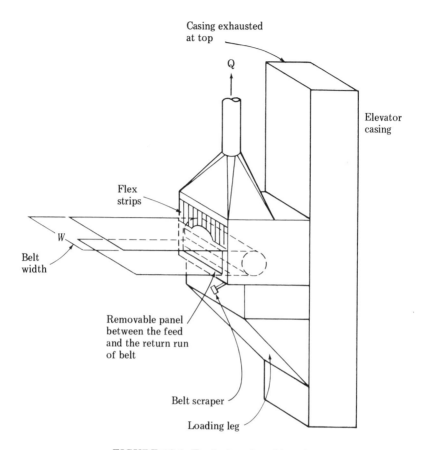

FIGURE 10.1. Typical enclosed hood.

Receiving Hoods

These are hoods specifically located for exhausting a stream of contaminated air from a process. They are similar to exterior hoods in that the contaminated air originates beyond the physical boundaries of the hood. However, the entire airflow is induced by receiving hoods, whereas with exterior hoods the airflow is freer.

Two common types of receiving hoods are canopy hoods and grinding hoods. Canopy hoods are frequently located directly above hot processes. They receive contaminated air, which rises into the hood primarily because of its own buoyancy. Canopy hoods are adversely affected by cross drafts and are less efficient than total enclosures. They cannot be used to capture toxic vapors if people must work between the source of contamination and the hood.

TABLE 26

Booth and Enclosure Hoods and Their Applications

Type	Application
Booth	Laboratory
	Paint and metal spraying
	Arc welding
	Bagging machines
Machine Enclosure	Bucket elevators
	Vibrating screens
	Storage bins
	Mullers, mixers, crushers
	Belt conveyers
	Packaging machines
	Abrasive blast cabinets

Grinding hoods are used where contaminants from a grinding or polishing wheel are too heavy to be captured by the conventional airflow patterns created by exhaust hoods. Heavy particles are released into the hood by inertial forces of the wheel. If hood space is limited by the process, baffles or shields may be placed across the line of flow of the particles to reduce their kinetic energy.

Figures 10.2 and 10.3 show a typical canopy hood and grinding wheel hood; Table 27 indicates some common receiving hoods and their applications.

TABLE 27

Receiving Hoods and Their Applications

Type	Application
Grinding	Surface grinders
	Stone and metal polishing
Woodworking	Shapers, stickers, saws
	Jointers, molders, planers
Stonecutting	Granite and marble cutters and grinders
	Granite surfacing
Sanding	Belt and drum sanding operations
Portable	Hand grinding, clipping
Canopy	Hot processes evolving fumes

Exterior Hoods

Such hoods must capture air contaminants being generated from a point outside the hood itself—sometimes relatively far away. They differ from the other types of hoods in that they must capture the contaminants without the aid of such natural phenomena as drafts, buoyancy, inertia. Exterior hoods must create directional air currents adjacent to the suction opening in order to provide exhausting action. They are sensitive to external conditions, and even

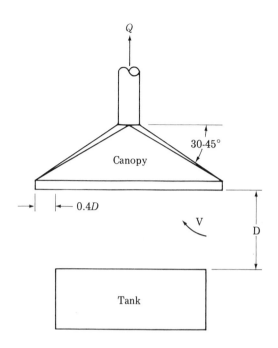

Q = 1.4 PDV

where Q = rate of air exhausted (cubic feet per minute)
       P = perimeter of source (feet)
       D = vertical distance between source and canopy (feet)
       V = required average air velocity through area between
             source and canopy (feet per minute)

FIGURE 10.2. Canopy hood.

a slight draft in the work area may make them entirely ineffective. They also require the most air to control a process.

Exterior hoods are used when the mechanical requirements of a process will not allow the obstruction that total or partial enclosure would entail. They include the numerous types of suction openings located next to sources of contamination that are not enclosed. Typical applications are shown in Table 28.

Fans

Fans are the mechanical air-moving devices that make most industrial exhaust systems operate. Although the industrial hygiene engineer is not generally concerned with the design of fans, it is important to understand enough of their theory and performance to be able to select, evaluate, and modify industrial exhaust fans.

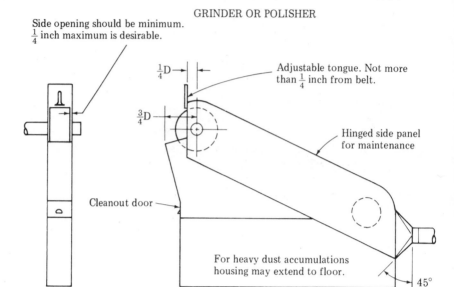

FIGURE 10.3. Grinding wheel hood.

Fans are usually divided into two groups—centrifugal and axial-flow. The distinction between the two is the direction of airflow produced. Air enters a centrifugal fan parallel to the fan shaft, and is discharged in a plane perpendicular to it. Axial-flow fans create a flow parallel to the fan shaft.

TABLE 28

Exterior Hoods and Their Applications

Type	Application
Slot	Open tanks
Push-pull	Plating tanks Cementing and lay-up tables
Down draft	Floor or bench-type welding, grinding, low-fog painting
Side draft	Some open-surface tanks Shakeout grates
Small canopy	Cool to warm processes
Wall fan (hood)	Some plastics operations Feed mills

Centrifugal Fans

These are commonly used for industrial supply and exhaust systems, pneumatic conveying, or forced draft applications. In these fans the air enters parallel to the fan shaft, turns 90°, passes between the blades of the fan wheel, and discharges at right angles to the inlet. The increase in velocity and static pressure is principally caused by centrifugal force. The spiral shape of the case or scroll determines the conversion of velocity pressure to static pressure. Centrifugal fans are classified as radial, forward curved, or backward curved, according to the shape of the blade.

Radial Blade or Paddle Wheel Fans. Paddle wheel fans were the first type developed, and are the only type that should be used for handling dirty air. Centrifugal force tends to keep the blades clean, permitting applications where suspended solids must pass through the fan. The rotors usually have between 5 and 12 blades, and the blades and housing are ruggedly fabricated. The pressure characteristic of this fan is stable, with only slight variations obtained. The horsepower required is almost a straight-line function of air volume. Maximum static efficiency is about 65 percent to 70 percent, lower than that of other types. Generally, the fan is used to handle air volumes up to about 100,000 cubic feet per minute at temperatures up to 1,000° F and at static pressures to 20 inches of water.

Forward-curved Blade Fans. These fans usually have a large number of closely spaced shallow blades. They are the standard "utility" or "squirrel cage" blower used for a number of general ventilation applications where erosion of the blades or the accumulation of solids is not a problem. The forward curve of the blade builds up a high velocity, which is partially converted to static pressure by the scroll. Usual operating static pressures in ventilating applications are from 2 to 4 inches of water, although forward-curved fans are available that will develop higher pressures. Maximum efficiencies are about 70 percent.

The horsepower requirement for this type of fan rises rapidly with increasing volume. Unless pressure requirements can be estimated accurately, the use of oversized motors is necessary.

Backward-curved Blade Fans. Sometimes called "power-limiting" or "non-overloading," these fans have a few flat, curved, or airfoil-shaped blades. The shape of the blades is such that the air tends to follow the blade even at high volumes. Consequently, the static pressure and efficiency are high at high volumes, and the horsepower decreases as the volume approaches wide open. This eliminates the necessity for oversized motors. Because of the backward curve of the blades, higher speeds are needed, and this necessitates more rugged construction. The noise will also be higher although this may not be a problem in industrial applications. The fan is a good choice for handling large volumes of clean air.

Airfoil Design. This unit is the latest development in centrifugal-type fans. It was developed to reduce noise levels; however, it is also backwardly inclined and has the nonoverloading feature. It too can deliver large volumes of air against high static pressures.

Axial-Flow Fans

These create airflow parallel to the air shaft. Within this group are low-pressure propeller and duct fans, and vane-axial fans that can develop higher pressures.

Propeller Fans. These fans consist of a simple propeller, usually with three or four blades mounted in a ring or cage. Since they have no way of confining the flow, conversion of velocity pressure to static pressure is minimal and usually they are not used against more than one inch of water static pressure. In addition, due to the waste of energy in the useless helical and unconverted radial motion of the air, they are inefficient. They are commonly used for general ventilation at low temperatures.

Tube-Axial Fans. If a propeller is placed in a duct, a part of the energy wasted in radial motion can be converted to velocity and static pressure. This tube-axial fan is able to operate against slightly higher resistances than a propeller fan but is still sensitive to variations in resistance. They are suitable for air containing materials that will not collect on fan blades.

Vane-Axial Fans. Tube-axial fans with guide vanes either ahead of or following the impeller are classed as vane-axial fans. These vanes straighten the helical flow pattern and convert velocity pressure into useful static pressure, increasing the efficiency. The impeller blades usually have an airfoil shape.

Vane-axial fans are best suited to clean-air applications. They are far more efficient and can operate at much higher pressures (up to 8 inches of static water pressure) with lower noise levels than propeller fans. Temperature applications up to 600°F are attainable in belt-driven units. Vane-axial fans are competitive with centrifugal fans for most industrial operations. At static pressures below about 4 inches of water they are quieter than centrifugal fans. At higher pressures they may be noisier, but are not necessarily so. Their efficiency may be as high or higher than that of centrifugal fans, and their compact design and simplicity of mounting and erection make them a good choice when conditions permit their use. In the very high-capacity range, vane-axial fans are superior to centrifugal fans, possibly except for the production of more noise.

Axial Centrifugal Fans. This is an axial fan with a centrifugal wheel. The wheel is available in both backwardly inclined and airfoil design. This fan is the latest improvement in axial/air-moving devices in that it combines the high efficiency of the axial unit with the quiet operation and high static-pressure level of the centrifugal fan. Although maximum operating temperatures are relatively low (under 200°F), static water pressures of up to 20 inches are attainable. In addition, the nonoverloading feature of the wheel is an important advantage.

Table 29 compares most of the fan types used in ventilation. Numerous exceptions will be found in individual cases, but this information serves as an indication of comparative characteristics.

TABLE 29

	Centrifugal Fans			Axial-Flow Fans	
	Backwardly Inclined	*Radial Blade*	*Forwardly Curved*	*Vane-Axial*	*Propeller and Tube-axial*
First Cost[1]	High[1]	Medium[1]	Low[1]	Low	Lowest
Efficiency	High	Medium	Medium	High	Lowest
Stability of Operation	Good	Good	Poor	Good	Poor
Space Required	Medium	Medium	Med-Small	Small	Small
Tip Speed (Noise)	High	Medium	Low	Medium	High
Resistance to Abrasion	Medium	Good	Poor	Medium	Medium
Ability to Handle Sticky Materials	Medium	Good	Poor	Medium	Medium

[1] On normal duty applications, first cost is essentially the same for all three types of centrifugal fans. Heavy-duty applications reflect the comparison listed.

Fan Selection

To select the proper fan for a given purpose, the following information is needed.*

1. Air volume to be moved.
2. Fan static pressure required.
3. Cleanliness of the air. A radial-bladed centrifugal fan will be needed if the air stream contains a high concentration of particulates.
4. Direct or belt drive. Belt drive can be changed for changes in air volume handled. Direct drive is inflexible but occupies less space and requires less maintenance.
5. Noise level. Noise level is a function of tip speed. It is not usually a limiting factor in industrial applications.

Industrial Ventilation: A Manual of Recommended Practice, 12th ed., American Conference of Governmental Industrial Hygienists, P.O. Box 453, Lansing, Michigan, 48902, 1972.

6. Special conditions such as high operating temperatures; corrosive, inflammable, or explosive materials; and unusual space limitations.

Fan Location and Maintenance

When air movers are used along with air-cleaning devices, good practice dictates that they be located downstream from the cleaner. This will minimize erosion, abrasion, and the building up of sticky deposits on the fan blades. The pressure drop through the air cleaner will be the same whether up or downstream of the fan. The fan should be located so that the inlet and exhaust ducts go through the minimum number of bends, and it should have an antivibration mounting. Another important consideration in the location of a fan is accessibility for inspection and servicing. However, all ducts inside the building should be on the inlet side of the fan under negative pressure to avoid leakage of contaminant into occupied spaces.

When a fan is installed it should be checked for proper mounting, grounding, and correct direction of rotation. The last is an easy error to make since a fan will move some air even when rotating backward. In addition to this initial inspection, a regular schedule of inspection should be established to include the following points:

1. Tightness of mounting bolts and ground connections.
2. Tension and condition of belts.
3. Lubrication of bearings and other parts.
4. Abrasion or accumulation of material on rotor.
5. Proper static pressure.
6. Vibration check.

Air Cleaning

Air cleaning is the operation of removing contaminants from an air or gas stream. The contaminants may be particulate matter or gases and vapors. This job can be simple and straightforward or complicated and difficult, depending on the number, nature, and concentration of the contaminants that must be removed, and the degree of removal required. Some air-cleaning mechanisms are effective on particular kinds of contaminants and less effective, or ineffective, on others. The selection of air-cleaning equipment to remove a given contaminant, or group of contaminants, must be based on an evaluation of the components of the air-stream to be cleaned, and on the efficiency of the various kinds of air-cleaning equipment for these components. Air-cleaning equipment is usually installed for one or more of the following reasons:

1. For compliance with an air pollution control ordinance.
2. For reclaiming or classifying product material.
3. To prevent a nuisance or damage to property.
4. To prevent re-entry of contaminants into workroom air.

Air Cleaning Equipment for Particulates

The choice of a particular type of air cleaner or collecting equipment depends upon several factors, including the amount of material to be collected, the nature of the material to be collected, and the efficiency of collection required. It must be kept in mind that particulates less than 10 microns in size are of hygienic significance and are more difficult to collect than particulates more than 10 microns, which create a nuisance problem only.

The basic type of collectors for particulates include settling chambers, inertial separators, fabric filters, and electronic precipitators. They are discussed below.

Settling Chambers. These devices, which are sometimes called chip traps, are enclosed spaces where the force of gravity is exploited to achieve separation. Retention is a function of the dimensions of the chamber and its flow rate.

Such a device is efficient only on particles larger than about 50 microns. If a series of horizontal trays is used instead of one large chamber, the device becomes a Howard Dust Chamber, which may remove particles as small as 10 microns. If a baffle is placed in the entrance of a settling chamber, it makes the chamber an inertial collector. Advantages include low initial cost and simple construction.

Inertial Separators. Inertial separators operate on the principle that when the path of an aerosol is changed, inertia tends to throw it from the airstream. There are two basic types of inertial separators— wet inertial collectors and cyclonic scrubbers. In wet inertial collectors, particulates impinge on wetted surfaces to facilitate removal. The spray admixes with the particulate material, agglomerates it, and causes it to coalesce to aid in its removal. Advantages of this type of collector are:

1. There is no secondary dust disposal problem.

2. It has the ability to clean high-temperature gases.
3. Moisture-laden gases cause no problem.
4. The efficiency curve drop-off indicates that about 5 micron minimum size particles are collected.
5. This collector can be used with sticky or linty materials.

In a cyclonic scrubber, contaminated gas enters the collector and is accelerated at high velocity where it impinges upon the liquid stream introduced at the throat of the Venturi. This results in the atomization of the liquid into a tremendous number of fine droplets. The high differential velocity between the gas and atomized droplets results in the impaction of the particles with the fine droplets. As the gas decelerates, further impaction and agglomeration of the droplets take place. After the particles have been trapped within the droplets, the resulting agglomerates are readily removed from the gas stream in the Cyclonic Separator. The scrubbed gas enters the Cyclonic Separator tangentially, spinning the liquid droplets against the wall, and leaves at the upper part of the unit. The particle-laden liquid drains by gravity from the bottom and may be recycled, sewered, or reclaimed.

The advantages of inertial separation include low initial cost and simple construction. Disadvantages include low efficiency and erosion due to impingement of particles.

Fabric Filters. These take advantage of inertial forces, diffusion, electrical charging, and straining. The effect of inertial forces and diffusion can be predicted with some accuracy from theoretical considerations. Electrical forces on the particles and/or the filter surface play an important part in filtration, but their effect in a given case is difficult to predict. Straining plays an insignificant role for dusts of hygienic significance.

The two most common types of commercial cloth filters are the bag type and the screen or envelope type. The difference between the two is in the manner of support for the filter. The sleeves or bags are usually contained in an enclosure for bag filters known as a "bag house." The bag house has an inlet for the dusty air, a manifold to distribute the air to the bags, an outlet, and a hopper to catch the collected dust. Most also contain some kind of shaking device to dislodge the collected dust when the critical resistance is reached.

The efficiency of the cloth filter increases with dust load since most of the filtering is done by the accumulated dust.

High-efficiency filters must be used for air-cleaning tasks where efficiencies by count greater than 99.5 percent are required. One type is a cellulose-asbestos paper. The asbestos does most of the filtering, and the cellulose gives the necessary support. Crushed flint, sand, and coke in deep beds make high-efficiency filters. Glass and resin wool fibers also are used for this purpose. High-efficiency filters are expensive and take up large amounts of space.

While high collecting efficiencies (about 99.9 percent) can be obtained with the use of certain types of filter media, it should be stressed that air containing toxic materials such as lead, beryllium, or radioactive materials should not be recirculated. Most states prohibit this practice because of possible failure in the collecting equipment.

Major problems may be high temperature, moisture, acidity, or alkalinity of the gas stream. In addition, both initial and maintenance costs are high and space requirements large. Many of these difficulties can be overcome by selecting the proper filter media or a combination of types.

Electronic Precipitators. When a high voltage is placed across two electrodes in air, an ionizing current will begin to flow between the electrodes. The stresses established against the molecules of air surrounding the ionizing electrode literally strip negative ions or electrons from the air molecule, and the ions migrate to the collecting (positive or grounded) electrode. Too great a voltage across the electrodes produces complete ionization or "arcing," but somewhere between the two extremes a condition known as "corona discharge" exists.

Industrial uses include the collection of fly ash, metallurgical fume, acid mists, cement dust, and other materials. Two types are in common use—the Cottrell and the two-stage.

Cottrell precipitators find use in metallurgical and chemical industries to collect oxide fumes and acid mists. Utilities use them to collect fly ash. Two-stage precipitators are used to remove oil and tobacco smoke and in collecting dusts and smokes from foundries. They have separate ionizing and collecting fields and are used for light dust loadings. Both types run on rectified alternating current: Cottrell at 25,000 volts and over, and the two-stage with 12,000 volts ionizing stage and 6,000 volts collecting. Cottrells have visible coronas and present a shock hazard.

Electrostatic precipitation is the most sophisticated, expensive, and yet the most efficient air-cleaning device known. Its high efficiency is maintained regardless of the size of particles in the gas stream. Furthermore, it can handle gases that are hot (up to about 1,500°F), cold, wet or dry. Maintenance and operating costs are low.

Some of the disadvantages of this method are: high initial cost; cannot be used for combustibles or in explosive atmospheres; a precleaner is often required to prevent overloading the precipitator; is not economical for low airflows; and large space requirements. Table 30 summarizes some of the important characteristics of collectors for particulates.

TABLE 30

Some Characteristics of Dust Collectors

Type of Collector	Particle Size Range Readily Collected (Microns)	Pressure Drop (Inches Water)
A. Settling Chambers		
1. Simple	40-100	0.1-0.5
2. Multiple Tray	10-50	0.1-0.5
B. Centrifugal (Dry)		
1. Baffle Chamber	20-50	1-3
2. Louvre	15-50	1-3
3. Single Cyclone	15-50	1-2
4. Multiple Cyclone (High efficiency)	5-20	3-6
5. Dynamic (Mechanical)	5-20	—
C. Scrubbers		
1. Spray Tower	5-10	0.5-1.5
2. Wet Centrifugal	2-5	2-6
3. Wet Dynamic	2-5	—
4. Wet Orifice	2-5	3-10
5. Venturi	<1-5	10-30
6. Packed Tower	3-10	1.5-4
(Trays of solid packing)		
7. Jet	1-5	—
D. Fabric Filter		
1. Tubular and Envelope	<1-5	2.5-6
2. Reverse Jet	<1-5	2.5-8
E. Electrostatic		
1. Single Stage	1-3	0.25-0.5
2. Two-Stage	1-3	0.25-0.5

Collection of Gases and Vapors

Removal of these from the airstream may be accomplished by absorption in liquids or solids, by adsorption on solids, and for vapors by condensation, and by catalytic combustion and incineration. In absorption the gas or vapor becomes distributed in the collecting liquid or solid. Adsorption is a surface phenomena. Both processes are diffusional operations. A theoretical discussion of these processes is beyond the scope of this book, and belongs in the study of chemical engineering. However, a brief discussion is given below.

Absorption. Equipment for absorption includes absorption towers, such as bubble-cap plate columns, sieve plate columns, packed towers, spray towers; washers, such as the wet-cell washers described for dust collection; and special devices such as Venturi injectors. Selection of equipment for a particular task is dependent on operational variables, such as vapor pressures, solubilities, and efficiency of removal required. Successful methods for the removal of sulfur dioxide, hydrogen sulfide, fluorides, and nitrogen oxides have been developed.

Adsorption. For the adsorption of gases and vapors, solid adsorbents have been developed with an affinity for certain substances. Various clays, charcoals, activated carbons, gels, aluminas, and silicates have been used. They are usually granular in form and are made up in beds or columns through which the gas passes. Adsorption may be practically complete even with very low vapor content, so the procedure is readily adaptable to solvent recovery operations. Adsorbents can collect from 8 percent to 25 percent of their weight in vapors. The vapors can be removed from the beds and the beds reused.

Charcoal, for example, will adsorb the following:

acetic acid	carbon disulfide
benzene	diethyl ether
ethyl alcohol	ammonia
carbon tetrachloride	hydrochloric acid
methyl alcohol	nitrous oxide
chloroform	carbon dioxide
acetone	acetaldehyde

The advantages of adsorbents include high efficiency and the fact that the materials adsorbed can be recovered. Disadvantages include high initial and operating costs, along with corrosion.

Catalytic Combustion and Incineration. In catalytic combustion, a platinum alloy-alumina catalyst is used to burn hydrocarbons, reducing them to less noxious compounds. Catalytic combustion is a low-temperature oxidation process by which many gases and vapors from industrial processing are converted to an odor-free, color-free gas. The catalyst simply provides an activated surface on which the reaction proceeds more readily than in its absence. Where large amounts of particulate are present, these unburnable solids must be removed by dust collectors prior to catalytic incineration. Minimum catalytic ignition temperatures may vary from 350°F to 600°F depending upon the character of the gas or vapor.

Many types of incinerators or boilers may be used to control emissions of noxious gases or vapors. Oil refineries usually "flare" hydrogen sulfide by burning it with other waste gases. In this case, toxic hydrogen sulfide is converted to less toxic sulfur dioxide. Where carbon monoxide is generated as a by-product in a process, it may be burned in waste heat boilers to convert it to carbon dioxide and thus gain some heat, usually in the form of process steam. Efficiencies of 98 percent or higher can be obtained in a well-designed incinerator.

Design of Ventilating System

The actual design of a ventilating system is highly complex and beyond the scope of this text. Persons wishing specific design information should consult specialized texts such as *Industrial Ventilation—A Manual of Recommended Practice*, American Conference of Governmental Hygienists, P.O. Box 453, Lansing, Michigan 48902 or *Fundamentals Governing the Design and Operation of Local Exhaust Systems Z9.2*, American National Standards Association, 1430 Broadway, New York, New York 10018.

MEASURING AIRFLOW AND QUANTITY

To determine the effectiveness of environmental control by any ventilation system, evaluations of airflow and air quantity are often necessary. In some instances, a qualitative determination may be sufficient to evaluate the effectiveness of control. In most cases, however, quantitative measurements are needed.

Airflow Measuring Devices

These may be divided into pressure-sensing instruments and thermal devices. The different types are discussed below.

Pressure-Sensing Instruments

These include Pitot tubes and vane anemometers.

Pitot Tube. The Pitot tube is the standard instrument for measuring the velocity of air. A standard Pitot tube, carefully made, will need no calibration. It consists of two concentric tubes—an impact tube whose opening faces axially into the flow, and the static pressure tube, a larger tube with circumferential openings. The difference between the impact and the static pressures is the velocity pressure. The Pitot measures pressure in the location in which it is placed. One measurement is usually insufficient to define the rate of airflow through a duct. However, if a circular duct is straight for at least 10 diameters upstream from the point of measurement, a single centerline Pitot reading may give a fairly accurate estimate of the flow rate.

The Pitot tube is used largely for measuring air velocity in ducts, or at high-velocity supply and exhaust openings. Its accuracy in the low-velocity range is limited by the sensitivity of the manometer or other pressure-sensing device used with it. Therefore, a vertical water manometer should not be used for velocities below about 2,000 lfm (linear feet/min.). Various accurate low-pressure measuring devices may be used with a Pitot tube, but their uses are usually limited to the laboratory, where they can be carefully set up and balanced.

Pitot tubes can be used under a wide range of conditions. There are no moving parts to get out of order, and if made of stainless steel, they are resistant to corrosion and high temperatures. Standard Pitot tubes cannot be used in dusty atmospheres because they will become plugged, but modified Pitot tubes have been made for use in such atmospheres. This type of Pitot tube should always be calibrated before use.

Deflecting Vane Anemometer (Velometer). This type of instrument is widely used for airflow and static-pressure measurement by industrial hygienists and heating and air-conditioning engineers.

Commercial velometers are available with one pressure and one, two, three, or four velocity ranges, with a variety of jets for each range. The velometer has the advantage of being a direct-reading instrument with ranges as low as 50 lfm or as high as 24,000 lfm, although the optimum range is 100 to 10,000 lfm.

The velometer operates by the pressure of an airstream against a spring-loaded swinging vane. It is fairly rugged, and if calibrated regularly the readings are sufficiently accurate for most field work. If the temperature of the airstream is outside the range $70^\circ F \pm 30^\circ$ or the altitude is greater than 1,000 feet, density corrections should be made. Dust or corrosive gases should not be permitted to enter the instrument. Velometers can be ordered with dust filters that will retain light dust loadings. If the instrument is calibrated with a dust filter in place, it must always be used in this manner.

To measure airflow or total pressure in a duct with a velometer, a hole is required in the duct large enough to accommodate the appropriate probe. Such a hole is larger than that required for a Pitot tube and may be difficult to provide.

Rotating Vane Anemometer. This instrument consists of a propeller connected through gears to a dial that counts rotations. The dial reads in linear feet, which when divided by the time interval of the measurement, gives linear feet per unit time. The instrument is available with 3-, 4-, or 6-inch wheels.

The rotating vane anemometer is useful for measuring air velocities in the range of 200 to 2,000 lfm. It is largely self-averaging for fluctuating flows and for traversing large openings. Readings are little affected by deviations in alignment with the direction of airflow up to about 20°. The instrument is delicate and should be calibrated frequently. It cannot be used in dusty or corrosive air. Since this type of instrument is rarely used in temperatures or pressures far from standard, pressure-temperature corrections are seldom needed.

Other Pressure-Actuated Anemometers. One type of commercially available anemometer is a direct reading instrument designed primarily for measuring air velocity at discharge grilles in the range of 30 to 2,700 lfm. It has a rotating impeller whose motion is opposed by a spring. When the velocity pressure on the impeller and the spring tension balance, the velocity can be read from the dial indicator. Another type of anemometer available is the cup anemometer used in meteorological work. It consists of four hemispherical cups mounted on light arms from a hub that turns freely on a vertical axis. This type of instrument has little application to ventilation work.

Thermal Devices

These include heated thermometer anemometers, heated thermocouple anemometers, hot-wire anemometers, and Kata thermometers.

Heated Thermometer Anemometer. The heated thermometer anemometer is based on the principle that the rate of heat loss from an object at elevated temperature is a function of air movement over the object. The instrument consists of a pair of thermometers, one with an electric resistance coil about the bulb to which a known voltage may be applied. The other is an ordinary thermometer. In use, the two thermometers are placed in the airstream to be measured, and the applied voltage adjusted to give a difference of 15° to 30° in the two thermometer readings. The thermometers are then allowed to come to equilibrium. Velocity of the airstream can be determined from a calibration chart using the temperature difference between the two thermometers, or by use of an equation.

The instrument indicates the cooling power rather than the velocity of an airstream. It is affected by radiant heat and convection. In the absence of radiant heat sources, it is very accurate as a velocity measuring device when carefully used. The instrument is nondirectional and has a range of 15 to 500 lfm. It is fragile and its use is time-consuming.

Heated Thermocouple Anemometer. This instrument works in the same manner as the heated thermometer anemometer, but uses thermocouples in place of thermometers. The advantages gained are a reduction in size, increased ruggedness of the sensing elements, faster response time, and it can be readily calibrated as a direct reading instrument. Several commercial instruments of this type are available, with varying airflow velocity and static-pressure use ranges. Most of these instruments use batteries that must be replaced periodically.

Hot-Wire Anemometer. A hot-wire anemometer consists of a fine, electrically heated wire that is placed in the airstream to be measured. The cooling effect on the wire will depend on the velocity of the airstream. As the wire cools, its resistance changes, and this change can be measured with a bridge circuit. If calibrated properly, this type of instrument can be used to measure a wide range of air velocities.

Kata Thermometer. The Kata thermometer is a special instrument with a large bulb containing alcohol and a stem with marks at 95 to 100°F. It is heated above 100°F, and the time required for it to cool from 100 to 95°F is a measure of the nondirectional air velocity in the room. It was designed for comfort ventilation measurement, and its surface-to-volume ratio is similar to that of the human body. The useful velocity range is 25 to 500 lfm. It has the disadvantage of being fragile and having large radiation and convection errors.

Air-Quantity Measuring Devices

These include orifice meters, flow meters, Venturi meters, rotameters, and thermal meters. The different types are discussed below.

Orifice Meter

This type of meter is simply a restriction in a pipe between two pressure taps. There are several types of orifice meters, but the simplest and most common is the square-edged orifice, which is a very short cylindrical passage in a thin metal plate. If it is properly constructed, the orifice plate will be at right angles to the flow, and the surface will be carefully smoothed to remove burrs and other irregularities. Since the square-edged orifice is easily constructed and the meter as a whole is simple and comparatively inexpensive, it is widely used as an accurate flow meter. Orifice meters are seldom used as permanent flow meters in ventilation systems because of their high permanent-pressure loss. They are more typically used in the ventilation laboratory for calibration purposes. Permanent head loss will vary from 40 to 90 percent of the static pressure drop across the orifice as the ratio of orifice diameter to pipe diameter varies from 0.8 to 0.3. Orifices have been intensively studied and their performance characteristics can be predicted if they are constructed to standard proportions.

Flow Meter

This term is often used to designate any restricted opening through which the rate of flow has been determined by calibration. One can be made by inserting a restriction in a pipe section, or for smaller flows by using a piece of glass capillary tubing as the restriction. A typical laboratory flow meter consists of a U-tube manometer with a capillary connected across its legs.

Venturi Meter

This consists of a $25°$ contraction to a throat, and a $7°$ re-expansion to the original size. It differs from the orifice meter, where the changes in cross section are sudden. The great advantage of the Venturi meter over the standard orifice is that the permanent reduction in static pressure is small, because the velocity head in the throat is largely reconverted to static pressure by the gradual re-enlargement. A well-designed and well-constructed Venturi will have a permanent static pressure loss of only 0.1 to 0.2 of the Venturi reading, as compared to 0.4 to 0.9 with a square-edged orifice.

Rotameter

This consists of a vertical transparent tube, increasing in cross-sectional area from bottom to top, in which an upward-flowing fluid stream supports a float at a level determined by the flow rate. The underlying principle of operation is the same as that of an orifice or Venturi meter. The float is the restriction, but instead of a variable pressure drop and a constant area of opening, there is a constant head (the weight of the float) and a variable area.

Rotameters are available for measuring a wide range of airflow rates, from as little as 10 cubic centimeters per minute of air, to as much as 350 cubic feet per minute. They are rarely used to measure ventilation airflow, but find wide use in laboratory calibration and metering.

Thermal Meter

This measures mass of air or gas flow with negligible pressure loss. It consists of a heating element in a duct section between two points at which the temperature of the air or gas stream is measured. The temperature difference between the two points is dependent on the mass rate of flow and the heat input.

OSHA STANDARDS FOR LOCAL VENTILATION

Regulations resulting from the Occupational Safety and Health Act of 1970 include several standards for local ventilation. In late 1972, such regulations included specific ventilation requirements for:

1. Abrasive blasting.
2. Grinding, polishing, and buffing.
3. Spray-finishing operations.
4. Open-surface tanks.
5. Welding, cutting, and brazing.
6. Gaseous hydrogen.
7. Oxygen.
8. Flammable and combustible liquids in storage rooms and enclosures.
9. Dip tanks containing flammable or combustible liquids.

These standards are based on the consensus-type standards developed by organizations such as the American National Standards Institute and the National Fire Protection Association. It is likely that the number and specifications of ventilation standards will increase with time.

EXERCISES: CHAPTER 10

1. Name the conditions under which dilution ventilation is successful.

2. Indicate the types of air contaminants that can and cannot be controlled successfully with dilution ventilation.

3. Name the components in a local exhaust ventilation system and discuss the advantages of this type of ventilation over dilution ventilation.

4. Indicate the design features that provide uniform airflow into a hood throughout its cross section.

5. Define the term "hood," name the four basic types, and list some typical applications for each.

6. Distinguish between centrifugal fans and axial-flow fans.

7. Describe the mode of operation, and the advantages and disadvantages, of each of the following types of particulate collectors:
   a. settling chambers
   b. inertial separators
   c. fabric filters
   d. electronic precipitators

8. Describe the mode of operation, and the advantages and disadvantages, of each of the following pressure-sensing devices for measuring airflow:
   a. Pitot tubes
   b. deflecting vane anemometers
   c. rotating vane anemometers

9. Describe the mode of operation, and the advantages and disadvantages, of each of the following thermal devices for measuring airflow:
   a. heated thermometer anemometers
   b. heated thermocouple anemometers
   c. hot-wire anemometers
   d. Kata thermometers

Appendix

PRE-EMPLOYMENT AND FOLLOW-UP
PHYSICAL EXAMINATIONS

When an employee's mental and physical attributes do not match the requirements of his or her job, stresses can arise. These so-called ergonomic stresses* can be prevented or combated by giving every new employee a pre-employment physical examination, matching the employee with a suitable job, and providing periodic follow-up examinations. Some of the factors involved in matching the employee with the job have already been touched on in the medical monitoring sections of various chapters. This section discusses pre-employment and follow-up examinations.

Pre-Employment Physical Examination

Every person being considered for employment by a business or industrial organization should be given a pre-employment physical examination. The same examination should also be administered to persons who are transferred to jobs that differ greatly from those previously held. An examination helps the employer to evaluate the risk of future health problems, assures safe and appropriate job placement, and establishes baseline information for further reference

*See J. B. Olishifski and F. E. McElroy, *Fundamentals of Industrial Hygiene* (Chicago, National Safety Council, 1971), pp. 301-42, for an excellent discussion of the general field of ergonomics.

so that changes in health can be properly assessed. The examination should be carried out by the company physician, if there is one. Small and medium-sized organizations without physicians should preferably send prospective employees to a local doctor rather than having them examined by a company nurse as is sometimes done. The local physician should, of course, be briefed thoroughly concerning the types of tests required. The examination routine outlined below has proved satisfactory for pre-employment physicals and is offered as suggested practice.

Whenever possible, the applicants should be examined singly rather than in groups. The heart and lungs should be checked first to avoid the stimulation caused by exercises carried out during other portions of the examination. Specific tests include recording the pulse, respiration rate, and blood pressure, and checking the heart and lungs with a stethoscope. If a question arises concerning cardiac capacity, the applicant should be asked to step up and down an 8-inch step for one minute, then pulse and blood pressure should be recorded immediately and after 3 minutes have passed.

Examination of the locomotor system involves having the applicant flex fingers, wrists, and elbows; raise arms above the head and lower them to the sides; do a deep-knee bend; flex the spine backward, forward, and laterally; rotate the neck and bend it forward and backward; and stand with eyes closed and feet together so the examiner can check for swaying. In addition to watching for improper functioning of the parts checked, the examiner should be alert for such conditions as finger tremor, muscular atrophy, missing members, deformities, and signs of recent serious injury.

For the first portion of the digestive system examination, the applicant is asked to assume a supine position on the examination table. The abdomen is palpated to detect enlargement of the liver or spleen, abdominal masses, or tender spots. Palpation is followed by a check for various types of hernia. If external inguinal rings are enlarged, having the applicant jump up and down on his or her heels 10 times will often bring down a hernia if one is present. If other types of hernias are suspected, having the applicant rise from a supine to a sitting position without using the hands may disclose them clearly. Once the foregoing checks have been completed, the applicant should be asked to stand, and the examiner should palpate the applicant's back for tender spots. The final portion of this phase consists of a rectal examination for external hemorrhoids, fissures, cysts, and sinuses. If there is rectal bleeding not caused by external hemorrhoids, the examiner should suggest that the applicant have his physician arrange for a proctoscopic examination. Following the rectal examination, the thighs, calves, and ankles of the applicant should be checked for varicose veins, ulcers, and other abnormalities, and the feet for fallen arches.

The hearing evaluation portion of the examination is best carried out by means of an audiometer (see Chapter 8). If an audiometric test cannot be performed, the whispered voice test, or a similar one, may be employed to detect pronounced defects in hearing acuity. The whispered voice test is carried out with the applicant standing 10 feet from the examiner and facing away from him or her. The applicant then plugs each ear in turn with a finger while the examiner asks a number of simple personal questions in a whispered voice. Following the check for hearing acuity, the ears should be examined with an otoscope for perforated eardrums or infections.

The eye examination should include a check for inflammation and pupillary reflexes, the nose exam should check for obstructions to breathing, the throat for inflamed tonsils and ulcers, and the teeth for decay. The absence of teeth should be noted, as should any infections of the gums or mouth.

The check for swaying that was part of the locomotor system examination and the pupillary reflex check made during the eye examination provide some evidence of the health of the applicant's nervous system. Additional evidence is obtained by carrying out a knee reflex text. If this test or either of the others is abnormal, further tests are indicated. As an example, the applicant may be asked to close his or her eyes, extend the arms, and bring the hands together so that the finger tips touch. The applicant should be checked for ataxia by walking a line with eyes closed. The examiner should also be alert for evidence of such other undesirable conditions, such as alcoholism, overaggressiveness, and psychoneurotic or psychotic reactions.

When considered for employment, persons who are greatly overweight, underweight, or overly tall should be restricted to work for which such a condition is not a handicap. A grossly obese person should be rejected until he or she has brought his or her weight down to acceptable standards or should be restricted to jobs not involving physical labor. Similarly, markedly underweight persons should be restricted from physical activities, such as heavy riveting, that would subject their bodies to continuous shock stress.

Under certain conditions, special tests may be necessary. For example, as discussed in Chapter 4, a person being considered for a job with lasers should have a complete blood count and urinalysis, and a detailed fundoscopic examination and tonometry by an ophthalmologist. Likewise, prospective operators of grinders and sand blasters will require chest X-rays; molders and finishers will require chest X-rays, red blood counts, and hemoglobin checks; and cafeteria workers will require chest X-rays and blood tests for syphilis.

In obtaining the applicant's personal medical history, the examiner should not pass over any of the questions asked on the medical application form. Where the applicant answers any question "yes," the examiner should require a further explanation and make the appropriate notations on the form. A detailed description of any abnormalities uncovered during the examination should be provided.

The procedure for female applicants is much the same as for men, except that no female applicant should be examined by a male physician without a female nurse in attendance. Depending upon the job being applied for, it is also often possible to omit an abdominal, hernial, or rectal examination.

The forms used by Dow Chemical for the pre-employment check of persons who will be working in laboratories and production plants are shown on the following pages. These forms are representative of those utilized in many other large industrial organizations.

Follow-Up Examinations

Although it is highly desirable that employees be given periodic complete physical examinations, this is impractical in many organizations because of the high cost and the demands on the plant physicians' time. As a result, where periodic examinations are given, they are usually confined to specialized groups of employees, such as those working with lasers or in microbiological laboratories.

Within the last decade, health screening has been adopted by an increasing number of companies as a practical means of worker health surveillance. Screening permits a large number of employees to be served at reasonable rates and without a greatly increased commitment of physician time.

The scope of the screening examination can be relatively narrow or very broad. A narrow examination checks only for those factors which experience indicates are likely to be affected by the particular employee's work exposure or exposures. This approach is deficient in that unforeseen job effects occasionally will be missed.

A much better approach is to provide the same broad basic examination for all individuals and to add specialized tests based on work exposure. This approach insures the early discovery of otherwise unforeseen effects. It also provides a better basis for future

FIGURE A.1

HEALTH HISTORY

☐ PRE-EMP ☐ RETIREMENT
☐ PERIODIC ☐ OTHER ☐
☐ PROFILE ☐

DOW CHEMICAL U.S.A.

Dow

DATE EXAM	DATE HIRED		

NAME

1.

	MASTER NUMBER	AGE	BIRTHDATE	SEX

ADDRESS

	BIRTHPLACE		CHURCH AFFILIATION

WORK APPLIED FOR

	BLDG. NO.	MARITAL STATUS

NOTIFY IN CASE OF AN EMERGENCY   LOCATION

☐ SINGLE ☐ MARRIED ☐ DIVORCED ☐ WIDOW(ER)

PHONE

DOW EMPLOYED RELATIVES

	FAMILY DOCTOR AND LOCATION

No. OF   (CIRCLE)
CHILDREN 0   1   2   3   4   5   6

2. PREVIOUS EMPLOYMENT HISTORY

NAME OF COMPANY	LOCATION OF WORK	TYPE OF WORK	DATE STARTED	DATE LEFT
				to
				to
				to
				to
				to
WHAT IS YOUR USUAL OCCUPATION	WHAT IS THE LONGEST PERIOD YOU HELD ANY JOB		THE SHORTEST PERIOD	to

3. FAMILY HISTORY

4. HAS ANY BLOOD RELATION (grandparents, parents, brothers, sisters, children, aunts, uncles, cousins) HAD:

RELATION	AGE	STATE OF HEALTH	IF DECEASED– CAUSE OF DEATH	AGE AT DEATH
FATHER				
MOTHER				
SPOUSE				
BROTHERS				
AND				
SISTERS				
CHILDREN				

CHECK EACH ITEM	YES	NO	RELATION
TUBERCULOSIS			
NERVE OR MUSCLE DISEASE			
DIABETES			
CANCER			
KIDNEY TROUBLE			
HEART TROUBLE			
BIRTH DEFECT			
RHEUMATISM (Arthritis)			
ASTHMA, HAY FEVER, HIVES			
EPILEPSY (Fits, Siezures)			
GLAUCOMA			
ANY MENTAL ILLNESS			

5. HAVE YOU EVER HAD OR HAVE YOU NOW (Please check each item)

ITEM		YES	NO	ITEM		YES	NO	ITEM		YES	NO	ITEM		YES	NO
RHEUMATIC FEVER	1			COUGH UP SPUTUM	19			SKIN TROUBLE	37			DEPRESSION OR EXCESSIVE WORRY	55		
SWOLLEN OR PAINFUL JOINTS	2			HEART MURMUR	20			VENEREAL DISEASE	38			LOSS OF MEMORY OR AMNESIA	56		
FREQUENT OR SEVERE HEADACHES	3			PALPITATION OR POUNDING HEART	21			RECENT GAIN OR LOSS OF WEIGHT	39			FEAR OF HIGH PLACES	57		
DIZZINESS OR FAINTING SPELLS	4			SWELLING OF ANKLES	22			ARTHRITIS OR RHEUMATISM	40			NERVOUS TROUBLE OF ANY SORT	58		
EYE TROUBLE	5			HIGH BLOOD PRESSURE	23			BONE, JOINT OR OTHER DEFORMITY	41			ANY DRUG OR NARCOTIC HABIT	59		
EAR, NOSE OR THROAT TROUBLE	6			CRAMPS IN YOUR LEGS	24			LAMENESS	42			USED ALCOHOLIC BEVERAGES	60		
RUNNING EARS	7			FREQUENT INDIGESTION	25			LOSS OF ARM, LEG, FINGER, TOE	43			OVER 2 DRINKS PER DAY	61		
CHRONIC OR FREQUENT COLDS	8			LOSS OF APPETITE	26			PAINFUL OR TRICK SHOULDER, ELBOW	44			EXCESSIVE DRINKING HABIT	62		
SEVERE TOOTH OR GUM TROUBLE	9			STOMACH, LIVER, INTESTINAL TROUBLE	27			BACK TROUBLE	45			ANY REACTION TO SERUM, DRUG, OR MEDICINE	63		
SINUSITIS	10			GALL BLADDER TROUBLE, GALLSTONES	28			"TRICK" OR LOCK KNEE	46			RIGHT HANDED	64		
HAY FEVER	11			JAUNDICE	29			FOOT TROUBLE	47			ANY OTHER PROBLEMS	65		
GOITER	12			TUMOR, GROWTH, CYST, CANCER	30			NEURITIS	48						
TUBERCULOSIS	13			RUPTURE	31			PARALYSIS (Inc. Infantile)	49						
SOAKING OR NIGHT SWEATS	14			BLEEDING FROM RECTUM	32			EPILEPSY OR FITS	50						
ASTHMA	15			PILES OR RECTAL DISEASE	33			CAR, TRAIN, SEA, AIR SICKNESS	51						
SHORTNESS OR BREATH	16			FREQUENT OR PAINFUL URINATION	34			FREQUENT TROUBLE SLEEPING	52						
PAIN OR PRESSURE IN CHEST	17			KIDNEY STONES-BLOODY URINE	35			LOSS OF SEXUAL POTENCY	53						
COUGH FOR MORE THAN 3 MOS.	18			DIABETES OR SUGAR IN URINE	36			FREQUENT NIGHTMARES	54						

FORM 10740 PRINTED R-2-73

323

HEALTH HISTORY PAGE 2

			SMOKING HISTORY		
6. HAVE YOU EVER (Please check each item)			**7.** ITEM	YES	NO

ITEM		YES	NO	ITEM		YES	NO	ITEM		YES	NO
WORN GLASSES	1.			ATTEMPTED SUICIDE	15.			DID YOU EVER SMOKE CIGARETTES	1.		
WORN AN ARTIFICIAL EYE	2.			BEEN A SLEEP WALKER	16.			DO YOU SMOKE CIGARETTES NOW	2.		
WORN HEARING AIDS	3.			LIVED WITH ANYONE WHO HAD TUBERCULOSIS	17.			HOW MANY PACKS PER DAY	3.		
STUTTERED OR STAMMERED	4.			COUGHED UP BLOOD	18.			HOW MANY YEARS (total) HAVE YOU SMOKED CIGARETTES	4.		
WORN A BRACE OR BACK SUPPORT	5.			BLED EXCESSIVELY AFTER INJURY OR TOOTH EXTRACTION	19.			IF YOU HAVE QUIT; HOW MANY YEARS AGO	5.		
WORKED WITH RADIOACTIVE SUBSTANCE	6.			APPLIED FOR OR COLLECTED WORKMEN'S COMPENSATION	20.			DO YOU SMOKE A PIPE OR CIGAR	6.		

Do you feel that exposure to plant chemicals has had
8. any effect on your healtn in the past year

☐ YES   ☐ NO

IF YES, WHY:

ITEM		YES	NO	ITEM		YES	NO
BEEN REFUSED EMPLOYMENT BECAUSE OF YOUR HEALTH	7.			BEEN DENIED LIFE INSURANCE	21.		
ENTERED LITIGATION OR CLAIMED DAMAGE BECAUSE OF INJURY TO OR EFFECT ON YOUR HEALTH	8.			HAD ANY ILLNESS OR INJURY OTHER THAN THOSE ALREADY NOTED IN THIS REPORT	22.		
BEEN A PATIENT IN A HOSPITAL	9.						

ITEM		YES	NO
HAVE YOU BEEN UNABLE TO HOLD A JOB BECAUSE OF:			
A. Allergy to chemicals, dust, sunlight, etc.	23.		

9. FEMALES ONLY (Please check each item)

ITEM		YES	NO
BEEN A PATIENT (either committed or voluntary) IN A MENTAL HOSPITAL OR SANITORIUM	10.		
HAD OR BEEN ADVISED TO HAVE ANY OPERATION	11.		
CONSULTED OR BEEN TREATED BY CLINICS, PHYSICIANS, HEALERS, OR OTHER PRACTITIONERS WITHIN THE PAST 5 YEARS	12.		
BEEN REJECTED FOR MILITARY SERVICE BECAUSE OF PHYSICAL, MENTAL, OR OTHER REASONS	13.		
BEEN DISCHARGED FROM MILITARY SERVICE BECAUSE OF PHYSICAL, MENTAL, OR OTHER REASONS OTHER THAN HONORABLE DISCHARGE	14.		

ITEM		YES	NO
B. Inability to perform certain motions	24.		
C. Inability to assume certain positions	25.		
D. Other medical reasons	26.		
ARE YOU TAKING ANY DRUGS OR MEDICATIONS	27.		
ARE YOU NOW OR HAVE YOU EVER TAKEN TRANQUILIZERS	28.		

ITEM		YES	NO
ARE YOU PREGNANT	1.		
HAVE YOU BEEN TREATED FOR FEMALE DISORDER	2.		
DO YOU HAVE PAINFUL MENSTRUATION	3.		
HAS ANY PREGNANCY NOT GONE TO TERMINATION	4.		
DO YOU NOW OR HAVE YOU EVER TAKEN ORAL CONTRACEPTIVE PILLS	5.		

10. I certify that I have reviewed the foregoing information supplied by me and that it is true to the best of my knowledge. I authorize any of the doctors, hospitals, or clinics mentioned above to furnish to The Dow Chemical Company a complete transcript of my medical records for purposes of processing my application for this employment.

SIGNATURE_____   DATE SIGNED _____

WITNESS_____

11. PHYSICIAN'S SUMMARY:

REPORT OF MEDICAL EXAMINATION
(FOR USE OF EXAMINING PHYSICIAN ONLY)

	CLINICAL EVALUATION	
NORMAL	ABNORMAL	CHECK EACH ITEM IN APPROPRIATE COLUMN. "NE" IF NOT EVALUATED

DESCRIBE EVERY ABNORMALITY IN DETAIL. ENTER PERTINENT ITEM NUMBER BEFORE EACH COMMENT. ATTACH ADDITIONAL PLAIN SHEETS IF NECESSARY.

NORMAL	ABNORMAL	ITEM
		12. HEAD, FACE, NECK, AND SCALP
		13. NOSE
		14. SINUSES
		15. MOUTH AND THROAT
		16. EARS–GENERAL
		17. DRUMS (Perforation)
		18. EYES–GENERAL
		19. OPHTHALMOSCOPIC
		20. PUPILS (Equality and reaction)
		21. OCULAR MOTILITY (Associated parallel movements, nystagmus.)
		22. LUNGS AND CHEST (Include breasts)
		23. HEART (Thrust, size, rhythum, sounds)
		24. VASCULAR SYSTEM (Varicosities, etc.)
		25. ABDOMEN AND VISCERA (Include hernia)
		26. ANUS AND RECTUM (Hemorrhoids, fistulae prostate if indicated)
		27. ENDOCRINE SYSTEM
		28. G.U. SYSTEM
		29. UPPER EXTREMITIES (Strength, range of motion)
		30. FEET
		31. LOWER EXTREMITIES
		32. SPINE, OTHER MUSCULOSKELETAL
		33. IDENTIFYING BODY MARKS, SCARS, TATTOOS
		34. SKIN, LYMPHATICS
		35. NEUROLOGIC
		36. PSYCHIATRIC (Specify any personality deviation)
		37. PELVIC

MEASUREMENTS AND OTHER FINDINGS

HEIGHT	WEIGHT	STANDARD WT. FOR HEIGHT	40. CHEST MEASUREMENTS	INSPIRATION	EXPIRATION
38.	39.				

41. ABDOMINAL MEASUREMENT	TEMPERATURE	42. BUILD	☐ SLENDER ☐ MEDIUM ☐ HEAVY ☐ OBESE

43. BLOOD PRESSURE Arm at Heart level	SYS. DIAS.	SYS. DIAS.	44. PULSE Arm at Heart level	SITTING	AFTER EXERCISE	2 MIN. AFTER	RECUMBENT

CODE
1. NORMAL OR NEGATIVE
2. BORDER LINE
3. ABNORMAL

48. VITAL CAPACITY				50. VISION	1. FAR	2. NEAR
EXPECTED FVC	OBSERVED FVC	FEV₁	TIME FOR FVC	BOTH		

FILM NO.

X-RAY
45. CHEST ☐1 ☐2 ☐3
46. LUMBAR ☐1 ☐2 ☐3

			ACUITY	RIGHT		
				LEFT		

47. E.K.G. ☐1 ☐2 ☐3

49. HEARING		DEPTH		
RIGHT	LEFT	COLOR		

54. URINALYSIS	SPECIFIC GRAVITY	MICROSCOPIC
	SUGAR	ALBUMIN

PSEUDO-ISOCHROMATIC 51. COLOR SCALE
VERHOEFF 52. STEREOPTOR

55. SEROLOGY	SPECIFY TEST USED AND RESULTS

53. INTRAOCULAR TENSION

RIGHT	LEFT

56. BLOOD TYPE AND Rh FACTOR	SPECIFY TEST USED AND RESULTS

57. OTHER TESTS

FIGURE A.3

PROFILE EVALUATION

Age_____
Ht. _____
Wt. _____

PROFILE FOR_____

JOB_____

P	E	S	R	E	P	X

P _____
E _____
S _____
R _____
E _____
P _____
X _____

_____ M.D.

DATE_____

Approved
By _____

58. PHYSICIAN'S SUMMARY: _____

_____

_____

_____

_____

_____

_____

_____

_____

_____

_____

_____

_____

_____

_____

_____

_____

_____

_____

_____

_____

_____

_____

_____

_____

_____

_____

_____

EXAMINING PHYSICIAN | DATE

59.   M.D.

documentation of the effect or lack of effect of work exposures to given agents and stresses. Finally, it makes possible the early discovery of certain personal illnesses, leading to early treatment and diminishing the chances for the untimely disablement or death of an employee.

A health inventory can be performed by two nurses, an audiometrician, and an X-ray technician, utilizing a specially designed trailer housing all the testing facilities. Such a group can carry out 20 to 25 inventories each day.

The following procedures should be performed during a broad inventory.

Interval Health History Plus Work History, Family Hisotry, and Smoking History (Pack-Years) Height and Weight	Blood Pressure and Pulse Rate Spirometry Sight Screen Skin Survey Audiometric Evaluation Skinfold Measurements to Determine True Obesity

Urinalysis

Protein Specific Gravity Ketone Sugar pH	Occult Blood Microscopic (if indicated by positive findings on basic urinalysis)

Hematology

Hemoglobin RBC Differential (if WBC is below 5,000 or above 10,000)	Hematocrit WBC Sedimentation Rate MCH, MCV, MCHC

Chemistry

Glucose (1 – 2 post prandial) Lipoprotein Phenotyping (in selected cases) Cholesterol	SGP-T, and SGOT in selected cases BUN LDH LDH Isoenzymes

Tonometry

X-ray

ECG

12 Lead

The findings should be recorded on a form such as that shown in Figure A.4.

A physician should review each inventory and carry out a complete examination at the central medical department on those whom testing has shown to have suspected abnormalities. The employee who is found to have a nonoccupational disease is referred

FIGURE A.4

MEDICAL EXAM DATE			
TRAILER EXAM DATE	[2] PERIODIC [6] HEALTH INVENTORY	**MEDICAL EXAMINATION**	◆DOW

NAME

	MASTER NO.	AGE	BIRTH DATE	SEX
1.				

NAME OF DEPARTMENT	LOCATION OF WORK	JOB CLASSIFICATION	START DATE-DEPT.	DATE LEFT DEPT.

SINCE YOUR LAST DOW PHYSICAL EXAM, have you had:

ITEM		YES	NO	ITEM		YES	NO	ITEM		YES	NO
FREQUENT OR SEVERE HEADACHES	03			FREQUENT INDIGESTION	14			EPILEPSY OR FITS	30		
DIZZINESS OR FAINTING SPELLS	04			STOMACH, LIVER, INTESTINAL TRBL.	15			DEPRESSION OR EXCESSIVE WORRY	31		
VISION OR HEARING DIFFICULTY				GALLBLADDER TROUBLE	16			NERVOUS TROUBLE OF ANY SORT	32		
EAR, NOSE OR THROAT TROUBLE	05			YELLOW JAUNDICE	17			EXCESSIVE DRINKING HABIT	34		
HAY FEVER	06			FREQUENT OR PAINFUL URINATION	19			REACTION TO SERUM OR DRUG	35		
ASTHMA	08			KIDNEY STONE, BLOODY URINE	20			ANY SERIOUS INJURIES OR ANY OPERATIONS			
SHORTNESS OF BREATH	09			SUGAR OR ALBUMIN IN URINE	21			ANY SERIOUS ILLNESSES			
PAIN OR PRESSURE IN CHEST	10			SKIN TROUBLE	22			CHANGE IN BOWEL MOVEMENTS			
CHRONIC COUGH	11			RECENT GAIN OR LOSS OF WEIGHT	23			DO YOU FEEL THAT EXPOSURE TO PLANT CHEMICALS HAS CONTRIBUTED TO ANY POOR HEALTH YOU MAY HAVE HAD THE PAST YEAR.			
COUGH UP BLOOD				ARTHRITIS OR RHEUMATISM							
SWELLING OF ANKLES	12			BACK TROUBLE	26						
HIGH OR LOW BLOOD PRESSURE	13			NEURITIS	28						

	DIABETES		CANCER		HEART TROUBLE		MENTAL ILLNESS	
4. HAVE YOU OR ANY BLOOD RELATION HAD ▶	[ ] YES [ ] NO		[ ] YES [ ] NO		[ ] YES [ ] NO		[ ] YES [ ] NO	

SMOKING HISTORY

		HOME ADDRESS:	
1. DID YOU EVER SMOKE?	[ ] YES [ ] NO	HOME PHONE:	PLANT PHONE:
2. HOW MANY YEARS?		FAMILY PHYSICIAN:	
3. DO YOU SMOKE NOW?	[ ] YES [ ] NO	ADDRESS:	
4. [ ] PIPE [ ] CIGARS [ ] CIGARETTES			
5. HOW MANY PER DAY?			

PHYSICIAN'S EXAM

DO NOT WRITE IN SPACE BELOW

NORM.	ABNORM.	✓ EACH ITEM OR WRITE "N.E." NOT EVAL.
		26. HEAD, FACE, NECK, AND SCALP
		27. NOSE
		28. SINUSES
		29. MOUTH AND THROAT
		30. EARS – GENERAL
		31. DRUMS (Perforation)
		32. EYES – GENERAL
		33. OPHTHALMOSCOPIC
		34. PUPILS (Equality and reaction)
		35. OCULAR MOTILITY (Associated parallel movements, nystagmus.)
		36. LUNGS AND CHEST (Include breasts)
		37. HEART (Thrust, size, rhythm, sounds)
		38. VASCULAR SYSTEM (Varicosities, etc.)
		39. ABDOMEN AND VISCERA (Include hernia)
		40. ANUS AND RECTUM (Hemorrhoids, fistulae, Prostate if indicated)
		41. ENDOCRINE SYSTEM
		42. G-U SYSTEM
		43. UPPER EXTREMITIES (Strength, range of motion)
		44. FEET
		45. LOWER EXTREMITIES
		46. SPINE, OTHER MUSCULOSKELETAL
		47. IDENTIFYING BODY MARKS, SCARS, TATOOS
		48. SKIN, LYMPHATICS
		49. NEUROLOGIC
		50. PSYCHIATRIC (Specify any personality deviation)

FORM 33272 PRINTED IN U.S.A. R10-68   SEE REVERSE   (FRONT)

FIGURE A.4 (*cont.*)

71. HEARING- I.S.O.			RIGHT					LEFT								70. VISION- ORTHORATER			1 FAR
500	1000	1500	2000	3000	4000	6000	8000	500	1000	1500	2000	3000	4000	6000	8000				

1 PHORIA	VERTICAL	
	LATERAL	

DOW NOISE HISTORY

INDEX	DEPARTMENT	JOB CLASSIFICATION	CODE	PROFILE	YEARS

2 ACUITY	BOTH	
	RIGHT	
	LEFT	

PRE-DOW EXPOSURE TO NOISE	YEARS
INDUSTRIAL	
MILITARY	
CIVILIAN GUN-FIRE	

| 3 | DEPTH |
| 4 | COLOR |

2 NEAR

66. X-RAY

1 ACUITY	BOTH	
	RIGHT	
	LEFT	

68. E.K.G.

2 PHORIA	VERTICAL	
	LATERAL	

72. INTRAOCULAR TENSION	MEDICATIONS BEING TAKEN

54. HEIGHT	55. WEIGHT	STD. FOR HEIGHT	60. VITAL CAPACITY	61. BLOOD PRESSURE	SYS.	DIAS.	SYS.	DIAS.	PULSE

CHEMISTRY	CODE	RESULTS		URINALYSIS	CODE	RESULTS		HEMATOLOGY	CODE	RESULTS	
BILIRUB: TOTAL	315		3	PROTEIN	710		5	HEMOGLOBIN	410		4
DIRECT	316		3	SUGAR	715		5	HEMATOCRIT	415		4
INDIRECT	317		3	MICRO:				RBC	420		9
CREATININE	370		3	LEUKOCYTES	720		5	WBC	425		6
GLUCOSE - P.P.	387		5	ERYTHROCYTES	721		5	DIFF. COUNT			
GLUCOSE - FAST	385		4	CASTS	722		5	NEUT.	430		4
PHOSPHATASE-ALK	341		3	EPITHELIALCEUS	723		5	LYMPH.	431		4
PROTEIN: ALBUMIN	350		3	SPECIFIC GRAVITY	750		2	MONO.	432		3
TOTAL	352		3	pH	755		3	STABS	433		3
S. G. P. T.	310		4	UROBILINOGEN, OIL	760		4	EOS	434		3
URIC ACID	325		3	UROBILINOGEN, COLOR	761		3	BAS	435		3
CHOLESTEROL	346		5	VOLUME	765		6	ATYP. LYMPH.	436		3
BUN	324		4	KETONE	800		5	JUVENILES	437		3
				OCCULT BLOOD	801		5				

* PREP DIET ☐ YES ☐ NO

75. PHYSICIAN'S SUMMARY

			YES	NO		FOLLOW UP REQUIRED	
EVIDENCE OF OCCUPATIONAL DISEASE							
POSITIVE FINDINGS OF OCCUPATIONAL INTEREST					☐ DOW	☐ FAMILY PHYSICIAN	

EXAMINING PHYSICIAN		DATE							
	M.D.			P	E	S	R	E	P	X

FORM 33272 PRINTED IN U.S.A. R10-68 (BACK)

329

to his or her personal physician. Follow-ups on occupational diseases are carried out by the industrial medical department. A report is made to each employee of the work completed.

A physician should review the examinations of personnel in each production department for any evidence of occupational disease and report the findings to the department head. The findings should also be correlated with those gathered by industrial hygienists on plant conditions and exposure levels. The findings may be processed for computer storage and retrieval.

NOTE ON INDUSTRIAL MEDICAL FACILITIES

Despite an extremely close link between the disciplines of industrial hygiene and industrial medicine, an extended discussion of the latter is beyond the scope of this book. Readers wishing a concise yet reasonably detailed overview of industrial health should consult J. S. Felton, *Organization and Operation of an Occupational Health Program*, Occupational Health Institute, Industrial Medical Association, Chicago, 1964.

This 72-page publication is divided into three main sections. The first section begins by defining occupational medicine, occupational physician, occupational disease, occupational injury, and occupational health program. These definitions are followed by an outline of the objectives of an industrial health program, and this in turn by discussions of what is involved in a complete in-plant medical program, a partial in-plant medical program, and an out-plant medical program. Following a brief consideration of the position of the medical department in the company organization, the section takes up in detail the functions and activities of the medical group. These include services to the work applicant, the healthy employee, and the sick or injured employee; record maintenance, budget development; occupational nursing services; and illness-absenteeism control. Program costs, the relationship between the occupational physician and the private practitioner, medical department manuals, small plant occupational medicine, and contractual agreements are the next topics of concern. The section ends with a list of organizations concerned with occupational medicine and an extensive bibliography.

The second section is concerned exclusively with the design of dispensary facilities. Site, size, configuration, component areas, and general design criteria are specifically discussed. Perhaps the most interesting and valuable feature of the section is the many floor plans presented. These include plans for small, medium, and large dispensaries, as well as a series of complete medical facility plans, each specifically tailored for a particular size plant and together encompassing the entire industrial size spectrum, from plants employing a few hundred persons to those employing as many as 15,000.

The final section, which has to do with special programs in industrial medicine, considers alcoholism, medical disaster preparedness, programs at airports, and programs in the construction industry and in hospitals.

A briefer write-up of similar but somewhat narrower scope is found in F. E. McElroy et al., *Accident Prevention Manual for Industrial Operations*, 6th ed., Chicago, National Safety Council, 1969, pp. 1222-42. In addition, the American Medical Association, Council on Occupational Health, has issued a number of brief publications dealing with particular aspects of industrial medicine, as have various insurance companies and industrial associations.

Where more detailed information is required, textbooks such as W. P. Shepard, *The Physician in Industry*, New York, McGraw-Hill Book Co., 1961, and H. L. Herschensohn, *Medical Forms and Procedures in Industry*, Springfield, Mass., Thomas Publishing Co., 1961, can be consulted. In addition, dozens of detailed articles are published annually in the various journals devoted to industrial safety, hygiene, and medicine. A listing of the more important ones can be found on pages 370 and 371 of the McElroy book cited above.

BASIC PRINCIPLES FOR TREATING
EXPOSURE TO HAZARDOUS CHEMICALS

This information is presented because of the general lack of knowledge, even among physicians, of the proper methods for dealing with exposure to hazardous chemicals. The suggestions are based on general principles. The specific approach will depend on the situation, degree of knowledge of the treating personnel, and many other factors. While there are several approaches, the one chosen here is the one that has proven best in Dow Chemical's experience. Although much of this material is elementary, a continuous review is desirable to assure prompt and proper action in an emergency. Anyone who may deal with this type of emergency must remember that, in many cases, immediate action is essential and even the slightest delay can result in a significantly more serious injury. Remember, a secondary material labeled as "solvent," "inert ingredient," or the like may be more hazardous than the "active" constituent. Such terminology is frequently based on intended use, not on biological activity.

Training

Persons who may have to help patients who have suffered chemical exposure often will have to deal with other problems. They should be trained and retrained at intervals in all first-aid procedures, in addition to those specific to chemical exposures.

Protection and Decontamination

No one should work with hazardous chemicals without proper instruction or in the absence of proper protective equipment and adequate decontamination facilities. Eye showers and overhead showers should be readily at hand and periodically checked for proper function. Every employee should be well acquainted with the locations of all eye showers, showers, protective equipment, and rescue supplies.

Removal from Exposure

Remove the patient to an uncontaminated area. This applies in all instances of exposure to a chemical, but is particularly important when there is continuing exposure to an atmospheric contaminant.

The rescuer must protect himself or herself. If the rescuer is disabled, he or she can be of little help. Use protective equipment to avoid self-injury by inhalation or skin contamination.

Eye Contamination

At the Site of Injury

If the eye is contaminated, irrigate it with water immediately and thoroughly. Time is of the essence. Seconds can make a difference. Continue irrigation for at least 15 minutes. Hold the eyelids open with the hands and move the eyes around to make sure all areas under the eyelid are irrigated. If water is not immediately at hand, use any bland aqueous fluid, such as milk, soda, or cool coffee.

At a Medical Facility

After the initial irrigation, the patient may be transferred to a medical facility where additional irrigation may be carried out. The additional irrigation will be facilitated by the use of a topical ophthalmic anesthetic. The irrigation should cover all areas of the eye, including those under the eyelids.

Long experience indicates that clean tap water is safe for eye irrigation and that the use of special preparations or "antidotes" for irrigation can be hazardous. The use of any substance other than water is not recommended, and such substance should be employed only when specifically ordered by the responsible physician. Examination always should be done with adequate illumination and magnification. The duration of irrigation at a medical facility will depend on the particular potential for eye injury of the involved chemical.

The extent of injury can often best be assessed after a fluorescein stain. If the cornea takes a stain, the patient must be seen by a physician qualified to judge the extent of injury. In certain severe injuries where corneal epithelium is completely lost, the stain will not be an adequate guide in assessing the extent or depth of injury.

Commonly, after irrigation is completed, a corticoid preparation will be prescribed by a physician.

Skin Contamination

If the skin is contaminated, shower immediately and thoroughly. If no medical facility is at hand, showering should be continued for one-half hour with abundant soap and water. If a medical facility is readily at hand, a common practice is to shower at the site of injury for 15 minutes and at the medical facility for an additional one-half hour. If a shower is not available, low-pressure hoses, pails, or any other method of thoroughly washing with flowing water should be used.

Clothing should be removed while the patient is showering. This is extremely important. Contaminated clothing should be decontaminated or discarded.

Where burns have occurred or the chemical has the potential for producing systemic effects, the patient should be seen by a physician. In certain instances, the showering cannot be continued because of other injuries to the patient or systemic effects. In this case, provisions must be made for decontamination by other means (such as using a low-pressure hose or spray) while necessary treatment is given.

The use of substances other than tap water for decontamination is to be discouraged. Experience indicates this can occasionally result in tragedy. Even if the contaminating chemical has a low solubility in water, large quantities of readily available water make up for this deficiency. After the victim's arrival at a medical facility, other substances may be used for decontamination or treatment only under guidance of the responsible medical attendant.

Ingested Chemicals

If a chemical is taken by mouth, *and the patient is conscious*, give copious bland fluid suitable for drinking, such as water or milk, and then induce vomiting *unless the chemical is a petroleum distillate or a corrosive.*

In all instances of ingested poisons, a physician should be contacted as soon as possible.

Never try to give fluids to an unconscious patient. Never cause vomiting in an unconscious patient. Never cause vomiting in a patient who has taken petroleum distillates or corrosive poisons unless specifically directed to do so by a physician.

Reports indicate that giving a slurry of activated charcoal, *freshly prepared*, can be of great help. It should be given immediately. Its effectiveness varies with different poisons but it does no harm. Dosage of activated charcoal for adults and teen-agers is 50 grams in about a cup of water. For children, 20 grams in a half cup of water may be sufficient. *Caution*—After giving activated charcoal, syrup of ipecac may be completely ineffective, and apomorphine should be given parenterally (intramuscularly) instead.

Syrup of ipecac is not as effective as apomorphine and may be ineffective, especially if the patient has previously been given activated charcoal. The recommended dosage of apomorphine is 0.02 milligrams per pound of body weight. (Since apomorphine is a narcotic, an antagonist should be kept on hand in case of an unexpectedly severe depressant reaction.)

Vomiting may be induced by tickling the back of the throat with a finger. Giving 2 tablespoons of salt in a glass of water may help to induce vomiting. Emetics have been reported to be more effective than gastric lavage.

In certain instances where, as an example, a potent poison dissolved in a petroleum distillate has been taken by mouth, the physician may wish to perform gastric lavage. This can be accomplished more safely by first placing and properly inflating a cuffed endotracheal tube. Before the cuff is deflated, thorough irrigation and suctioning should be performed.

Artificial Breathing

If breathing has stopped, start artificial respiration. Mouth-to-mouth breathing is generally easiest and most effective. Oxygen is of use only when the patient is breathing.

Make sure the patient has no airway obstruction.

If the patient has swallowed a potent poison, mouth-to-mouth breathing may be a hazard to the rescuer, and alternative manual methods should be considered.

Once artificial respiration has been started, do not discontinue until a physician has ordered it discontinued or the patient is breathing adequately on his or her own.

Mechanical resuscitators are to be discouraged unless the individual using such a device is continuously retrained in its use. Except with simple devices such as the hand-pressure bag and mask, much time can be lost trying to remember how to operate these devices. Remember—the more complicated the device, the more likely it is to fail at the crucial moment.

Cardiac Resuscitation

If the heart has stopped, give closed chest cardiac resuscitation, but only if trained to recognize the signs of such an occurrence and in the proper procedure.

Medical Help

Get medical help as soon as possible. In general, it is best to get the patient to an adequately equipped facility. Even well-trained personnel can be limited in their ability to help if adequate equipment is not available.

An unconscious patient should be transported on his or her side to diminish the chance of aspirating vomited material. Occasionally, other injuries or the requirement of resuscitative measures contraindicate such positioning.

Delegate someone to convey all known information on the occurrence to the personnel at the medical facility.

THE OCCUPATIONAL SAFETY AND HEALTH ACT: A SUMMARY

The stipulations of the Occupational Safety and Health Act (OSHA) of 1970 that pertain to the specific industrial stresses have been discussed in preceding chapters of this book. The following discussion summarizes the general impact of the law on employers. The law is designed to assure safe and healthful working conditions for nearly every man or woman in the United States. In writing this legislation, Congress relied for guidance on numerous earlier laws and regulations, such as state laws; special laws enacted for the coal mining, construction, and maritime industries; and the regulations applying to organizations fulfilling government contracts. The scope of OSHA, however, far exceeds that of any other law. Coverage includes some 57 million workers in 4.1 million business establishments, and its impact will be felt from the gigantic industrial complexes to the small garage operator cleaning dirty car parts in a bucket of solvents.

Obligations of Employers

As defined by OSHA, an employer is "a person engaged in a business affecting commerce who has employees. . ." The definition, however, excludes the United States, individual states, and the political subdivisions of a state. *Any business whose raw materials, power, or communications cross state lines is subject to the act, even though its products are not sold out-of-state.*

The duties of the employer are numerous. First of all, he or she must "furnish to each of his employees employment and a place of employment that are free from recognized hazards that are likely to cause death or serious physical harm to his employees" and "comply with occupational and health standards promulgated under the act."

Each employer must keep a log and a supplementary record of all recordable job injuries and illnesses, as well as compile an annual summary of illnesses and injuries at the end of each calendar year.

Three types of injuries are recorded in the log: fatalities, non fatal lost-time mishaps, and no-lost-time mishaps. An injury or illness causing death is regarded as a fatality regardless of the time that elapses before death occurs. Fatalities must be reported within 48 hours to the nearest OSHA area office. Lost-time mishaps take in accidents and illnesses that cause lost workdays. In addition, they include cases in which workers are transferred to temporary jobs or remain at their present jobs but work less than full time or have reduced duties. Any nonfatal accident in which five or more workers are hospitalized must be reported to the OSHA area office within 48 hours. No-lost-time mishaps are those which involve loss of consciousness, transfer to another job, restriction of duties, or medical treatment beyond first aid, but do not result in lost workdays.

The supplementary record of injuries and illnesses must list the employer's name and mailing address; the name, social security number, home address, sex, occupation, and department of the employee; the place of the accident and exposure and what the employee was doing when it occurred; a description of the illness or injury, and the name and address of the treating physician or hospital. Each accident must be recorded within six days.

An annual summary must be prepared for each location of the employer. This summary must be posted by February 1 and remain posted for 30 days for the employees' inspection.

OSHA has available forms for keeping these three types of records. The forms are shown on the following pages. Detailed record-keeping instructions may be obtained from OSHA.

Besides the foregoing records, the employer may be required to keep records of worker exposure to substances monitored under the law as potentially toxic or harmful. To protect the health of employees exposed to toxic substances or potentially harmful physical agents, employers must provide physical examinations to determine if exposure has exceeded permissible limits. If standards require, employers must conduct periodic inspections of safety and health hazards and provide employees with protective equipment. Finally, the employer is obliged to inform the worker of any hazards or potential hazards that exist in his or her environment and to post notices informing workers of their rights and obligations under the act.

Obligations and Rights of Employees

OSHA imposes duties on employees as well as employers. According to the act:

> Each employee shall comply with OSHA standards and all rules, regulations, and orders issued pursuant to this act which are applicable to his own actions and conduct.

Along with this obligation, certain rights are conferred on employees. Thus, an employee or his or her representative may request an inspection of an establishment if he or she feels that a violation of an OSHA standard exists or there is "imminent danger" of one. If an inspection is made, an employee or employee representative must be given a chance to accompany and assist the inspector. If no violation is found, the inspector is required to notify the employee or the representative. No employee can be discriminated against for making an inspection request.

Enforcing Agency

The Department of Labor is charged with the responsibility for enforcing OSHA. To expedite this task, the Occupational Safety and Health Administration has been created. This group, headed by an Assistant Secretary of Labor, should number 2,000 at full strength.

Inspections

Inspections may be carried out by the Labor Department upon its own initiative or at the request of an employee or representative. In the latter case, the request must be in writing and state "with reasonable particularity" the grounds for the request. The following priorities have been established for conducting investigations:

Investigation of deaths and catastrophes.
Investigation of employee complaints.
Investigation of target industries.
Other industries chosen at random.

Target industries, which receive special inspection priority because of their high injury frequency rates, include the roofing and sheet metal, longshoring, meat and meat products, mobile home, transportation equipment, and lumber and wood products industries.

Inspectors, called compliance officers, have the right to enter an establishment without delay and at reasonable times. Ordinarily, inspections will be conducted during working hours, but "after hours" inspections are permitted provided they do not violate the "reasonable times" criteria. Ordinarily, no advance notice is given of inspections. However, in cases of employee complaints, the OSHA area director may notify the establishment 24 hours before the inspection.

Upon arriving for an inspection, the compliance official presents his or her credentials to the "highest official of the company available," explains the general nature and purpose of the visit, and names the employees, if any, that he or she wishes to question. In the event of a special inspection, the officer provides the official with a copy of the request for inspection, with the employee's name omitted.

A representative of the employer and a representative authorized by the employees may accompany the compliance officer during any inspection. Each is subject to approval by the officer. The employee representative may or may not be an employee, depending upon the attitude of the union. Where employees are represented by a number of unions, there may be a different representative for

FIGURE A.5

LOG OF OCCUPATIONAL INJURIES AND ILLNESSES

Form Approved
OMB NUMBER 44R 1453

Case or file no.	Date of injury or initial diagnosis of illness. If diagnosis of illness was made after first day of absence enter first day of absence. (mo./day/yr.)	Employee's Name (First name, middle initial, last name)	Occupation of injured employee at time of injury or illness	Department to which employee was assigned at time of injury or illness	DESCRIPTION OF INJURY OR ILLNESS		EXTENT OF AND OUTCOME OF INJURY OR ILLNESS					
					Nature of injury or illness and part(s) of body affected (Typical entries for this column might be: Amputation of 1st joint right forefinger; Strain of lower back; Contact dermatitis on both hands; Electrocution—body)	Injury or illness code See codes at bottom of page.	Fatalities	Lost Workday Cases		Nonfatal Cases Without Lost Workdays		
							Enter date of death (mo./day/yr.)	Enter workdays lost due to injury or illness (see instructions on back.)	If, after lost workdays, the employee was permanently transferred to another job or was terminated, enter a check in the column below	If no entry was made in columns 8 or 9, but the injury or illness did result in: Transfer to another job or termination, or; medical treatment, other than first aid, or; diagnosis of occupational illness, or; loss of consciousness, or; restriction of work or motion; Enter a check in the column below	If a check in columns 8 or 9, but the injury or illness represented a transfer or termination, enter another check in column 12
1	2	3	4	5	6	7	8	9	10	11	12

Company Name _____

Establishment Name _____

Establishment Location _____

Injury Code
10 All occupational injuries

Illness Codes

21 Occupational skin diseases or disorders
22 Dust diseases of the lungs (pneumoconioses)
23 Respiratory conditions due to toxic agents
24 Poisoning (Systemic effects of toxic materials)
25 Disorders due to physical agents (other than toxic materials)
26 Disorders due to repeated trauma
29 All other occupational illnesses

FIGURE A.6

OSHA No. 101
Case or File No. _____

Form approved
OMB No. 44R 1453

Supplementary Record of Occupational Injuries and Illnesses

EMPLOYER

1. Name _____

2. Mail address _____
   (No. and street)   (City or town)   (State)

3. Location, if different from mail address _____

INJURED OR ILL EMPLOYEE

4. Name _____ Social Security No. _____
   (First name)   (Middle name)   (Last name)

5. Home address _____
   (No. and street)   (City or town)   (State)

6. Age _____   7. Sex: Male_____ Female_____ (Check one)

8. Occupation _____
   (Enter regular job title, *not* the specific activity he was performing at time of injury.)

9. Department _____
   (Enter name of department or division in which the injured person is regularly employed, even
   though he may have been temporarily working in another department at the time of injury.)

THE ACCIDENT OR EXPOSURE TO OCCUPATIONAL ILLNESS

10. Place of accident or exposure _____
    (No. and street)   (City or town)   (State)
    If accident or exposure occurred on employer's premises, give address of plant or establishment in which
    it occurred. Do not indicate department or division within the plant or establishment. If accident oc-
    curred outside employer's premises at an identifiable address, give that address. If it occurred on a pub-
    lic highway or at any other place which cannot be identified by number and street, please provide place
    references locating the place of injury as accurately as possible.

11. Was place of accident or exposure on employer's premises? _____ (Yes or No)

12. What was the employee doing when injured? _____
    (Be specific. If he was using tools or equipment or handling material,
    _____
    name them and tell what he was doing with them.)

13. How did the accident occur? _____
    (Describe fully the events which resulted in the injury or occupational illness. Tell what
    _____
    happened and how it happened. Name any objects or substances involved and tell how they were involved. Give
    _____
    full details on all factors which led or contributed to the accident. Use separate sheet for additional space.)

OCCUPATIONAL INJURY OR OCCUPATIONAL ILLNESS

14. Describe the injury or illness in detail and indicate the part of body affected. _____
    (e.g.: amputation of right index finger
    _____
    at second joint; fracture of ribs; lead poisoning; dermatitis of left hand, etc.)

15. Name the object or substance which directly injured the employee. (For example, the machine or thing
    he struck against or which struck him; the vapor or poison he inhaled or swallowed; the chemical or ra-
    diation which irritated his skin; or in cases of strains, hernias, etc., the thing he was lifting, pulling, etc.)
    _____
    _____

16. Date of injury or initial diagnosis of occupational illness _____
    (Date)

17. Did employee die? _____ (Yes or No)

OTHER

18. Name and address of physician _____

19. If hospitalized, name and address of hospital _____
    _____

    Date of report _____ Prepared by _____
    Official position _____

FIGURE A.7

Summary: Occupational Injuries and Illnesses

Establishment Name and Address:

Code 1	Category 2	Fatalities 3	Lost Workday Cases				Nonfatal Cases without Lost Workdays*	
			Number of Cases 4	Number of Cases Involving Permanent Transfer to Another Job or Termination of Employment 5	Number of Lost Workdays 6		Number of Cases 7	Number of Cases Involving Transfer to Another Job or Termination of Employment 8
	Injury and Illness Category							
10	Occupational Injuries							
	Occupational Illnesses							
21	Occupational skin diseases or disorders							
22	Dust diseases of the lungs (pneumoconioses)							
23	Respiratory conditions due to toxic agents							
24	Poisioning (systemic effects of toxic materials)							
25	Disorders due to physical agents (other than toxic materials)							
26	Disorders due to repeated trauma							
29	All other occupational illnesses							
	Total—occupational illnesses (21-29)							
	Total—occupational injuries and illnesses							

*Nonfatal Cases without Lost Workdays—Cases resulting in: Medical treatment beyond first aid, diagnosis of occupational illness, loss of consciousness, restriction of work or motion, or transfer to another job (without lost workdays).

FIGURE A.7

different phases of the inspection. During an inspection, any employee must be afforded a reasonable opportunity to consult with the compliance officer in private. The officer may also talk with a reasonable number of employees concerning health and safety at their work places.

Regardless of the reason for the inspection, compliance officers may examine any conditions, structures, machines, apparatus, devices, and equipment. They may also review any records required under OSHA as well as other pertinent records relating to occupational safety and health. In addition, they have the authority to take photographs and to employ other reasonable investigative techniques. Information that contains or might reveal a trade secret must be treated as confidential, and the employer representative has the right to ask the compliance officer for written acknowledgement that the latter has received confidential information.

A little advance planning will enable an employer to handle inspections expeditiously and minimize the likelihood of violations. First, appropriate personnel should be assigned to become knowledgeable about OSHA, to keep up-to-date on new developments, and to suggest actions to insure compliance. As part of its safety program, each location or department should formulate a policy outlining responsibilities for compliance. Procedures should be established for insuring that the proper records are kept and other duties carried out. Receptionists and internal security personnel should be briefed on how to deal with the compliance officer when he or she arrives. This briefing should include the names of alternates for the company official to whom the compliance officer would ordinarily be sent. Needless to say, any company official involved in any phase of an inspection should be familiar with the regulations on inspections and on the act itself.

Citations for Violations

After completing an inspection, the compliance officer discusses with the plant management representative any violations that have been uncovered, and asks the representative what, in his or her opinion, would be a reasonable time in which to correct the violation. He or she then consults with the area director of the Occupational Safety and Health Administration, who sets abatement deadlines, issues citations, or imposes penalties, as the case warrants. The establishment is notified of the decision by registered mail. Citations must be issued within 6 months of the alleged violation. The employer then has 15 days to contest any penalty.

The law recognizes three different types of violations:

Imminent danger: a situation "which could reasonably be expected to cause death or serious physical harm immediately before the . . . danger can be eliminated."

Serious: a situation which could result in severe harm or death.

Other types: a situation which could result in less severe injury.

If the compliance officer discovers that an imminent danger situation exists, he or she will ask management to correct it before leaving the premises. If this can be done, the case is closed. If it cannot, the compliance officer will call the area director, with his or her permission post a notice telling employees of the situation, and return to the local office for legal consultation. Where such action is deemed desirable, the Secretary of Labor can seek an injunction in the United States District Court to restrain the violation. The court can prohibit employees from entering the premises where the imminent danger exists.

Penalties for Violations

A wide range of penalties has been established for employers cited for violating standards, rules, orders, or regulations issued under OSHA. Some of the penalties are discretionary while others are mandatory.

Discretionary penalties include:

A civil penalty of up to $1,000 for each nonserious violation.

A civil penalty of up to $1,000 per day for failure to correct a violation within the set deadline.

A civil penalty of not more than $10,000 for each violation when the employer has willfully or repeatedly violated standards under the act.

Mandatory penalties include:

A civil penalty of up to $1,000 for each violation of posting requirements.

A civil penalty of up to $1,000 for each serious violation.

A $10,000 fine and/or imprisonment for not more than six months upon conviction for any willful violation that causes death to an employee. If, however, one or more previous convictions have occurred, the penalty is increased to a $20,000 fine and/or one-year imprisonment.

A $10,000 fine and/or imprisonment for not more than one year upon conviction for knowingly issuing any false statement or falsifying any document required under OSHA.

Appealing an Alleged Violation

An employer who wishes to contest an adverse finding may do so by notifying the Secretary of Labor of his or her intention. The notification must be in writing and filed within 15 working days of the time when notice of the citation or penalty was received. The secretary then notifies the Occupational Safety and Health Review Commission, a three-person committee appointed by the president, and this commission appoints a hearing examiner to hear the dispute and make a ruling. If the examiner's ruling is unfavorable, the employer has 30 days to ask the full commission for a review. Such a review is granted if one commission member feels it to be in order. Following the review, the commission may uphold or modify the hearing examiner's decision. Orders of the commission can be appealed to the United States Court of Appeals for review.

Keeping Abreast of Changes in Standards

On May 29, 1971, the Department of Labor, in accordance with the provisions of OSHA, published in the *Federal Register* a basic, comprehensive 248-page set of OSHA compliance standards. All changes and new standards are likewise published in the *Federal Register*. This journal can be found in most public libraries or can be subscribed to for about $25 per year.

The Occupational Safety and Health Reporter Services of the Bureau of National Affairs, Inc., 1231 25th Street, N.W., Washington, D.C. 20037, will for a fee keep subscribers up-to-date on all revisions and additions.

Regional OSHA offices or state health departments can be a source of pertinent information. A listing follows of regional and area OSHA offices, up-to-date as of spring 1975.

Regional Offices

REGION 1—Boston
Donald E. Mackenzie
John F. Kennedy Federal Building
Government Center 308 E
Boston, Massachusetts 02203
617-223-6712

REGION II—New York
Alfred Barden
1515 Broadway
Room 3445
New York, New York 10036
212-971-5754

REGION III—Philadelphia
David H. Rhone
Penn Square Building   Room 623
1317 Filbert Street
Philadelphia, Pennsylvania 19107
215-597-4102

REGION IV—Atlanta
Basil Needham
1375 Peachtree Street, N.E.
Suite 587
Atlanta, Georgia 30309
404-526-3573

REGION V—Chicago
Edward E. Estkowski
300 South Wacker Drive
Room 1201
Chicago, Illinois 60606
312-353-4716

REGION VI—Dallas
John Barto
Texaco Building, Suite 600
1512 Commerce Street
Dallas, Texas 75201
214-749-2477

REGION VII—Kansas City
Joseph Reidinger
Waltower Building, Room 300
823 Walnut Street
Kansas City, Missouri 64106
816-374-5249

REGION VIII—Denver
Howard J. Schulte
Federal Building, Room 15010
1961 Stout Street
Box 3588
Denver, Colorado 80202
303-837-3883

REGION IX—San Francisco
Warren Fuller
9470 Federal Building
450 Golden Gate Avenue
Box 36017
San Francisco, California 94102
415-556-0584

REGION X—Seattle
James W. Lake
1808 Smith Tower Building
506 Second Avenue
Seattle, Washington 98104
206-442-5930

Area and District Offices

REGION I—Boston
Area Offices
John V. Fiatarone
Custom House Building, Room 703
State Street
Boston, Massachusetts 02109
FTS and commercial phone:
617-223-4511/12

Harold R. Smith
Federal Building, Room 617B
450 Main Street
Hartford, Connecticut 06103
FTS and commercial phone:
203-244-2294

Francis R. Amirault
Federal Building, Room 425
55 Pleasant Street
Concord, New Hampshire 03301
FTS phone:
603-224-7725
Commercial:
603-224-1995/6

District Office
Steven J. Simms
503-A Federal Building
Providence, Rhode Island 02903
FTS and commercial phone:
401-528-4466

REGION II—New York
Area Offices
Nicholas A. DiArchangel
90 Church Street
Room 1405
New York, New York 10007
FTS and commercial phone:
212-264-9840/1/2

William J. Dreeland
Midtown Plaza, Room 203
700 East Water Street
Syracuse, New York 13210
FTS and commercial phone:
315-473-2700/1

James H. Epps
370 Old Country Road
Garden City
Long Island, New York 11530
FTS and commercial phone:
516-294-0400

Thomas W. Fullam, Jr.
Federal Office Building, Room 635
970 Broad Street
Newark, New Jersey 07102
FTS and commercial phone:
201-645-5930/1/2

Louis Jacob
Condominium San Alberto Building
Room 328
605 Condado Avenue
Santurce, Puerto Rico 00907
FTS and commercial phone:
809-724-1059

REGION III—Philadelphia
Area Offices
Harry Sachkar
1317 Filbert Street
Suite 1010
Philadelphia, Pennsylvania 19017
FTS and commercial phone:
215-597-4955

Byron R. Chadwick
Stanwick Building, Room 111
3661 Virginia Beach Boulevard
Norfolk, Virginia 23502
FTS and commercial phone:
703-441-6381/2

Lapsley C. Ewing, Jr.
Federal Building, Room 8015
P.O. Box 10186
400 North 8th Street
Richmond, Virginia 23240
FTS and commercial phone:
703-782-2241/2

Maurice R. Daly
Federal Building, Room 1110A
31 Hopkins Plaza, Charles Center
Baltimore, Maryland 21201
FTS and commercial phone:
301-962-2840

Harry G. Lacey
Federal Building, Room 445D
1000 Liberty Avenue
Pittsburgh, Pennsylvania 15222
FTS and commercial phone:
412-644-2905/6

REGION IV—Atlanta
Area Offices
William F. Moerlins
1371 Peachtree Street, N.E.
Room 723
Atlanta, Georgia 30309
FTS and commercial phone:
404-526-5806/7 or 5883/4

James E. Blount
Bridge Building, Room 204
3200 East Oakland Park Boulevard
Fort Lauderdale, Florida 33308
FTS phone:
305-350-7331
Commercial:
305-525-0611 x 331

William W. Gordon
U.S. Federal Office Building
Box 35062
400 West Bay Street
Jacksonville, Florida 32202
FTS and commercial phone:
904-791-2895

Frank P. Flanagan
600 Federal Place, Room 561
Louisville, Kentucky 40202
FTS and commercial phone:
502-582-6111/12

Harold J. Monegue
Commerce Building, Room 801
118 North Royal Street
Mobile, Alabama 36602
FTS phone:
205-433-4382
Commercial:
205-433-3581 x 482

Quentin F. Haskins
1361 East Morehead Street
Charlotte, North Carolina 28204
FTS phone:
704-372-7495
Commercial:
704-372-0711 x 495

Eugene E. Light
1600 Hayes Street
Suite 302
Nashville, Tennessee 37203
FTS and commercial phone:
615-749-5313

Joseph L. Camp
Todd Mall
2047 Canyon Road
Birmingham, Alabama 35216
FTS phone:
205-325-6081
Commercial:
205-822-7100

Bernard E. Addy
Enterprise Building, Suite 201
6605 Abercorn Street
Savannah, Georgia 31405
FTS phone:
912-232-4393
Commercial:
912-354-0733

REGION V—Chicago
Area Offices
William Funcheon
300 South Wacker Drive
Room 1200
Chicago, Illinois 60606
FTS and commercial phone:
312-353-1390

Peter Schmitt
Bryson Building, Room 224
700 Bryden Road
Columbus, Ohio 43215
FTS and commercial phone:
614-469-5582

Robert B. Hanna
Clark Building, Room 400
633 West Wisconsin Avenue
Milwaukee, Wisconsin 53203
FTS and commercial phone:
414-224-3315/6

J. Fred Keppler
U.S. Post Office and Courthouse
Room 423
46 East Ohio Street
Indianapolis, Indiana 46204
FTS and commercial phone:
317-633-7384

Kenneth Bowman
847 Federal Office Building
1240 East Ninth Street
Cleveland, Ohio 44199
FTS and commercial phone:
216-522-3818

Earl J. Krotzer
Michigan Theatre Building
Room 626
220 Bagley Avenue
Detroit, Michigan 48226
FTS and commercial phone:
313-226-6720

Vernon Fern
110 South Fourth Street
Room 437
Minneapolis, Minnesota 55401
FTS and commercial phone:
612-725-2571

Ronald McCann
Federal Office Building, Room 5522
550 Main Street
Cincinnati, Ohio 45202
FTS and commercial phone:
513-684-2355

Glenn Butler
Federal Office Building, Room 734
234 North Summit Street
Toledo, Ohio 43604
FTS and commercial phone:
419-259-7542

REGION VI—Dallas
Area Offices
Charles J. Adams
Federal Building, Room 6B1
1100 Commerce Street
Dallas, Texas 75202
FTS and commercial phone:
214-749-1786/7/8

Robert B. Simmons
Federal Building, Room 421
1205 Texas Avenue
Lubbock, Texas 79401
FTS phone:
806-747-3681
Commercial:
806-747-3711 x 3681

James T. Knorpp
Petroleum Building, Room 512
420 South Boulder
Tulsa, Oklahoma 74103
FTS and commercial phone:
918-584-7151 x 7676

Thomas T. Curry
Old Federal Office Building
Room 802
201 Fannin Street
Houston, Texas 77002
FTS and commercial phone:
713-226-5431

John K. Parsons
Federal Building, Room 1036
600 South Street
New Orleans, Louisiana 70130
FTS and commercial phone:
504-527-2451/2 or 6166/7

District Office
Burtrand C. Lindquist
U.S. Custom House Building
Room 325
Galveston, Texas 77550
FTS and commercial phone:
713-763-1472/4

REGION VII—Kansas City
Area Offices
Robert J. Borchardt
1627 Main Street
Room 1100
Kansas City, Missouri 64108
FTS and commercial phone:
816-374-2756

A. F. Castranova
210 North 12th Boulevard
Room 554
St. Louis, Missouri 63101
FTS and commercial phone:
314-622-5461/2

Warren P. Wright
City National Bank Building
Room 630
Harney and 16th Streets
Omaha, Nebraska 68102
FTS and commercial phone:
402-221-3276/7

REGION VIII—Denver
Area Offices
Jerome J. Williams
Squire Plaza Building
8527 West Colfax Avenue
Lakewood, Colorado 80215
FTS and commercial phone:
303-234-4471

Charles F. Hines
Executive Building, Suite 309
455 East Fourth South
Salt Lake City, Utah 84111
FTS and commercial phone:
801-524-5080

Vernon A. Strahm
Petroleum Building, Suite 225
2812 First Avenue, North
Billings, Montana 59101
FTS phone:
406-245-6640/6649
Commercial:
406-245-6711, x 6640-6649

REGION IX—San Francisco
Area Offices
Donald T. Pickford
100 McAllister Street
Room 1706
San Francisco, California 94102
FTS and commercial phone:
415-556-0536

Lawrence E. Gromachey
Amerco Towers, Suite 910
2721 North Central Avenue
Phoenix, Arizona 85004
FTS and commercial phone:
602-261-4857/8

Anthony Mignano
Hartwell Building, Room 514
19 Pine Avenue
Long Beach, California 90802
FTS phone: 213-831-9281, then ask
Long Beach FTS operator for
432-3434
Commercial:
213-432-3434

Paul F. Haygood
333 Queen Street
Suite 505
Honolulu, Hawaii 96813
FTS and commercial phone:
808-546-3157/8

REGION X—Seattle
Area Offices
Richard L. Beeston
1906 Smith Tower Building
506 Second Avenue
Seattle, Washington 98104
FTS and commercial phone:
206-442-7520/27

Darrell Miller
Willholth Building, Room 217
610 C Street
Anchorage, Alaska 99501
FTS phone: dial local FTS operator
and ask for 907-272-5661 x 851
Commercial:
907-272-5561 x 851

Eugene Harrower
Pittock Block, Room 526
921 Southwest Washington Street
Portland, Oregon 97205
FTS and commercial phone:
503-221-2251

ORGANIZATIONS AND AGENCIES CONCERNED WITH INDUSTRIAL HYGIENE

Professional Organizations

ACGIH
American Conference of Governmental
Industrial Hygienists
P. O. Box 1937
Cincinnati, Ohio 45201

AIHA
American Industrial Hygiene Association
25711 Southfield Road
Southfield, Michigan 48075

Federal Governmental Agencies

NIOSH
National Institute for Occupational
Safety and Health. Headquarters:
Public Health Service
Department of Health, Education
and Welfare
Danac Building
5600 Fisher's Lane
Rockville, Maryland 20852

Main Laboratory:
NIOSH
U.S. Post Office Bldg.
Fifth and Walnut Streets
Cincinnati, Ohio 45202

ACOSH
Appalachian Center for Occupational
Safety and Health
DHEW/PHS/CDC/NIOSH
944 Chestnut Ridge Road
Morgantown, West Virginia 26505

OSHA
Occupational Safety and Health
Administration
U.S. Department of Labor
Fourteenth Street and
Constitution Avenue, N.W.
Washington, D.C. 20210

State Agencies

Alabama
Alabama State Department of Labor
600 Administrative Building
64 North Union Street
Montgomery, Alabama 36104

Alaska
Department of Labor
P.O. Box 1149
Juneau, Alaska 99801

Arizona
Industrial Commission of Arizona
1601 W. Jefferson
P.O. Box 19070
Phoenix, Arizona 85005

Arkansas
Department of Labor
Capitol Hill Building
Little Rock, Arkansas 72201

California
  Occupational Health Section
  State Department of Health
  2151 Berkeley Way
  Berkeley, California 94704

Colorado
  Department of Labor & Employment
  200 E. Ninth Avenue
  Denver, Colorado 80203

Connecticut
  Connecticut Department of Labor
  200 Folly Brook Boulevard
  Wethersfield, Connecticut 06109

Delaware
  Department of Labor & Industrial Relatic
  801 West Street
  Wilmington, Delaware 19801

District of Columbia
  Government of the District of Columbia
  Minimum Wage & Industrial Safety Board
  615 Eye Street, N.W.
  Washington, D.C. 20001

Florida
  Division of Labor
  Department of Commerce
  Caldwell Building
  Tallahassee, Florida 32301

Georgia
  State Board of Workmen's Compensation
  1182 W. Peachtree Street, N.E.
  Atlanta, Georgia 30309

Guam
  Department of Labor
  Government of Guam
  P.O. Box 884 (JSMP)
  Agana, Guam 96910

Hawaii
  Department of Labor and
  Industrial Relstions
  825 Mililani Street
  Honolulu, Hawaii 96813

Idaho
  Department of Labor
  Industrial Administration Bldg.
  317 Main Street
  Boise, Idaho 83702

Illinois
  Department of Labor
  160 N. LaSalle Street
  Chicago, Illinois 60601

Indiana
  Indiana Division of Labor
  Indiana State Office Bldg., Room 101
  100 North Senate Avenue
  Indianapolis, Indiana 46204

Iowa
  Bureau of Labor, State House
  East 7th & Court Avenue
  Des Moines, Iowa 50319

Kansas
  Department of Labor
  401 Topeka Avenue
  Topeka, Kansas 66603

Kentucky
  Kentucky Department of Labor
  Capitol Plaza
  Frankfort, Kentucky 40601

Louisiana
  Louisiana Department of Labor
  P.O. Box 44063
  Baton Rouge, Louisiana 70804

  Louisiana Department of Health
  P.O. Box 60630
  New Orleans, Louisiana 70160

Maine
  Office of the Governor
  State House
  Augusta, Maine 04330

Maryland
  Department of Licensing & Regulation
  Division of Labor & Industry
  203 E. Baltimore Street
  Baltimore, Maryland 21202

Michigan
  Michigan Department of Labor
  300 East Michigan Avenue
  Lansing, Michigan 48933

  Michigan Department of Public Health
  3500 North Logan Street
  Lansing, Michigan 48914

Minnesota
Department of Labor & Industry
Space Center Building, Fifth Floor
444 Lafayette Road
St. Paul, Minnesota 55101

Mississippi
Mississippi Board of Health
P.O. Box 1700
Jackson, Mississippi 39205

Missouri
Industrial Commission
Department of Labor & Industrial Rela.
Box 599
Jefferson City, Missouri 65101

Montana
Division of Workmen's Compensation
815 Front Street
Helena, Montana 59601

Nebraska
Department of Labor
Box 94600, State House Station
Lincoln, Nebraska 68509

Nevada
Nevada Industrial Commission
Department of Occupational
Safety & Health
515 E. Musser Street
Carson City, Nevada 89701

New Hampshire
Department of Labor
1 Pillsbury Street
Concord, New Hampsire 03301

New Jersey
Department of Labor & Industry
John Fitch Plaza
Trenton, New Jersey 08625

New Mexico
Health and Social Services Department
P.O. Box 2348
Santa Fe, New Mexico 87501

New York
Department of Labor
State Office Building
Campus
Albany, New York 12226

North Carolina
Department of Labor
P.O. Box 27407
Raleigh, North Carolina 27611

North Dakota
State Department of Health
Capitol Building
Bismarck, North Dakota 58501

Ohio
Department of Industrial Relations
Division of Occupational
Safety & Health
220 Parsons Avenue
Columbus, Ohio 43215

Oklahoma
Oklahoma State Department of Labor
State Capitol
Oklahoma City, Oklahoma 73105

Oregon
Workmen's Compensation Board
Labor & Industries Building
Salem, Oregon 97301

Pennsylvania
Department of Labor & Industry
1700 Labor and Industry Building
Harrisburg, Pennsylvania 17120

Puerto Rico
Department of Labor
Commonwealth of Puerto Rico
414 Barbosa Avenue
San Juan, Puerto Rico 00917

Rhode Island
Division of Occupational Safety
State Department of Labor
235 Promenade Street
Providence, Rhode Island 02908

South Carolina
Department of Labor
P.O. Box 11329
Columbia, South Carolina 29211

South Dakota
State Department of Health
Pierre, South Dakota 57501

Tennessee
Tennessee Department of Labor
Cordell Hull Building
Nashville, Tennessee 37219

Texas
  State Department of Health
  and State Safety Engineer
  Texas Occupational Safety Board
  1100 West 49th Street
  Austin, Texas 78756

Utah
  Industrial Commission of Utah
  Occupational Safety &
  Health Division
  560 East 500 South
  Salt Lake City, Utah 84111

Vermont
  Department of Labor & Industry
  State Office Building
  Montpelier, Vermont 05602

Virginia
  State Department of Health
  109 Governor Street
  Richmond, Virginia 23219

Washington
  Department of Labor and Industries
  General Administration Building
  Olympia, Washington 98504

West Virginia
  State Department of Health
  Rm. 535, 1800 E. Washington St.
  Charleston, West Virginia 25305

Wisconsin
  Department of Labor, Industry and
  Human Relations
  310 Price Place
  P.O. Box 2209
  Madison, Wisconsin 53701

Wyoming
  Occupational Health and
  Safety Department
  200 E. 8th Avenue
  P.O. Box 2186
  Cheyenne, Wyoming 82002

SELECTED BIBLIOGRAPHY

This section lists some of the more important, relatively recent publications in the field of industrial hygiene. The selections represent only a few of the many hundreds of publications on this subject. Persons seeking additional references should consult the bibliography sections of the publications listed below.

General

    Accident Prevention Manual for Industrial Operations (7th ed.). Chicago: National Safety Council, 1974.

    Factory Mutual System. *Handbook of Industrial Loss Prevention.* New York: McGraw-Hill Book Co., 1967.

    The Industrial Environment—Its Evaluation and Control, U.S. Department of Health, Education, and Welfare, Public Health Services, 1974.

    Olishifski, J.B., and F.E. McElroy, *Fundamentals of Industrial Hygiene.* Chicago: National Safety Council, 1971.

    Patty, F.A., ed., *Industrial Hygiene and Toxicology* (2nd ed.) vols. 1 and 2. New York: Interscience Publishers, 1958, 1963.

    Sax, Irving, *Dangerous Properties of Industrial Materials* (3rd ed.). New York: Van Nostrand Reinhold Co., 1968.

    Steere, N.V., *Handbook of Laboratory Safety.* Cleveland: Chemical Rubber Co., 1971.

    Threshold Limit Values, Cincinnati, American Conference of Governmental Industrial Hygienists, published annually.

Air Contaminants

Air Sampling Instruments Manual, American Conference of Governmental Industrial Hygienists, Cincinnati (1967).

Boyland, E., and R. Gooding, *Modern Trends in Toxicology.* New York: Appleton-Century-Crofts, 1968.

Browning, E., *Toxicity and Metabolism of Industrial Solvents.* New York: Elsevier Publishing Co., 1965.

Drinker, P., and T.F. Hatch, *Industrial Dust.* New York: McGraw-Hill Book Co., 1954.

Fisher, A.A., *Contact Dermatitis.* Philadelphia: Lea and Febiger, 1967.

Gerarde, H.W., *Toxicology and Biochemistry of Aromatic Hydrocarbons.* New York: Elsevier Publishing Co., 1960.

Gleason, M.N., Robert E. Gosselin, and Harold C. Hodge, *Clinical Toxicology of Commercial Products.* Baltimore: The Williams and Wilkins Co., 1963.

Handbook of Organic Industrial Solvents (3rd ed.). Chicago: American Mutual Insurance Alliance, 1966.

Hemeon, W., *Plant and Process Ventilation* (2nd ed.). New York: Industrial Press, Inc., 1963.

Industrial Ventilation Manual (10th ed.). American Conference of Governmental Industrial Hygienists, Lansing, Mich. (1968).

NIOSH, *Toxic Substance List.* Cincinnati, Ohio (1971).

Patty, F.A., ed., *Industrial Hygiene and Toxicology* (2nd ed.) vols 1 and 2. New York: Interscience Publishers, 1958, 1963.

Sunshine, I., ed., *Handbook of Analytic Toxicology.* Cleveland: Chemical Rubber Co., 1971.

Nonionizing Radiation

Boysen, J.E., "Microwave Radiation," Proceedings Sanitary and Industrial Hygiene Engineering A.S.T.I.A., no. AD 287799 (1961), 65-80.

Brotherton, M., *Masers and Lasers: How They Work, What They Do.* New York: McGraw-Hill Book Co., 1964.

Cleary, S.R., "The Biological Effects of Microwave and Radiofrequency Radiation," CRC Critical Review in Environmental Control, 1 (2) (1970), 257.

Kallen, Lewis R., *Ultraviolet Radiation* (2nd ed.). New York: John Wiley and Sons, Inc., 1965.

Schwan, H.P., *Biological Effects and Health Implications of Microwave Radiation,* U.S. Government Printing Office (1970).

Sperling, H.G., ed., *Laser Eye Effects,* A Report of the Armed Forces NRC Committee on Vision, Washington, D.C. (1968).

Wacker, P., *Biological Effects and Health Implications of Microwave Radiation,* U.S. Government Printing Office (1970).

Weiss, M.M., and W.W. Munford, "Microwave Radiation Hazards," *Health Physics,* 5 (1961), 168.

Wilkening, G.M., "The Potential Hazards of Laser Radiation," Proceedings of Symposium on Ergonomics and Physical Environmental Factors, Rome, Italy, (September 1968), International Labor Office, Geneva.

Ionizing Radiation

Barns, D.E., and D. Taylor, *Radiation Hazards and Protection.* New York: Pitman Publishing Corp., 1959.

Blatz, H., *Radiation Hygiene Handbook.* New York: McGraw-Hill Book Co., 1959.

Drummer, J.E., Jr., *General Handbook for Radiation Monitoring* (3rd ed.). LA-1835. Superintendent of Documents, Washington, D.C. (1958).

Glueckhauf, E., *Atomic Energy Waste, Its Nature, Use, and Disposal.* New York: Interscience Publishers, 1961.

Henry, H.F., *Fundamentals of Radiation Protection.* New York: John Wiley and Sons, Inc., Interscience Publishers, 1969.

Regulations for the Safe Transport of Radioactive Materials, IAEA Safety Series, no. 6 and 7, New York, National Agency for International Publications (1961).

Light

Foote, F.M., "Ophthalmic Aspects of Light and Vision," *Industrial Medicine and Surgery,* 27 (1958), 546.

IES Lighting Handbook (5th ed.). New York: Illuminating Engineering Society, 1972.

Lighting Design Committee of the IES, "General Procedure for Calculating Maintained Illumination," *Illuminating Engineering,* 57 (1962), 517.

Lighting Survey Committee of the IES, "How to Make a Lighting Survey," *Illuminating Engineering,* 57 (1962), 87.

Kahler, W.H., "Visual Comfort in the Plant," *Industrial Medicine and Surgery,* 27 (1958), 556.

Niedhart, J.J., "Visual Comfort in the Office," *Industrial Medicine and Surgery,* 27 (1958), 558.

Practice for Industrial Lighting. New York: American National Standards Institute, 1965.

Sell, Floyd, "Application of the Four Factors of Seeing to a Variety of Industrial Operations," *Industrial Medicine and Surgery,* 27 (1958), 578.

Heat

Belding, H.S., and T.F. Hatch, "Index for Evaluating Heat Stress in Terms of Resulting Physiological Strains," *Journal of the American Society of Heating and Ventilating Engineers,* 27 (1955), 129.

Burton, A.C., and O.G. Edholm, *Man in a Cold Environment.* London: Arnold, 1955.

"Cold and Its Effect on the Worker," Occupational Health Bulletin, Occupational Health Division, Department of Health and Welfare, Ontario, Canada, vol. 15 (1960).

Guide and Data Book, Fundamentals and Equipment (current ed.), American Society of Heating, Refrigerating, and Air Conditioning Engineers, New York.

Hardy, J.D., ed., *Temperature, Its Measurement and Control in Science and Industry*, vol. 3. New York: Reinhold Publishing Corporation, 1963.

Hardy, J.D., A.P. Gagge, and J.A.J. Stolwijk, *Physiological and Behavioral Temperature Regulation*. Springfield, Ill.: Charles C. Thomas, 1970.

Leitland, C.S., and A.R. Lind, *Heat Stress and Heat Disorders*. Philadelphia: F.A. Davis, Co., 1964.

Noise

Baron, R.A., *The Tyranny of Noise*. New York: St. Martins Press, 1970.

Beranek, L.L., *Noise Reduction*. New York: McGraw-Hill Book Co., 1960.

Glorig, A., *Noise and Your Ear*. New York: Grune and Stratton, 1958.

Guide for Conservation of Hearing in Noise, Subcommittee on Noise of the Committee on Conservation of Hearing of the American Academy of Ophthalmology and Otolaryngology, Dallas, Texas (1964).

Harris, C.M., *Handbook of Noise Control*. New York: McGraw-Hill Book Co., 1962.

Industrial Noise: A Guide to Its Evaluation and Control, U.S. Department of Health, Education, and Welfare, Public Health Service, Publication no. 1572 (1967).

Industrial Noise Manual (2nd ed.). Detroit: American Industrial Hygiene Association, 1966.

Kryter, K.D., *The Effects of Noise on Man*. New York: Academic Press, 1970.

Peterson, A.P.G., and E.E. Gross, *Handbook of Noise Measurement*. West Concord, Mass.: General Radio Co., 1963.

Welch, B.L., and A.S. Welch, eds., *Physiological Effects of Noise*. New York: Plenum Press, 1970.

Microbiological Hazards

Akers, R.L., R.J. Walker, and F.L. Sabel, "Development of a Laminar Air Flow Biological Cabinet," *American Industrial Hygiene Association Journal*, 30 (1969), 177.

Jenski, J.V., and G.B. Phillips, "Microbiological Safety Equipment," *Laboratory Animal Care*, 13 (1963), 2.

Phillips, G.B., and R.S. Runkle, "Laboratory Design for Microbiological Safety," *Applied Microbiology*, 15 (1967), 378.

Pike, R.M., S.E. Sulkin, and M.L. Schulze, et al., "Continuing Importance of Laboratory-Acquired Infections," *American Journal of Public Health*, 55 (1965), 190.

Wedum, A.G., "Laboratory Safety in Research with Infectious Aerosols," *Public Health Reports*, U.S. Department of Health, Education, and Welfare, Public Health Services, 79 (1964), 619.

Wedum, A.G., "Prevention of Laboratory-Acquired Infections," *American Journal of Medical Technology*, 22 (1956), 311.

Wolf, H.W., and others, *Sampling Microbiological Aerosols*, U.S. Department of Health, Education, and Welfare, Public Health Services, Public Health Monograph no. 60 (1964).

Ergonomic Stresses

Dreyfuss, H., *Designing for People.* New York: Paragraphic Books, Division of Grossman Publishers, 1967.

Fitts, P.M., and M.I. Posner, *Human Performance.* Belmont, Calif.: Wadsworth Publishers, 1967.

Hunter, D., *The Diseases of Occupation.* Boston: Little, Brown and Co., 1969.

McCormick, E.J., *Human Factors Engineering* (3rd ed.). New York: McGraw-Hill Book Co., 1970.

Morgan, C.T., and others, *Human Engineering Guide to Equipment Design.* New York: McGraw-Hill Book Co., 1963.

Morrell, K.F.H., *Ergonomics—Man in His Working Environment.* London: Chapman and Hall, 1970.

Muviell, K.F., *Human Performance in Industry.* New York: Reinhold Publishing Corp., 1965.

Woodson, W.E., and D.W. Conover, *Human Engineering Guide for Equipment Designers.* Berkeley, Calif.: University of California Press, 1964.

Index